Electron–Phonon Interaction and Lattice Dynamics in High T_c Superconductors

Peking University–World Scientific Advanced Physics Series

ISSN: 2382-5960

Vol. 10 *Advances in Theoretical and Experimental Research of High Temperature Cuprate Superconductivity*
edited by Rushan Han

Vol. 9 *Electron–Phonon Interaction and Lattice Dynamics in High T_c Superconductors*
edited by Han Zhang

Vol. 8 *Physics on Ultracold Quantum Gases*
by Yongjian Han, Wei Yi and Wei Zhang

Vol. 7 *Endless Quests: Theory, Experiments and Applications of Frontiers of Superconductivity*
edited by Jiang-Di Fan

Vol. 6 *Superconductivity Centennial*
edited by Rushan Han

Vol. 5 *Quasi-One-Dimensional Organic Superconductors*
by Wei Zhang and Carlos A. R. Sá de Melo

Vol. 4 *Applied Symbolic Dynamics and Chaos (Second Edition)*
by Bailin Hao and Weimou Zheng

Vol. 3 *Computer Simulations of Molecules and Condensed Matter: From Electronic Structures to Molecular Dynamics*
by Xin-Zheng Li and En-Ge Wang

Vol. 2 *Geometric Methods in Elastic Theory of Membranes in Liquid Crystal Phases (Second Edition)*
by Zhanchun Tu, Zhongcan Ou-Yang, Jixing Liu and Yuzhang Xie

Vol. 1 *Dark Energy*
by Miao Li, Xiao-Dong Li, Shuang Wang and Yi Wang

Peking University-World Scientific Advanced Physics Series

Electron–Phonon Interaction and Lattice Dynamics in High T_c Superconductors

Editor

Han Zhang

Peking University, China

Published by

World Scientific Publishing Co. Pte. Ltd.

5 Toh Tuck Link, Singapore 596224

USA office: 27 Warren Street, Suite 401-402, Hackensack, NJ 07601

UK office: 57 Shelton Street, Covent Garden, London WC2H 9HE

Library of Congress Cataloging-in-Publication Data
Names: Zhang, Han, 1954– editor.
Title: Electron-phonon interaction and lattice dynamics in high T_c superconductors / editor, Han Zhang (Peking University, China).
Other titles: Peking University-World Scientific advance physics series ; v. 9.
Description: Singapore ; Hackensack, NJ : World Scientific Publishing Co. Pte. Ltd., [2019] | Series: Peking University-World Scientific advanced physics series, ISSN 2382-5960 ; vol. 9
Identifiers: LCCN 2018050989| ISBN 9789813271135 (hardcover ; alk. paper) | ISBN 9813271132 (hardcover ; alk. paper)
Subjects: LCSH: High temperature superconductors. | Electron-phonon interactions. | Lattice dynamics.
Classification: LCC QC611.98.H54 E44 2019 | DDC 537.6/233--dc23
LC record available at https://lccn.loc.gov/2018050989

British Library Cataloguing-in-Publication Data
A catalogue record for this book is available from the British Library.

B&R Book Program

The Work is originally published by Peking University Press in 2013.
This edition is published by World Scientific Publishing Company Pte Ltd by arrangement with Peking University Press, Beijing, China.

For any available supplementary material, please visit
https://www.worldscientific.com/worldscibooks/10.1142/11010#t=suppl

Typeset by Stallion Press
Email: enquiries@stallionpress.com

Contents

Part II. Isotopic Effect 171

Preface

In early studies of high-T_c superconductivity (HTSC), the role of electron–phonon interaction (EPI) in cuprates was questioned by researchers because some important properties of the cuprates could not be explained by conventional BCS theory [1]. For instance, the isotope effect in the cuprates is not as evident as in conventional superconductors. Many theories were proposed to account for the mechanism of HTSC, such as resonating valence bond (RVB) theory [2], strong coupling theory [3], spin exciton model [4], $t - J$ model [5] and so on. But it was soon realized that these new theories were inadequate to explain the mechanism of HTSC. At the same time, substantial electron–phonon interaction became visible in the cuprates. In 2001, Lanzara et al., using angle-resolved photoemission spectroscopy (ARPES), observed for the first time that there was an abrupt change of the electron velocity at 50-80 meV in the different families of the cuprates which was referred to as the "kink" in the electronic dispersion [6]. Interestingly, however, the most likely candidate responsible for this "kink" was really the strong interaction between electrons and phonons. Meevasana interpreted this "kink" as the result of the coupling between electrons and special phonons with some collective behavior [7]. Subsequently, it was found that the isotope effect in high-T_c superconductors was not negligible, as previously thought, but instead, it was nontrivial. For example, Khasanov directly observed evident oxygen isotope ($^{16}O/^{18}O$) effect on the in-plane penetration depth of YBCO film [8]. And Iwasava found a distinct oxygen isotope shift near the electron–phonon coupling "kink" in the electronic dispersion of BSCCO system, which demonstrated the dominant role of the EPI in the cuprates [9].

Of course, a lot of researchers did not agree with the EPI scenario. For instance, Allen thought the EPI in the cuprates could not induce their *d*-wave symmetry [10]. This was just the difference between the high T_c superconductors and conventional superconductors, whose wave function was of *s*-wave symmetry. Another important fact was that the pseudogap existed universally in the high T_c cuprate superconductors [11]. Further, it was shown that there are two distinct energy scales and quasi-particles in the cuprates, one with larger energy corresponding to the pseudogap and the other one with smaller energy corresponding to the superconducting gap [12, 13]. This has raised an important question whether the pseudogap and the *d*-wave symmetry are induced by the EPI? Although the experiments mentioned above indicated the important significance of the EPI, the relationship between the pseudogap and the EPI and that between the *d*-wave symmetry and the EPI were not clearly clarified. In order to explain the mechanism of HTSC, one must answer these two questions inevitably.

Polarionic or bipolaronic models have been proposed to explain how the EPI or lattice dynamics in the cuprates determines the HTSC [14–16]. In conventional superconductors, the energy of phonons is just considered, and the symmetry of wave function is s-type. With the strong anisotropy of HTSC, the anisotropy of phonons should be taken into account which may induce *d*-wave symmetry. It was suggested that quasi-2D charge carriers weakly coupled with the anisotropic phonons undergo a quantum phase transition from conventional *s*-wave symmetry to unconventional *d*-wave symmetry. Therefore, the anisotropic phonons and thereby the anisotropic EPI are responsible for the *d*-wave pairing in the cuprates.

In spite of this, the strong anisotropy of phonons must arise from the anisotropy of the crystalline structure. It may be from some special local structure in the cuprates [17]. What is the particularity of the phonons in the cuprates? One must find the answers from the crystalline structure.

As more and more important results on the EPI and lattice dynamics have been discovered, it is felt most timely to edit a book that includes these issues. Because the field has grown indeed so vastly there is always the risk that I may be missing a number of important publications for which, I hope, the reader will generously excuse me.

The book contains the experimental and theoretical studies about the EPI. The experimental part covers the results of ARPES, isotopic effect, elastic neutron scattering study of electron–phonon and lattice role and so on. The theoretical part covers the electron–phonon, polaron and bipolaron aspects, also includes a number of other related papers supporting lattice role.

I would like to thank all the contributors for their support and encouragement to create this book. They are (in alphabetical order): Prof. Alexandrov at Loughborough University UK, Dr. Aiura at Tokyo University of Science, Prof. Kresin at Lawrence Berkeley, Dr. Koikegami at Japan and Nanoelectronics Research Institute, Prof. Lanzara at University of California, Berkeley, Prof. Pintschovius at Forschungszentrum Karlsruhe, Germany, Prof. Reznik at University of Colorado Boulder, Prof. Wolf at University of Virginia, Dr. Weyeneth and Prof. Mueller at Universität Zürich, and Prof. Zhao at UCLA.

Han Zhang
Peking University, China
Sept 26, 2012

References

[1] J. Bardeen, L. N. Cooper, and J. R. Schrieffer, *Phys. Rev.* **108** (1957), 1175.
[2] P. W. Anderson, *Science* **235** (1987), 1196.
[3] R. Zeyher and G. Zwicknagl, *Z. Phys. B* **78** (1990), 175.
[4] for example, S. M. Hayden, H. A. Mook, Pengcheng Dai, T. G. Perring, and F. Dogan, *Nature* **429** (2004), 531.
[5] F. C. Zhang and T. M. Rice, *Phys. Rev. B* **37** (1988), 3759.
[6] A. Lanzara, P. V. Bogdanov, X. J. Zhou, S. A. Kellar, D. L. Feng, E. D. Lu, T. Yoshida, H. Eisaki, A. Fujimori, K. Kishio, J. I. Shimoyama, T. Noda, S. Uchida, Z. Hussain, and Z. X. Shen, *Nature* **412** (2001), 510.
[7] W. Meevasana, N. J. C. Ingle, D. H. Lu, J. R. Shi, F. Baumberger, K. M. Shen, W. S. Lee, T. Cuk, H. Eisaki, T. P. Devereaux, N. Nagaosa, J. Zaanen, and Z. X. Shen, *Phys. Rev. Lett.* **96** (2006), 157003.
[8] R. Khasanov, D. G. Eshchenko, H. Luetkens, E. Morenzoni, T. Prokscha, A. Suter, N. Garifianov, M. Mali, J. Roos, K. Conder, and H. Keller, *Phys. Rev. Lett.* **92** (2004), 057602.
[9] H. Iwasawa, J. F. Douglas, K. Sato, T. Masui, Y. Yoshida, Z. Sun, H. Eisaki, H. Bando, A. Ino, M. Arita, K. Shimada, H. Namatame, M. Taniguchi, S. Tajima, S. Uchida, T. Saitoh, D. S. Dessau, and Y. Aiura, *Phys. Rev. Lett.* **101** (2008), 157005.
[10] P. B. Allen, *Nature* **412** (2001), 494.
[11] S. Hüfner, M. A. Hossain, A. Damascelli, and G. A. Sawatzky, *Rep. Prog. Phys.* **71** (2008), 062501.

[12] C. W. Luo, C. C. Hsieh, Y. J. Chen, P. T. Shih, M. H. Chen, K. H. Wu, J. Y. Juang, J. Y. Lin, T. M. Uen, and Y. S. Gou, *Phys. Rev. B* **74** (2006), 184525.
[13] T. Kondo, T. Takeuchi, A. Kaminski, S. Tsuda, and S. Shin, *Phys. Rev. Lett.* **98** (2007), 267004.
[14] A. S. Alexandrov, *Phys. Scr.* **83** (2011), 038301.
[15] A. S. Alexandrov, *Europe Phys. Lett.* **95** (2011), 27004.
[16] V. Kresin, *J Supercond. Nov. Magn.* **23** (2010), 179.
[17] W. T. Jin, S. J. Hao, and H. Zhang, *New J. Phys.* **11** (2009), 113036.

Part I

Reviews

1

Colloquium: Electron–Lattice Interaction and Its Impact on High-T_c Superconductivity*

V. Z. Kresin[†] and S. A. Wolf[‡]

[†]*Lawrence Berkeley Laboratory, University of California, Berkeley, California 94720, USA*

[‡]*Department of Materials Science and Engineering and Department of Physics, University of Virginia, Charlottesville, Virginia 22904, USA*

In this colloquium, the main features of the electron–lattice interaction are discussed and high values of the critical temperature up to room temperature could be provided. While the issue of the mechanism of superconductivity in the high-T_c cuprates continues to be controversial, one can state that there have been many experimental results demonstrating that the lattice makes a strong impact on the pairing of electrons. The polaronic nature of the carriers is also a manifestation of strong electron–lattice interaction. One can propose an experiment that allows an unambiguous determination of the intermediate boson (phonon, magnon, exciton, etc.) which provides the pairing. The electron–lattice interaction increases for nanosystems, and this is due to an effective increase in the density of states.

1.1. Introduction

This colloquium addresses the current experimental and theoretical situation concerning the importance of the interaction between electrons

*Reprinted with the permission from V. Z. Kresin, S. A. Wolf. Original published in *Rev. Mod. Phys.* **81**, 481–501 (2009).

and the crystal lattice in novel superconducting systems, especially in high-T_c cuprates. It will be demonstrated that the electronlattice interaction is an important factor underlying the nature of high-T_c superconductivity.

The phenomenon of superconductivity was discovered by Kamerlingh-Onnes in 1911 [1], and presently we are approaching the 100th anniversary of this event. The phenomenon was explained only in 1957 by Bardeen, Cooper, and Schrieffer (BCS). According to the classical BCS theory, the key phenomenon occurring in superconductors is the pairing of electrons. The system of conducting electrons in a superconducting metal forms pairs of bound electrons ("Cooper" pairs). There is still the fundamental problem of the mechanism of superconductivity, i.e., the origin of the pairing should be explained. Indeed, pairing means that there is an attraction between the paired electrons; as a result, they can form a bound state. What is the origin of such a force? As was shown in the BCS theory, and later supported by experimental and theoretical studies of many superconducting materials, this attraction is provided by the electron–lattice interaction.

According to the quantum theory of solids, the lattice excitations in bulk metals, which correspond to small ionic vibrations ($a/d \ll 1$, where a is the amplitude of vibrations and d is the lattice period), can be described as acoustic quanta (phonons) with energies $\varepsilon^i_{\mathrm{ph}} = \hbar\Omega_i(\vec{q}); \vec{q} = \hbar\vec{k}$ (in the following we set $\hbar = 1$), with momentum $\vec{q}$ and wave number $\vec{k}(k = 2\pi/\lambda$, where λ is the phonon's wavelength), and with i corresponding to the various phonon branches (longitudinal, transverse, optical). For such systems, the electron–lattice interaction, e.g., the energy exchange between the electrons and lattice, can be described as radiation and adsorption of phonons and is denoted as the electron–phonon interaction.

In the Debye model [2], all acoustic branches are described by the linear law $\Omega = uq$, where u is the average sound velocity. The value of the so-called Debye frequency, which is the maximum frequency of vibrations ($\Omega_D \equiv \Omega_{\mathrm{max}}$), is determined by the condition that the total number of vibrations $V\Omega^3_{\mathrm{max}}/2\pi^2u^3$ is equal to the total number of vibrational degrees of freedom $3N$(N is the number of ions). One can estimate $\Omega_D \approx uq_{\mathrm{max}} \approx u/d$, where d is the lattice period.

According to the BCS theory of superconductivity, pairing is provided by the electron–phonon interaction, or more specifically by the exchange of phonons between the electrons forming the pair. This exchange means the emission of a phonon by an electron moving through the lattice and the subsequent absorption of the phonon by another electron.

In 1986, the 75th anniversary of superconductivity was marked by the discovery of a new class of superconducting materials, namely, high-T_c copper oxides (usually called cuprates). Bednorz and Müller [3] discovered that the $La_{1.85}Ba_{0.15}CuO_4$ compound became superconducting with a critical temperature $T_c \approx 30$ K, which noticeably exceeded the previous record ($T_c \approx 23.2$ K for Nb_3Ge). Optimization of the synthesis of the similar compound (La–Sr–Cu–O) moved the transition temperature close to 40 K. This achievement was quickly followed by discoveries of other high-T_c copper oxides (cuprates). The most studied is the $YBa_2Cu_3O_{6+x}$ (YBCO) compound with $T_c \approx 93$ K at $x \approx 0.9$ [4]. At present, the highest observed value of T_c is about 150 K and is for the $HgBa_2Ca_2Cu_3O_{8+x}$ compound under pressure. The discoveries of new cuprates were accompanied by intensive studies of their structure and properties [5, 6]. It turns out that all cuprates have a layered structure. The main structural unit that is typical for the whole family is the Cu–O plane (see Fig. 1.1). One should distinguish between the main layer (Cu–O plane) where pairing originates and the charge reservoir. For example, in addition to the Cu–O planes, the YBCO compound contains Cu–O chains, and the change in the oxygen content in the chain layers leads to charge transfer between these two subsystems. The charge transfer occurs through the apical oxygen ion located between the planes and chains (Fig. 1.1).

Another important property of these novel superconductors is that they are doped materials. The doping is provided either by chemical substitution (e.g., by the La→ Sr substitution in the $La_{2-x}Sr_xCuO_4$ compound; the value $T_c \approx 40$ K corresponds to $x \approx 0.15$) or by changing the oxygen content. Doping leads to the appearance of carriers in the Cu–O planes. There are two types of carriers [7]. One of them (electrons) is created by the dopants, which are called donors. The second type (holes; they have a positive charge) is produced by doping, which removes electrons. Some of the cuprates (e.g., Nd–Ce–Cu–O) contain electrons as the carriers. Such an important material as YBCO contains carriers that are holes. It is important that the value of T_c depends strongly on the in-plane carrier concentration. The undoped parent compounds are insulators. Doping leads to conductivity and then, for larger carrier concentration, to superconductivity. There is some characteristic value of the carrier concentration n_m which corresponds to the maximum value of $T_c \equiv T_c^{\max}$. The underdoped

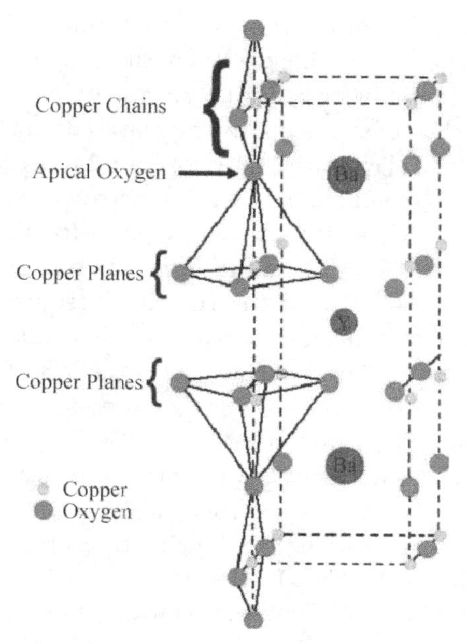

Fig. 1.1. Structure of the Y–Ba–Cu–O (YBCO) compound. One can see the apical, in-plane, and in-chain oxygen ions.

($n < n_m$) and overdoped ($n > n_m$) regions are characterized by values of T_c lower than $T_c^{\max}$.

Since the discovery of high-T_c oxides, there has been an intensive and fruitful study of these novel materials. However, despite intensive research, the question of the mechanism for these materials is still open. There has been growing evidence, mainly from various experimental studies, that the electron–lattice interaction is important for understanding the nature of high-T_c superconductivity in the cuprates. This interaction provides a direct contribution to the pairing of electrons, and also is clearly manifested in polaronic effects. The polaronic effects appear as a result of the strong electron–lattice interaction. In this case, a moving electron polarizes the lattice, and a shift in positions of neighboring ions forms a potential "box" for the electron. A polaron is a unit containing an electron that is moving with the lattice polarization caused by the electron itself (see Refs. [7, 8] and Section 4).

This colloquium is not a review, but rather a systematic description of our view, reflected in many publications on the subject. This colloquium also contains an extensive list of references that are related to this subject. Our viewpoint is that the electron–lattice interaction is an important ingredient of the current scenario and can explain superconductivity in novel systems including high-temperature superconducting cuprates.

The structure of the paper is as follows. The general properties of superconductivity caused by the electron–phonon interaction are discussed in Section 2. Experimental data demonstrating the impact of this interaction are described in Section 3. Section 4 is concerned with the polaronic effect and the isotopic substitution. The phonon-plasmon mechanism for layered systems is described in Section 5. Section 6 contains a discussion of the electron–lattice interaction in the "pseudogap" state. A critical experiment to provide more insight into the mechanism is proposed in Section 7. High-T_c superconductivity in nanoclusters caused by the electronvibrational interaction is discussed in Section 8. Section 9 contains concluding remarks.

1.2. Electron–Lattice Interaction and the Upper Limit of T_c

The electron–lattice interaction can, in principle, lead to high values of T_c (see below). Of course, this statement alone does not provide the answer to the question about the nature of the superconducting state in cuprates, but it means that the electron–lattice interaction cannot be summarily ruled out as a potential mechanism.

One should note that immediately after the discovery of high-T_c oxides, many excluded the electron–lattice interaction from the list of potential mechanisms. For the most part, this was done because of the natural temptation to introduce something new and exciting into the field as opposed to relying on the important principle of Occam's razor, *pluralitas non est ponenda sine neccesitate* ("one should not increase, beyond what is necessary, the number of entities required to explain anything"). An additional key factor was the conviction that despite the electron–phonon interaction being successful as an explanation for superconductivity in conventional materials, this mechanism is not sufficient to explain the observed high values of T_c. We address this important second aspect.

The BCS theory [9] was developed in the weak-coupling approximation ($\lambda \ll 1$, where λ is the electron–phonon coupling constant at $T = 0$; its value reflects the strength of the electron–lattice interaction).

Electron–phonon coupling leads to attraction between electrons in a superconductor. More specifically, an electron polarizes the lattice, that is, it induces ionic motion which affects another electron. As mentioned, in the quantum picture the process can be visualized as an exchange of phonons; such an exchange leads to an attraction between electrons (a detailed description of this interaction has been given by Ashcroft and Mermin [7], Kresin and Wolf [5]). In superconductors, this attraction overcomes Coulomb electron repulsion.

The expression for the critical temperature derived in the BCS theory has the form

$$k_B T_c \approx \tilde{\Omega} e^{-1/(\lambda-\mu^*)}, \tag{1.2.1}$$

where $\tilde{\Omega}$ is the characteristic phonon frequency, $\tilde{\Omega} \approx \Omega_D, \Omega_D$ is the Debye frequency, and $\mu^* = V_c[1+V_c \ln(\varepsilon_0/\tilde{\Omega})]^{-1}$ describes the Coulomb repulsion; $\varepsilon_0 \approx E_F$, where E_F is the Fermi energy; usually $\mu^* \approx 0.1$.

As mentioned, Eq. (1.2.1) is valid in a weak-coupling approximation. Since $\lambda \ll 1$ (e.g., $0.5 \gtrsim \lambda$), one could easily come to the conclusion, which follows from Eq. (1.2.1), that T_c should be at least an order of magnitude below the Debye temperature (the Debye temperature θ_D is determined by the relation $k_B\theta_D = \hbar\Omega_D$; in the following discussion, we have set $k_B = \hbar = 1$ so that energy E, frequency Ω, and temperature T all have the same units).

In many superconductors, the condition $\lambda \ll 1$ is not satisfied and $\lambda \gtrsim 1$. For example, in lead, $\lambda = 1.4$; in mercury, $\lambda = 1.6$; and in the alloy $Pb_{0.65}Bi_{0.35}$, the coupling constant has the value $\lambda = 2.1$ (see, e.g., [10, 11]). To understand the consequences of these high values of λ, it is necessary to go beyond the limit of weak coupling. This more universal approach was developed shortly after the creation of the BCS theory [12, 13] and allows us to analyze the properties of superconductors with strong electron–phonon coupling.

Strong-coupling theory is a generalization of the theory of normal metals [14]. It is also based on the method developed by Gor'kov [15], which was initially applied for the weak-coupling case (see, e.g., [16]). A detailed description of the fundamentals of superconductivity with strong coupling can be found in a number of reviews and monographs (see, e.g., [17–19]) and is based on the Green's-function method of the many-body theory.

We introduce here the main quantities that enter the theory. The phonon spectrum contains a continuous distribution of phonon frequencies and it is described by the phonon density of states $F(\Omega)$, where Ω is the phonon frequency. An important material-dependent parameter is $\alpha^2(\Omega)F(\Omega)$, where $\alpha^2(\Omega)$ is a measure of the phonon-frequency-dependent electron–phonon interaction. The electron–phonon coupling constant λ, which determines the value of T_c (see Eqs. (1.2.1) and (1.2.8)–(1.2.12)), can be written as

$$\lambda = 2\int \alpha^2(\Omega)F(\Omega)\Omega^{-1}d\Omega. \tag{1.2.2}$$

One can introduce the characteristic phonon frequency $\tilde{\Omega}$, which is defined as an average over $\alpha^2(\Omega)F(\Omega)$,

$$\tilde{\Omega} = \langle\Omega^2\rangle^{1/2}. \tag{1.2.3}$$

The average is determined by $\langle f(\Omega)\rangle = (2/\lambda)\int d\Omega f(\Omega)\alpha^2(\Omega)F(\Omega)\Omega^{-1}$, so that $\langle\Omega^2\rangle = (2/\lambda)\int \Omega\alpha^2(\Omega)F(\Omega)d\Omega$; the coupling constant is defined by Eq. (1.2.2).

The main quantity of interest is the pairing order parameter $\Delta(\omega)$. The pairing energy gap can be determined as the root of the equation $\omega = \Delta(i\omega)$.

The equation for the pairing order parameter $\Delta(\omega)$ has the form (at $T = 0$ K)

$$\Delta(\omega) = [Z(\omega)]^{-1}\int_0^{\omega_c} d\omega' P(\omega')[K_+(\omega,\omega') - \mu^*], \tag{1.2.4}$$

where

$$[1 - Z(\omega)]\omega = \int_0^{\infty} d\omega' N(\omega')K_-(\omega,\omega'),$$

$$P(\omega) = \mathrm{Re}\{\Delta(\omega)[\omega^2 - \Delta^2(\omega)]^{-1/2}\},$$

$$N(\omega) = \mathrm{Re}\{|\omega|[\omega^2 - \Delta^2(\omega)]^{-1/2}\},$$

$$K_\pm(\omega,\omega') = \int d\Omega\alpha^2(\Omega)F(\Omega) \times \left(\frac{1}{\omega' + \omega + \Omega + i\delta} \pm \frac{1}{\omega' - \omega + \Omega - i\delta}\right).$$

Here, Ω is the phonon frequency and Z is the so-called renormalization function describing the "dressing" of electrons moving through the lattice. Equations (1.2.1) and (1.2.4) also contain the Coulomb pseudopotential μ^*.

The important aspect of pairing is the logarithmic weakening of the Coulomb repulsion (see Refs. [20, 21, 22]), which is related to the difference in the energy scales of the attractive and repulsive effects (see discussion following Eq. (1.2.1)). The attraction is important in an energy interval $\tilde{\Omega}$, whereas the repulsion is characterized by the energy scale $\varepsilon_0 \sim E_F$, where E_F is the Fermi energy. In usual metals $E_F \approx 10$ eV, the characteristic phonon frequency $\tilde{\Omega} \approx 20 - 50$ meV, so that $E_F \gg \tilde{\Omega}$. As a result, the Coulomb pseudopotential $\mu^* = V_c[1 + V_c \ln(\varepsilon_0/\tilde{\Omega})]^{-1}$ contains a large logarithmic factor that reduces the contribution of the Coulomb repulsion. For the cuprates, the electronic energy scale $\varepsilon_0 \sim 1$ eV, and although it is smaller than the corresponding energy scale in conventional superconductors it is still much larger than the scale of the lattice energy. For simplicity, we omit μ^* below in some equations.

At finite temperature it is convenient to use the thermodynamic Green's-function formalism (see, e.g., [16]). Then the major equation can be written in the form

$$\Delta(\omega_n)Z = T\sum_{\omega_{n'}} \int d\Omega\, \Omega^{-1}\alpha^2(\Omega)F(\Omega) \times D(\omega_n - \omega_{n'};\Omega)F^+(\omega_{n'}), \qquad (1.2.5)$$

where

$$D = \Omega^2[(\omega_n - \omega_{n'})^2 + \Omega^2]^{-1}$$

is the so-called phonon Green's function, $\omega_n = (2n+1)\pi T$. Equation (1.2.5) can be approximated to a high degree of accuracy using Eq. (1.2.2),

$$\Delta(\omega_n)Z = \lambda T \sum_{\omega_{n'}} D(\omega_n - \omega_{n'};\tilde{\Omega})F^+(\omega_{n'}), \qquad (1.2.6)$$

where $\tilde{\Omega}$ is the characteristic phonon frequency (see Eq. (1.2.3)), the coupling constant is defined by Eq. (1.2.2), and

$$F^+ = \Delta(\omega_n)/[\omega_n^2 + \xi^2 + \Delta^2(\omega_n)]$$

is the pairing Green's function, introduced by Gor'kov [15], ξ is the electron energy relative to the chemical potential. One can also write the equation for the renormalization function Z.

McMillan [23] introduced a convenient expression for the coupling constant λ,

$$\lambda = \nu \langle I^2 \rangle / M\tilde{\Omega}^2. \quad (1.2.7)$$

In Eq. (1.2.7), $\nu = m^* p_F / 2\pi^2$ is the bulk density of states, $\langle I^2 \rangle$ contains the average value of the electron–phonon matrix element I (see, e.g., [18]), and $\tilde{\Omega}$ is defined by Eq. (1.2.3). One can see from Eq. (1.2.7) that λ is not a universal constant, but a material-dependent parameter.

Equation (1.2.6) is especially convenient for evaluating T_c and for analyzing the thermodynamic properties. It is important that the strong-coupling superconductivity theory is valid if $\tilde{\Omega} \ll E_F$. This is the *only* condition for its applicability. It is important to stress also that there is no limit on the value of T_c and the theory even allows T_c to exceed the Debye temperature.

As noted above, the derivation of Eqs. (1.2.4) and (1.2.5) is based on a special method (Green's-function formalism), and its description is beyond the scope of this paper. It is worth noting that these equations are a generalization of the BCS theory. Indeed, if we assume that the electron–phonon coupling is weak ($T_c \ll \tilde{\Omega}$), one can neglect the dependence of the D function on $\omega_n = (2n+1)\pi T_c$, and then $\Delta(\omega_n) = \text{const}$. At $T = T_c$ one should put $\Delta = 0$ in the denominator of Eq. (1.2.6). Then one can calculate T_c. Performing a summation, we arrive at the usual BCS expression (1.2.1).

However, strong-coupling theory is based on one very important assumption. Namely, it assumes that the phonon spectrum is fixed, and this implies that the lattice is not affected by the pairing. Strictly speaking, this is not the case, and if the value of the coupling constant exceeds some value $\lambda_{\max}$, then the lattice could become unstable. This problem was studied by Browman and Kagan [24] and Geilikman [25, 26]. Based on rigorous adiabatic theory, one can prove that the change in phonon characteristic frequency caused by the electron–lattice interaction is small, and the lattice becomes unstable (that is, the characteristic frequency becomes imaginary) only at very large values of $\lambda (\lambda \gg 1)$. Therefore, high values of T_c are theoretically possible within this framework.

It is interesting that an explicit expression for T_c depends on the strength of the coupling. As noted above, the BCS expression (1.2.1) is valid for weak-coupling superconductors only. For larger values of $\lambda (1.5 \gtrsim \lambda \gtrsim)$ one can use the expression obtained by McMillan [27] and then modified by Dynes [27]. This expression has the form

$$T_c = \frac{\tilde{\Omega}}{1.2} \exp\left[-\frac{1.04(1+\lambda)}{\lambda - \mu^*(1+0.62\lambda)}\right]. \tag{1.2.8}$$

If the coupling constant λ is large ($\lambda > 1.5$), one should use a different expression for the critical temperature. We initially discuss the case of very strong coupling ($\lambda \gtrsim 5$; then $\pi T_c \gtrsim \tilde{\Omega}$). In this case, the dependence of T_c on λ differs drastically from the dependences given by Eqs. (1.2.1) and (1.2.8). As shown initially by Allen and Dynes [10] using numerical calculations, and later analytically by Kresin *et al.* [28], this dependence has the form (here we assume $\mu^* = 0$)

$$T_c = 0.18\lambda^{1/2}\tilde{\Omega}. \tag{1.2.9}$$

Kresin *et al.* [28] also obtained the expression for T_c when $\mu^* \neq 0$, that is,

$$T_c = 0.18\lambda_{\text{eff}}^{1/2}\tilde{\Omega}, \quad \lambda_{\text{eff}} = \lambda(1+2.6\mu^*)^{-1}. \tag{1.2.10}$$

This analytical expression was obtained using the matrix method [29]. The scaling behavior for T_c can be seen directly from Eq. (1.2.6). Indeed, if $\pi T_c \gg \tilde{\Omega}$, then one can neglect $\tilde{\Omega}^2$ in the denominator of the phonon Green's function, and then one can directly see the scaling behavior $T_c \propto \lambda^{1/2}\tilde{\Omega}$.

One can see from Eqs. (1.2.9) and (1.2.10) that for large λ the expression for T_c is very different from Eqs. (1.2.1) and (1.2.8). As mentioned earlier, Eq. (1.2.10) is valid for $\lambda \gtrsim 5$. For the intermediate case, one can use the general equation [30] that was obtained by solving Eq. (1.2.6) and is valid for any value of the coupling constant,

$$T_c = 0.25\tilde{\Omega}/(e^{2/\lambda_{\text{eff}}} - 1)^{1/2},$$
$$\lambda_{\text{eff}} = (\lambda - \mu^*)[1 + 2\mu^* + \lambda\mu^* t(\lambda)]^{-1}. \tag{1.2.11}$$

The universal function $t(\lambda)$ decreases exponentially with increasing λ; $t(\lambda)$ can be approximated quite accurately by $t(\lambda) = 1.5\ \exp(-0.28\lambda)$; such an approximation was proposed by Tewari and Gumber [31], see also Kresin and Wolf [5]. If we neglect μ^*, we obtain

$$T_c = 0.25\tilde{\Omega}/(e^{2/\lambda} - 1)^{1/2}. \tag{1.2.12}$$

As mentioned above, Eqs. (1.2.11) and (1.2.12) are valid for any strength of the coupling. One can easily see that for the weak-coupling case Eq. (1.2.11) reduces to Eq. (1.2.1), whereas for $\lambda \gg 1$ we obtain the dependence of Eq. (1.2.10).

Equation (1.2.12) was obtained by Kresin [30] as a solution of Eq. (1.2.6). The dependence of Eq. (1.2.12) was used as a trial function, and then it was demonstrated that it satisfied Eq. (1.2.6) with a high degree of precision. Later the same expression was obtained analytically by Bourne *et al.* [32] for the model case: $\alpha^2(\Omega)F(\Omega) = \text{const}$ for $0 < \Omega < \Omega_{\text{max}}$ and $\alpha^2(\Omega)F(\Omega) = 0$ for $\Omega > \Omega_{\text{max}}$.

One can see directly from Eqs. (1.2.9)–(1.2.11) that there is a large range of values for the coupling constant where the lattice is still stable, and the value of T_c is high. In principle, T_c can reach room temperature (e.g., for $\lambda_{\text{eff}} \approx 5$; $\tilde{\Omega} \approx 60$ meV). The values of T_c observed in the cuprates are even more realistic (e.g., for $\lambda_{\text{eff}} \approx 3 - 3.5$, $\tilde{\Omega} \approx 25$ meV).

The question about an upper limit of T_c for the phonon mechanism has some interesting history. Based on the so-called Froelich Hamiltonian, which is the sum of the electronic term, the phonon term with experimentally measured phonon frequency, and the electron–phonon interaction, one can obtain

$$\Omega = \Omega_0(1 - 2\lambda)^{1/2} \tag{1.2.13}$$

(Migdal, 1960). Based on this expression, one can conclude that the value of the coupling constant λ cannot exceed $\lambda_{\text{max}} = 0.5$, and this implies that the value $T_c \lesssim 0.1\tilde{\Omega}(\tilde{\Omega} \approx \Omega_D)$ is the upper limit of T_c. Indeed, such a point of view was almost generally accepted after the appearance of the BCS theory. However, it soon became clear that something is wrong with this criterion, since there were many superconductors discovered with $\lambda > 0.5$ (e.g., Sn, Pb, Hg). The problem was clarified later by Browman and Kagan [24] and Geilikman [25, 26]. As mentioned, the total Hamiltonian that leads to Eq. (1.2.13) contains terms describing free electronic and phonon fields and their interaction; the phonon term contains an experimentally observed phonon spectrum, including an acoustic branch. However, one can demonstrate that the formation of an acoustic dispersion law is also provided by the electron–ion interaction. In other words, in this model we are double-counting. This means that the analysis of the electron–phonon interaction has to be carried out with considerable care. This has been done based directly on the adiabatic approximation by Browman and Kagan [24] and Geilikman [25, 26]; see also the review by Kresin *et al.* [19]. The theory starts from the initial picture of electrons and ions, and the formation of the phonon branch and the residual electron–phonon interaction has been obtained by rigorous and self-consistent analysis. The conclusion is that the electron–phonon interaction does not lead to the dependence of Eq. (1.2.13).

Note that this conclusion does not mean the absence of lattice instabilities. In fact, the electron–phonon interaction can lead to various instabilities, especially for systems containing low-dimensional units. But this fact does not support the conclusion about the existence of an upper value of λ and, correspondingly, an upper limit for T_c. It is likely that such a limit exists for very large $\lambda(\lambda \gtrsim 10)$, but this is still an open question.

Another faulty restriction on T_c was later proposed and was based on the McMillan equation (1.2.8). Indeed, this equation taken at face value leads to an upper limit of T_c. If one neglects μ^* for simplicity, one can easily find that the maximum value of T_c corresponds to $\lambda = 2$; then $T_c^{\max} \approx \tilde{\Omega}/6$. This conclusion, however, assumes that Eq. (1.2.8) is valid for $\lambda > 1.5$. But the McMillan equation is valid only for $\lambda \lesssim 1.5$. Therefore, the value $\lambda = 2$ is outside of the range of its applicability.

The treatment based on Eqs. (1.2.10) and (1.2.11) leads to a very different conclusion, namely, to the absence of the upper limit for T_c. As noted before, experimentally large values of λ have been determined for several superconductors. For example, $\lambda \approx 2.1$ for $Pb_{0.65}Bi$, $\lambda \approx 2.6$ for Am-$Pb_{0.45}Bi_{0.55}$ [10, 11]. Also, if the material is characterized by relatively large values of both λ and $\tilde{\Omega}$, it might have a very high value of T_c.

As stressed above, this conclusion by itself does not mean that high-temperature superconductivity in the cuprates is provided by the electron–phonon interaction. This can be determined only by special and detailed experimental study (e.g., by tunneling spectroscopy), but such a mechanism cannot be ruled out on any theoretical grounds.

1.3. Experimental Data and Analysis

Only some selected experimental techniques can provide information about the pairing mechanism. Indeed, many experimental studies are not sensitive to the pairing interaction. For example, thermodynamic and electromagnetic properties contain the energy gap ε_0 as a parameter and, since the energy gap is directly proportional to the critical temperature (according to the BCS theory $\varepsilon_0 = 1.76k_BT_c$), T_c becomes the key parameter of the theory. As a result, all such properties are parametrized by the critical temperature, and they are not sensitive to the nature of the pairing interaction. Only selected methods are sensitive to the nature of the interaction which provides the observed values of T_c.

According to the BCS theory, the pairing is provided by the electron–phonon interaction, that is, by phonon exchange between the paired electrons. However, after Little's paper [33] it becomes clear that the pairing can be caused by other excitations as well. Among these excitations are electronic ones. This electronic mechanism can be important if the material contains two groups of electrons. Excitations within one of these groups serve as "gents" giving rise to pairing in the other group. Another electronic mechanism represents exchange through coupling to plasmons which are electronic collective excitations (see Section 5). Pairing can be provided also by exchange of magnetic excitations (magnons).

In principle, these and other mechanisms can provide pairing in novel materials. In addition, the superconducting state can be caused by the contributions of different excitations. Based on special experiments, one should be able to determine the key factors responsible for pairing in these novel materials. Below we discuss some of these techniques and their relevance to the problem of determining the mechanism of pairing in the cuprates.

1.3.1. *Tunneling Spectroscopy*

1.3.1.1. *McMillan–Rowell method*

It is only because of the special tunneling method developed by McMillan and Rowell [34, 35] that we have rigorous evidence that the phonon mechanism, that is, the mechanism based on the electron–phonon coupling is the dominant one for conventional superconductors.

Superconducting tunneling spectroscopy was developed by McMillan and Rowell and described in their review [35]; see also the review by Wolf [11]. Here are some key elements of this approach.

Tunneling spectroscopy (see, e.g., [36]) is based on observation of the tunneling of electrons through a typically very thin ($\approx$ 10 A) insulating barrier separating the superconductor that is being studied and some other metallic layer. One then measures the tunneling current as a function of the applied voltage and analyzes the result according to the McMillan–Rowell procedure [35]. There are several experimental techniques that have been used to generate tunneling spectra. The most widely used method requires the deposition of the superconducting electrode, the formation of a barrier, by either oxidation of the superconductor or depositing an insulating layer, and then final deposition of another metallic or superconducting electrode on top of the insulator.

The important quantities that need to be measured are the direct current–voltage characteristic $I{-}V$, the derivative of the $I{-}V, dI/dV$ as a function of the voltage, and the second derivative d^2I/dV^2 also as a function of the voltage. These data must be taken with the sample in both the normal and the superconducting state. If the counterelectrode is a normal metal, then measurements of the normalized conductance of the junction $\sigma = (dj/dV)_s/(dj/dV)_n$ allow us to determine the key quantity, the tunneling density of states $N_T(\omega)$,

$$N_T(\omega) = \mathrm{Re}[|\omega|/(\omega^2 - \Delta^2)^{1/2}], \tag{1.3.1}$$

where $N_T(\omega)$ is normalized by the density of states in the normal state. Note that the tunneling density of states contains the order parameter $\Delta \equiv \Delta(\omega)$. This analysis is based on Eq. (1.2.4), i.e., it is assumed that the superconducting state is provided by the electron–phonon interaction. It is important to note also that the peak position in the function $\alpha^2(\Omega)F(\Omega)$ corresponds to the point of the most negative slope in the tunneling conductance. Therefore, the second derivative d^2I/dV^2 allows one to locate the peaks in the phonon spectrum.

The key part of the method is the inversion procedure. The function $\Delta(\omega)$ determined from the tunneling conductance measurements can be used to evaluate the function $\alpha^2(\Omega)F(\Omega)$ and the Coulomb pseudopotential μ^* from Eqs. (1.3.1) and (1.2.4). Inverting Eq. (1.2.4) allows one to determine the function $\alpha^2(\Omega)F(\Omega)$ introduced in Section 2, and μ^*. The McMillan–Rowell method involves numerically solving the integral equations (1.2.4) for a given set of parameters, calculating $N_T(\omega)$ (Eq. (1.3.1)), comparing the calculated values to the measured values, adjusting the input parameters, and iterating the procedure until the calculated tunneling density of states matches the measured one. As mentioned above, a detailed description of the procedure and the application to Pb can be found in McMillan and Rowell [35]; see also Wolf [11].

The function $\alpha^2(\Omega)F(\Omega)$ contains two factors. One of them $[\alpha^2(\Omega)]$ depends weakly on frequency, whereas the phonon density of states $F(\Omega)$ usually contains two peaks, corresponding to transverse and longitudinal phonons. (The peaks occur in the short-wavelength region, which makes a major contribution to pairing; the dispersion law in this region deviates from the usual acoustic law dependence and is close to being rather flat; this leads directly to a peaked structure of $F(\Omega) \propto dq/d\Omega$). A further important check on this procedure can be provided using inelastic neutron scattering measurements to determine the phonon density of states $F(\Omega)$. These measurements are not related to superconductivity. Comparison of the tunneling and neutron scattering measurements can provide important information. In fact, one can compare the position of the peaks determined by these two different methods. The coincidence of these positions verifies the initial assumption that superconductivity in the material of interest is caused by the electron–phonon coupling, that is, pairing occurs by exchange of phonons. The dependence $\alpha^2(\Omega)F(\Omega)$ and the value of μ^* obtained by inverting the tunneling spectrum can be used to calculate T_c directly from Eqs. (1.2.4) and (1.2.6), or with the use of Eq. (1.2.3), and then Eqs. (1.2.8) and (1.2.11), which can then be compared with the experimental value.

This method was applied to many conventional superconducting elements (see Fig. 1.2) and compounds, and by virtue of the remarkable agreement between theory and experiment, the mechanism in most conventional superconductors has been proven to be the electron–phonon interaction, or as is usually stated the phonon mechanism.

1.3.1.2. *Tunneling studies of the cuprates*

It is very temping to use tunneling spectroscopy to study the nature of high-T_c superconductivity in the cuprates. However, there is a serious challenge. As we know the coherence length, which is defined as $\xi = \hbar v_F / 2\pi T_c$

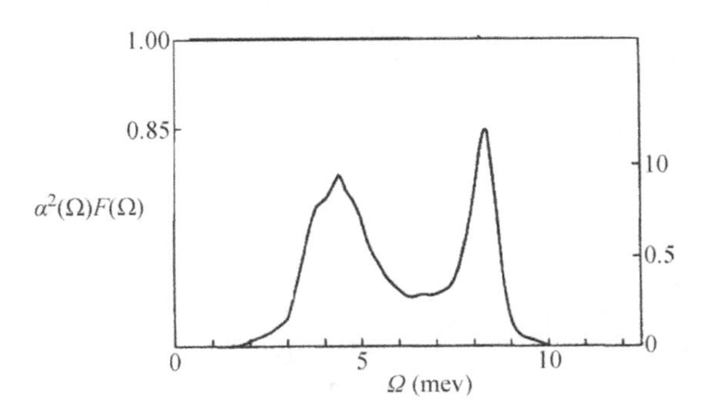

Fig. 1.2. Function $\alpha^2(\Omega)F(\Omega)$ for Pb.

(where v_F is the Fermi velocity), is an important parameter; its value characterizes the scale of pairing and can be visualized as the size of the pair. For usual superconductors, the value of ξ is rather large ($\sim 10^3$–10^4 Å), whereas for the cuprates it is quite small: $\xi \approx 15$–20Å. The length scale for providing the tunneling current at the interface between the superconductor and the insulator, that is, the depth over which the tunneling current originates is the pairing coherence length, and as noted above for conventional superconductors this is a large quantity that greatly exceeds the thickness of the surface layer. In the cuprates, the coherence length is very short, and this makes the measurements difficult. Nevertheless, such experiments were performed.

One of the first tunneling experiments [37] was carried out to study the yttrium–bariumcopper-oxide (YBCO) compound. Unfortunately, this paper has stayed mainly unnoticed by the high-T_c community. The inversion procedure carried out in this paper resulted in the dependence $\alpha^2(\Omega)F(\Omega)$, shown in Fig. 1.3. The calculated value of the critical temperature was $T_c \approx 60$ K. This value lies below the experimental one, but is still quite high. In addition, the experimentally measured (via neutron scattering) peak in the phonon density of states is somewhat below the peak position for the function $\alpha^2(\Omega)F(\Omega)$ obtained from the inversion procedure. Such a difference might reflect the presence of some additional mechanism, or perhaps is caused by a pair-breaking effect [38, 39] that has not been considered in the analysis. We believe that this effect is caused by magnetic

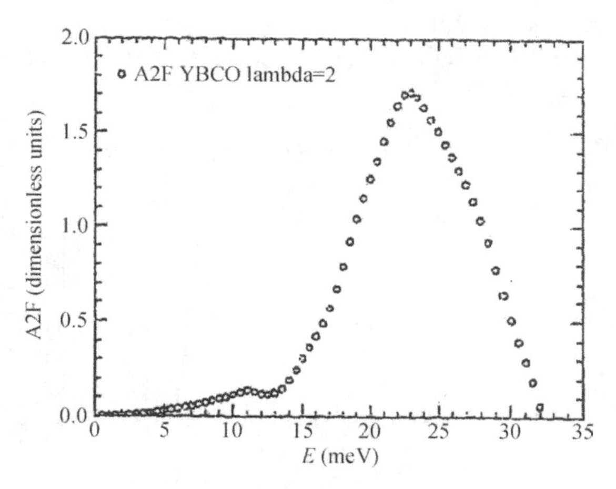

Fig. 1.3. Function $\alpha^2(\Omega)F(\Omega)$ for YBCO. From Ref. [37].

impurities. As we know, the pair is formed by two electrons with opposite momenta and opposite spin. Each localized magnetic moment (magnetic impurity) is trying to align the spins in the same direction, and this destroys the pairing. The presence of such broken pairs leads to the appearance of a gapless spectrum. Indeed, YBCO, contrary to conventional materials, does not display a sharp gap structure; its spectrum is rather gapless. In connection with this, it is interesting to note that Dynes *et al.* [37], by applying an external magnetic field, induced gaplessness in Pb. They carried out an analysis using the inversion procedure and observed a result (with proper scaling) similar to that observed for YBCO.

Break junction tunneling spectroscopy, which provides a high-quality contact, was employed by Aminov *et al.* [40] and Ponomarev *et al.* [41]. They demonstrated (Fig. 1.4) that the current-voltage characteristic for the $Bi_2Sr_2CaCa_2O_8$ compound contains an additional substructure that strongly correlates with the phonon density of states; the phonon density of states was obtained by Renker *et al.* [42, 43] using inelastic neutron scattering. Such a correlation is a strong indication of the importance of the electron–phonon interaction.

The tunneling conductance of $Bi_2Sr_2CaCa_2O_8$ was measured by Shiina *et al.* [44], Shimada *et al.* [45], and Tsuda *et al.* [46]. They also observed a correspondence of the peaks in d^2I/dV^2 and the phonon density of states. Moreover, the McMillan-Rowell inversion was performed, and the

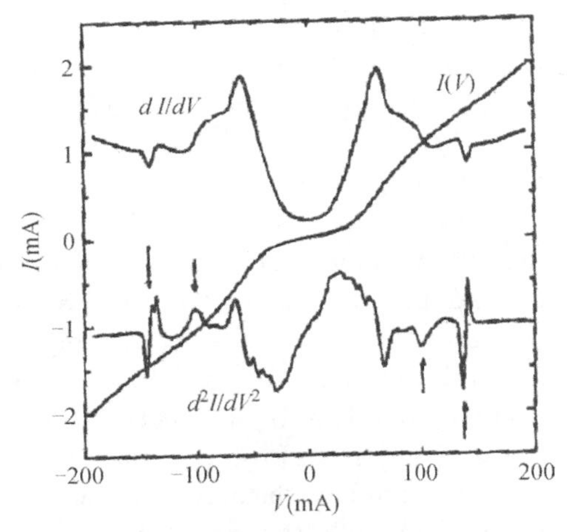

Fig. 1.4. $I(V), dI/dV$, and d^2I/dV^2 characteristics for a BSCCO break junction. From Ref. [40].

result was supportive of the electron–phonon scenario. The spectral function $\alpha^2(\Omega)F(\Omega)$ contains two groups of peaks at $\Omega \approx 15$–20 and ≈ 30–40 meV. The positions of the peaks corresponds with a high degree of accuracy to the structure of the phonon density of states. The coupling constant λ appears to be equal to about 3.5; such strong coupling is sufficient (see Eq. (1.2.11)) to provide the observed value of $T_c (T_c \approx 90$ K).

Large values of $2\Delta/T_c \gtrsim 10$, which greatly exceed the conventional values, observed for the underdoped sample [47] can be explained by the fact that the energy gap persists above T_c up to T_c^* ("pseudogap" region) and is in agreement with recent data [48]. This effect is caused by an intrinsic inhomogeneity of the sample (see Section 6 and [49]).

Another tunneling technique that appears to be a powerful tool in many studies is scanning tunneling microscopy (STM). This method is widely used in order to obtain information about the local structure of the order parameter, its inhomogeneity, etc. This type of tunneling (STM) can also be used to perform a study that can probe the mechanism of high T_c. For example, Lee *et al.* [50] carried out an STM analysis of the $Bi_2Sr_2CaCu_2O_{8+\delta}$ compound. As a part of the study, measurements of the tunneling current and its second derivative d^2I/dV^2 were performed. The locations of the peaks in the second derivative coincide with the position of specific phonon modes. This is a strong indication of the importance of electron–phonon coupling. Of course, a complete analysis requires the inversion procedure, which so far has not been carried out. Lee *et al.* stated that they are planning to perform this procedure; perhaps soon it will be done. Recently a similar correlation between the tunneling and Raman data for LaSrCuO was observed by Shim *et al.* [51].

A detailed STM study of the three-layer $Bi_2Sr_2CaCu_2O_{10+\delta}$ compound was performed recently by Levy de Castro *et al.* [52]. They concluded that it is necessary to take into account the band structure of the material, and especially the presence of the van Hove singularity, that is, the cusp in the electronic density of states that often appears in compounds with lower-dimensional substructural units (e.g., planes and/or chains). The interaction with some collective mode is also an essential factor in the analysis. These factors allow us to describe the observed features in the conductivity such as the dip asymmetry as well as the observed dip-hump structure [53]. Phonons could provide such a collective mode, but some magnetic excitations could do the same. Additional measurements can determine the exact nature of the mode. As a whole, tunneling spectroscopy continues to be a powerful and promising tool.

1.3.2. *Infrared Spectroscopy*

A new method based on precise infrared measurements can be used to reconstruct the function $\alpha^2(\Omega)F(\Omega)$. This method was proposed by Little and collaborators (for a description, see Ref. [54]) and is based on the so-called thermal-difference-reflectance spectroscopy. This method was demonstrated by Holcomb *et al.* [55–57] and allows one to determine the function $\alpha^2(\Omega)F(\Omega)$ for an energy interval that is larger than that in the tunneling method. The reflectivity of the sample was measured with a high degree of precision at different temperatures, and the ratio of the difference relative to their sum was determined. The theoretical method developed by Shaw and Swihart [58] was used in order to perform the inversion for thallium-cuprate (2212) samples. The larger extent of the accessible energy range allowed Little *et al.* [54] to take into consideration the electronic modes whose energy lies noticeably higher than typical phonon energies.

As we know, Little [33] introduced the electronic mechanism of superconductivity in his pioneering work; see also Gutfreund and Little [59]. Many interesting and novel aspects of various electronic mechanisms were also described by Ginzburg [60] (see also [61]) and Geilikman [62, 63]. In these papers, pairing is provided not by phonons but by electronic excitations, e.g., by excitons (while a usual electronic excitation corresponds to an appearance of an electron at $E > E_F$ and a hole at $E < E_F$, an exciton can be viewed as a bound electron–hole state; see, e.g., [64]). It is important to note that the superconducting state can benefit from the large energy scale characteristic of the electronic mechanism.

According to Little *et al.* [65], the superconducting state in the cuprates is caused by both phonon and electronic contributions, and each of them is of key importance. The phonon contribution is characterized by an intermediate coupling constant ($\lambda_{\text{phon}} \approx 0.9$). In addition, there are two electronic peaks at higher energies with the strengths $\lambda_{1.2\text{ eV}} \approx 0.1$ and $\lambda_{1.7\text{ eV}} \approx 0.3$. This combination can provide the observed high values of T_c. The excitoniclike excitations, namely, the d–d transitions of the Cu ions, are the electronic excitations of interest. Resonant inelastic X-ray emission spectroscopy was employed to confirm the presence of such excitations.

For our purpose it is important to note that although the electronic mechanism in this scenario is playing an important role, the contribution of the electron–phonon interaction is essential to obtain the high value of T_c observed.

1.3.3. *Photoemission and Ultrafast Electron Spectroscopy*

After the discovery of high-T_c cuprates, the photoemission technique was developed as a powerful tool used to obtain information about the energy spectrum and electronic structure of these novel materials. Photoemission experiments indicating the presence of substantial electron–phonon coupling were published by Lanzara *et al.* [66]. They studied different families of hole-doped cuprates, Bi2212, LSCO, and Pb-doped Bi2212, and they investigated the electronic quasi-particle dispersion relations. A kink in the dispersion around 50–80 meV was observed. This energy scale corresponds to the energy scale of some high-energy phonons; it is much higher than the energy scale for the pairing gap. Such a kink cannot be explained by the presence of a magnetic mode, because such a mode does not exist in LSCO, while the kink structure was also observed in this cuprate.

The structure observed by photoemission is consistent with the data on the phonon spectrum obtained by neutron spectroscopy. These measurements were also used in order to obtain a crude estimate of the electron–phonon coupling, since the quasi-particle velocity in the low-temperature region is renormalized by the electron–phonon coupling constant λ: $v = v_b(1 + \lambda)^{-1}$, where v_b is the bare (unrenormalized) velocity, which corresponds to the high-temperature region. This estimate indicates substantial electron–phonon coupling.

A different type of spectroscopy, so-called ultrafast electron crystallography, was employed by Gedik *et al.* [67]. The $La_2CuO_{4+\delta}$ compound was used and doping by photoexcitation was performed. It is interesting to note that the number of photon-induced carriers per copper site was close to the density of chemically doped carriers in the superconducting compound. The study of time-resolved relaxation dynamics demonstrated the presence of transitions to transient states which are characterized by structural changes (noticeable expansion of the c axis). Such a large effect on the lattice caused by electronic excitations is a strong signature of the electron–lattice interaction.

1.3.4. *Isotope Effect*

The isotope effect played an important role in understanding superconductivity. This effect manifests itself in the dependence of T_c on the ionic mass. This dependence has the form

$$T_c \propto M^{-\alpha}, \tag{1.3.2}$$

where M is the ionic mass and α is the so-called isotope coefficient. If we neglect μ^* and consider the simplest case of a monatomic lattice, then according to Eq. (1.2.1), $\alpha = 0.5$, since $T_c \propto \tilde{\Omega} \propto M^{-1/2}$. Note that the pure electronic or magnetic mechanisms of pairing do not involve participation of the lattice, and therefore do not contribute to the isotope effect.

The isotope effect [68, 69] provided strong evidence that the electron–lattice interaction is involved in the formation of Cooper pairs. However, the isotope effect is a very complex phenomenon, and it is difficult to carry out a quantitative analysis that determines the degree of involvement of the lattice in the formation of the superconducting state and/or its contribution relative to other possible mechanisms. Indeed, there are many other factors that can affect the value of the isotope coefficient α. Among them are the Coulomb pseudopotential μ^*, which depends explicitly on phonon frequency. Anharmonicity of the lattice is an another factor that can lead even to negative values of α. This situation becomes even more complicated if the material contains several varieties of ions, and this is exactly the situation for compounds and alloys [70]. Inhomogeneity of the sample, e.g., the coexistence of normal metal and superconducting regions (proximity effect), also strongly affects the isotopic dependence [71]. The presence of pair breakers, e.g., magnetic impurities (for the d-wave case even nonmagnetic impurities act as pair breakers), is another factor [71, 72]. A peculiar polaronic effect can also manifest itself in an isotopic dependence; this effect will be discussed in Section 4.

It is interesting to note that the isotope effect has been observed in the cuprates and its temperature dependence is a peculiar one (see, e.g., [73–75]). More specifically, the value of α is relatively small at optimum doping, but increases with decreasing doping up to values that are even larger than that in the BCS theory. We discuss this feature in Section 4. But as described above, it is hard to draw any quantitative conclusion based solely on the value of α.

It is interesting to note that not only T_c but other quantities can also display an isotopic dependence. Among them is the penetration depth (see Section 4). We mentioned above (Section 1.3.3) that the electron–lattice interaction manifests itself in a peculiar behavior of the phonon dispersion curve. This was detected using the photoemission technique. According to Gweon *et al.* [76], the isotope substitution $O^{16} \rightarrow O^{18}$ strongly affects this dispersion curve. However, the latest study by Douglass *et al.* [77] showed a much smaller impact of the isotopic substitution, which is more consistent with the STM data by Lee *et al.* [50]; see Section 1.3.1.2.

For our purposes, it is important to realize that the isotope effect strongly indicates that the ionic system and the electron–lattice interaction are involved in the formation of the superconducting state in the cuprates. As for a quantitative analysis, this should be carried out with considerable care, because there are many factors affecting the isotopic dependence. As a result, other techniques such as tunneling spectroscopy can provide more substantial information about the nature of the pairing and interplay of various contributions.

1.3.5. *Heat Capacity*

A study of thermodynamics properties can also provide information about the pairing mechanism. This is due to the fact that the effective mass and the electronic heat capacity are renormalized by the electron–phonon interaction (see, e.g., [18, 78]). Namely, $m^* = m_b[1 + \lambda(T)]$; here m^* and m_b are the values of the effective mass and band mass, respectively. Also, the Sommerfeld constant γ is given by $\gamma \equiv \gamma(T) = \gamma_0[1 + 2\int d\Omega\alpha^2(\Omega)F(\Omega)\Omega^{-1}g(T/\Omega)]$, where $g(x)$ is the universal function; $g \rightarrow o$ if $T \gg \tilde{\Omega}$, and $g = 1$ at $T = 0$ K, so that $\gamma(0) = \gamma_0(1+\lambda)$; see Eq. (1.2.2). The presence of the second term in the expression for $\gamma(T)$ reflects the fact that moving electrons become "dressed" by the phonon cloud. As the temperature increases, the "cloud" becomes weaker, so that $\gamma(T)$ decreases. As a result, the measurements of electronic heat capacity at high temperatures and in the low temperature region can be used to evaluate the value of the electron–phonon coupling constant which determines T_c. Such measurements were performed by Reeves *et al.* [79] for the YBCO compound. The main challenge was to evaluate the electronic contribution to the heat capacity at high temperatures where the heat capacity is dominated by the lattice. The lattice contribution was calculated using the phonon density of states obtained by neutron scattering. As a result, the value of $\lambda > 2.5$ was obtained, which means that there is strong electron–lattice coupling sufficient to provide high T_c.

1.4. Polaronic Effect

1.4.1. *Polarons and Isotope Effects*

A strong electron–lattice interaction could lead to specific polaronic effects. The concept of polarons was introduced and studied by Pekar [80] and Pekar and Landau [81]. A polaron can be created if an electron is added to the crystal with a small carrier concentration (see, e.g., [7]). Because of strong local electron–ion interactions, the electron appears to be trapped and can be viewed as being dressed in a "heavy" ionic "coat". In reality, we are dealing with a strong (nonlinear) manifestation of the electron–lattice interaction.

The concept of polarons is an essential ingredient of the physics of high-T_c oxides. In fact, the formation of a Jahn–Teller polaronic state was a main motivation for Bednorz and Muller to search for superconductivity in these systems, and this led to their breakthrough discovery. They gave a significant amount of credit to the paper on Jahn–Teller polarons by Hock *et al.* [82].

The formation of polaronic states is a strong nonadiabatic phenomenon. As we know, the usual adiabatic method (Born–Oppenheimer approximation [83]; see also [19, 84, 85]) allows us to separate electronic and ionic motions. Indeed, this approximation is based on the fact that in metals the ionic motion is much slower than the motion of electrons (the inequality $\tilde{\Omega}/E_F \ll 1$ is a condition of applicability of this approximation), and it allows us as a first step to neglect the kinetic energy of the ions and to study the electronic structure for a "frozen" lattice. The electronic energy (electronic terms) appears to be a function of the ionic positions $[\varepsilon_{\rm el} \equiv \varepsilon_n(\vec{R})]$. Next, one can study the ionic dynamics; it turns out that the electronic terms $\varepsilon_n(\vec{R})$ form the potential for the ionic motion. The total wave function Ψ can be written as a product: $\Psi = \psi_{\rm el}\phi_{\rm ionic}$. However, such a separation of electronic and ionic terms is impossible for polaronic states. Speaking of the high-T_c cuprates, it is important that oxygen ions actively participate in the formation of such states. Note that these ions play a unique role in the lattice dynamics, because they are the lightest elements

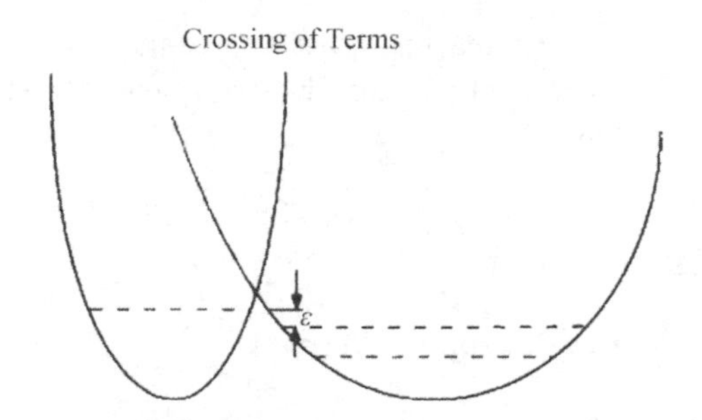

Fig. 1.5. Electronic terms (diabatic representation).

in the cuprates. Polaronic effect increases the phase space for pairing virtual transitions [86].

The implications for the isotope effect because of the presence of polaronic (bound electron–ionic) states was described by Kresin and Wolf [87].

One can assume that an oxygen ion is characterized not by the usual local minimum of the potential, but rather by two closely spaced minima (Fig. 1.5; "doublewell" structure). Note that the "double-well" structure is a characteristic feature for both the in-plane and apical ions (see Fig. 1.1). Such a double-well structure has been observed experimentally using the X-ray-absorption finestructure technique [88]; see Fig. 1.6.

Note that the double-well structure is a result of the crossing of electronic terms. The ionic configuration at this crossing corresponds to a degeneracy of the electronic states, which is a key ingredient of the Jahn–Teller effect (see, e.g., [89]).

We start with an apical oxygen. The dynamics of the apical oxygen ions plays an essential role in these compounds (see, e.g., [90]). The cuprates are doped materials, and because of it charge transfer through this ion is an important factor. One can show [87, 180] that the doping and therefore the carrier concentration are affected by an isotopic substitution. Since the value of T_c depends strongly on carrier concentration $[T_c \equiv T_c(n)]$, we are dealing with a peculiar isotopic dependence of T_c. If the charge transfer occurs in the framework of the usual adiabatic picture, so that only the carrier motion is involved, then the isotope substitution does not affect the forces and therefore does not change the charge-transfer dynamics. However, strong non-adiabaticity changes the picture rather dramatically. The electronic and nuclear motions are not separable, and in this case the charge transfer is a more complex phenomenon that does involve nuclear motion.

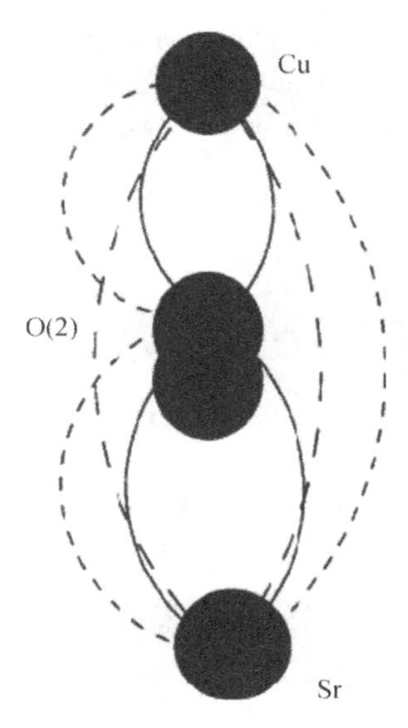

Fig. 1.6. "Double-well" structure for the apical oxygen. From Ref. [88].

The presence of two close minima means that the degree of freedom describing the ionic motion corresponds to electronic terms crossing (see Fig. 1.5). The charge transfer in this case is described by polaronic motion, that is, by the motion of the nearly bound electron–ionic unit (this can be described as a dynamic polaron). Note that a similar effect leads to the isotope effect in manganites [91].

Qualitatively, the charge transfer for such non-adiabaticity can be visualized as a multistep process: first the carrier makes a transition from the chain site to the apical oxygen, then the apical oxygen transfers to another term (see Fig. 1.5), and this is finally followed by the transition of the carrier to the plane. The second step is affected by the isotope substitution. For the entire crystal, it can be viewed as the motion of a polaron (dynamic polaron).

In order to describe this phenomenon, it is convenient to use a so-called "diabatic" representation (see, e.g., [92, 93]). In this representation, we are dealing directly with the crossing of electronic terms. The operator $\hat{H}_{\mathrm{el}} = \hat{T}_{\vec{r}} + V(\vec{r}, \vec{R})[\vec{T}_{\vec{r}}$ is a kinetic energy operator, $V(\vec{r}, \vec{R})$ is a total potential energy, and $\vec{r}$ and $\vec{R}$ are the electronic and nuclear coordinates, respectively] has non-diagonal terms (unlike the usual adiabatic picture when $\hat{H}_{\mathrm{el}}$ is diagonal). The charge transfer in this picture is accompanied by the transition to another electronic term. Such a process is analogous to the Landau–Zener effect (see Ref. [89]).

The total wave function can be written in the form

$$\Psi(\vec{r},\vec{R},t) = a(t)\Psi_1(\vec{r},\vec{R}) + b(t)\Psi_2(\vec{r},\vec{R}). \tag{1.4.1}$$

Here,

$$\Psi_i(\vec{r},\vec{R}) = \psi_i(\vec{r},\vec{R})\Phi_i(\vec{R}), \quad i = \{1,2\}.$$

$\Psi_i(\vec{r},\vec{R}), \Phi_i(\vec{R})$ are the electronic and vibrational wave functions that correspond to two different electronic terms (see Fig. 1.5).

In the diabatic representation, the transition between terms is described by the matrix element V_{12}, where $\hat{V} \equiv \hat{H}_{\vec{r}}$. One can show that

$$V_{12} \cong L_0 F_{12}, \tag{1.4.2}$$

where $L_0 = \int d\vec{r}\Psi_2^*(\vec{r},\vec{R})\hat{H}_{\vec{r}}\Psi_1(\vec{r},\vec{R})|_{R_0}$ is the electronic constant (R_0 corresponds to the crossing configuration), and $F_{12} \int \varphi_2^*(\vec{R})\varphi_1(\vec{R})d\vec{R}$ is the so-called Franck–Condon factor. The presence of the Franck–Condon factor is a key ingredient of our analysis. Its value depends strongly on the ionic mass and, therefore, is affected by the isotope substitution. The calculation [87] leads to the following expression for the isotope coefficient:

$$\alpha = \gamma(n/T_c)\partial T_c/\partial n, \tag{1.4.3}$$

where γ has a weak logarithmic dependence on ionic mass M. Therefore, the polaronic isotope effect ($\alpha \equiv \alpha_{\mathrm{ac}}$; α_{ac} corresponds to the apical oxygen ion) is determined by the dependence of T_c on n, where n is the carrier concentration. A strong non-adiabaticity (the apical oxygen in YBCO is in such a non-adiabatic state) results in a peculiar polaronic isotope effect.

The impact of the isotope substitution $O_{16} \rightarrow O_{18}$ on the in-plane oxygen ($\alpha \equiv \alpha_p$) looks different. The corresponding vibrational mode is directly affected by the isotopic substitution and thus makes a direct contribution to the pairing as in the normal isotope effect. In addition, the polaronic nature of the carriers in the planes also provides a novel isotope effect due to an increase ($O_{16} \rightarrow O_{18}$) in the carriers effective mass, which leads to a change in the value of T_c. Therefore, the polaronic effects are essential for both the in-plane and apical oxygen sites [94]. According to Eq. (1.4.3), at optimum doping $\partial T_c/\partial n = 0$ and, therefore, the apical oxygen ion does not make any contribution. In this case, the main contribution comes from the in-plane oxygen. This was confirmed by site-selected experiment [95]. One can expect that the value of α_{ac} increases for the region that is far from optimum T_c. It is important (see, e.g., Ref. [75]) for such experiments that the isotope effect should be measured on the same sample to guarantee that the doping level (oxygen concentration) is unchanged with the isotopic substitution.

Note that there is no one-to-one correspondence between the amount of oxygen and the in-plane carrier concentration. The in-plane carrier concentration can be affected by the isotopic substitution on the apical site. Because of the polaronic effect, the probability of tunneling becomes different and this leads to the redistribution of the total electronic wave function between the Cu–O plane and the charge reservoir.

The site-selected experiments have been performed for $Y_{1-x}Pr_xBa_2Cu_3O_{7-\delta}$ samples [96, 97]. The Pr substitution leads to a depression in T_c; the samples studied have $T_c \approx 44$ K. Both α_p and α_{ac} are large relative to their values at $x = 0$; the in-plane term α_p is larger than α_{ac}. This increase can be caused by mixed valence of the Pr ions as well as a pair breaking effect caused by magnetic moments on the Pr site. Indeed, pair breaking affects the value of the isotope coefficient [71, 72]. In connection with this, it would be interesting to carry out the site-selective experiments for samples with different oxygen contents.

The polaronic effect also leads to the possibility of observing an unusual isotopic dependence of the penetration depth, since this quantity also depends on the carrier concentration as well as on the effective mass. This effect was introduced theoretically by Kresin and Wolf [180] (see also [98]) and observed experimentally by Zech *et al.* [99] and Khasakov *et al.* [100]. The muon-spin rotation technique (see Ref. [101]) was employed; this method allows the direct determination of the penetration depth. According to recent experimental data [102, 103], the correlation between isotope effects on T_c and penetration depth can be explained by the interplay of both polaronic channels affecting the carrier concentration and effective mass. It is clear that these data demonstrate the importance of polaronic effects.

Another polaronic effect that also reflects the importance of the electron–lattice interaction was observed by Oyanagi *et al.* [104] using X-ray absorption spectroscopy. This method (see Refs. [104, 105]) reveals that doping leads to displacement of oxygen atoms, and this demonstrates the impact of the electron–phonon interaction. More specifically, Oyanagi *et al.* [104] measured the Cu–O radial distribution function. Upon cooling, a sharp decrease in this function at T_c was observed. Such a sharpening in the radial distribution function reflects the appearance of correlated motion of oxygen ions and is connected with the phase coherence of the electronic subsystem. Such a large impact of the pairing on the dynamics of the ions is caused by the fact that it is impossible to separate the electronic and ionic degrees of freedom, and again this corresponds to the propagation of a dynamic polaron.

1.4.2. *"Local" Pairs: Bipolarons, U Centers, and the BEC–BCS Scheme*

A bipolaron represents a local structure that can be viewed as a bound state of two polarons. This type of structure is supposedly caused by a very strong electron–lattice interaction. Therefore, the bipolaronic scenario represents an extreme case of electron–phonon (lattice) dynamics. It is interesting to note that a scenario of "local" pairs was proposed as an explanation of superconductivity even before the BCS theory [106]. A more rigorous concept of a bipolaron, which is a bound state of two polarons, was introduced by Vinetskii [107] and Eagles [108]. The qualitative picture of bipolaronic superconductivity is rather elegant and is very different from the conventional BCS concept. The main difference is the nature of the normal state. As we know, the starting point of the BCS picture is that in the normal state (above T_c or above the critical field) we are dealing with the usual fermions (delocalized electrons) and, correspondingly, with a Fermi surface. According to the bipolaronic picture, the normal state represents a Bose system formed by pairs of polarons: pairing occurs in real space. As a result, the nature of the phase transition at T_c is entirely different. According to the bipolaronic scenario, we are dealing with the Bose–Einstein condensation of bosons, whereas the formation of pairs (Cooper pairs) in usual superconductors occurs at T_c. The Cooper pair is formed by two electrons with opposite momenta, so that the pairs are formed in momentum, not real space.

A more detailed model of bipolaronic superconductivity, namely, the picture that the bosons (bipolarons) formed on a lattice could form a superconducting system, was proposed by Alexandrov and Ranninger [109]. A small value of the coherence length, along with a low carrier concentration, typical in the cuprates made the bipolaronic picture attractive. And, indeed, after the discovery of high-T_c cuprates several (see, e.g., Refs. [110–114] proposed such a picture and developed many of its aspects. However, Chakraverty *et al.* [115] later came to the conclusion that this scenario is not applicable to the cuprates, because of its incompatibility with experiments. The value of the effective mass that is required for the observed critical temperature appears to be drastically different from the observed one. Moreover, the bipolaronic picture requires a bosonic nature of the carriers. This factor is even more important, since it contradicts the existence of the Fermi surface that was established experimentally [116, 117]. Note that at present the evidence for the existence of a Fermi surface is even stronger (see, e.g., [118]).

Note also that the statement about the lattice instability leading to the formation of bipolarons at $\lambda \gtrsim 1$ was based on the usual Froelich Hamiltonian (see Section 1.2). According to rigorous adiabatic theory [26], this approach is valid only if $E_F \ll \tilde{\Omega}$, which is not the case for conventional superconductors, and is also not the case for the cuprates where $E_F \sim 1$ eV, $\tilde{\Omega} \sim 10 - 50$ meV.

A more general picture was described by Müller *et al.* [119]; see Ref. [120]. According to this approach, the high T_c compound contains two components: bipolarons and free fermions. The presence of free fermions explains the presence of the Fermi surface. As a whole, the model describes many experimental results.

A picture of negative U centers formed by two electrons localized on the same lattice site was introduced by Anderson [121] to study amorphous semiconductors. The appearance of such centers is caused by a strong local electron–lattice interaction. After the discovery of the high T_c cuprates, it was suggested [122] that the presence of U impurities can result in a large increase of T_c. The theoretical study by Oganesyan *et al.* [123] demonstrated that the U centers can provide the resonant tunneling channel between the CuO_2 layers (see Ref. [124]).

Another interesting approach is concerned with a scenario that is intermediate between the Bose–Einstein condensation (BEC) and Cooper pairing (BCS). Such a generalization was considered initially by Leggett [125] and later by Nozieres and Schmitt-Rink [126] and Nozieres [127]. The properties of a Fermi gas with an attractive potential have been studied as a function of the coupling strength. BEC and BCS cases correspond to two limits (strong and weak coupling). It is remarkable that the evolution between these two limits is smooth. Nozieres and Schmitt-Rink [126] studied not only the evolution of the ground state, but also the change in the transition temperature, and they stressed the importance of individual excitations for the Cooper pairing channel versus collective excitations for the BEC case.

All the examples described in this section are directly related to the impact of the lattice and the electronlattice interaction on the electronic subsystem and its superconducting state. The possibility of the appearance of local pairs and the impact of such factors as the presence of two components or U centers in the cuprates or other complex systems deserve additional theoretical and, especially, experimental study.

1.5. Phonon–Plasmon Mechanism

In this section, we discuss the phonon mechanism, which is combined with a peculiar plasmon contribution. Plasmons represent collective electronic modes; they can be visualized as collective electronic oscillations with respect to positive ionic background (see, e.g., Ref. [7]). For simple metals, the value of the plasmon frequency is rather high ($\approx 5{-}10\,\text{eV}$), and it depends weakly on momentum. Metals with a complex band structure display additional low-lying plasmon branches. The layered conductors also have a peculiar structure of their plasmon spectrum, and in this section we focus on this case.

The plasmon mechanism implies that pairing occurs via the exchange of plasmons; in other words, plasmons play a role similar to that of phonons. Here we discuss the situation when pairing is provided by contributions of both channels, that is, by phonons and plasmons.

The plasmon mechanism of superconductivity has been studied previously [129, 130, 179]. The interelectron coupling is provided by the acoustic plasmon branch; this mode corresponds to the collective motion of the light carriers with respect to the heavy ones (e.g., for the case of two overlapping different bands). For the cuprates such a channel was studied by Ruvalds [131]. Another possibility was studied by Ashkenazi *et al.* [132]. It has been proposed that a charge-density-wave instability will lead to softening of the plasmon branch, and this leads to strong pairing.

Here, we focus on the plasmon spectrum specific for layered conductors. This question is interesting not only for the study of the cuprates. Indeed, the past few years have witnessed the discovery of many new superconducting materials: high-temperature cuprates,

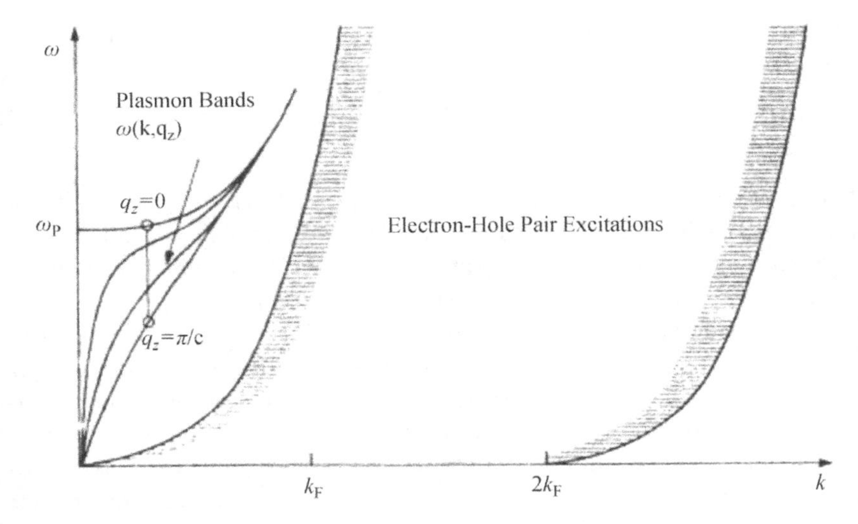

Fig. 1.7. Plasmon spectrum for a layered electron gas.

fullerides, borocarbides, ruthenates, MgB_2, metal-intercalated halide nitrides, intercalated Na_xCoO_2, etc. Systems such as organics, heavy fermions, and nanoparticles have also been studied intensively. Many novel systems belong to the family of layered (quasi-2D) conductors and are characterized by strongly anisotropic transport properties. One can raise an interesting question: Why is layering a favorable factor for superconductivity? One can show [133–135] that layering leads to a peculiar dynamic screening of the Coulomb interaction. Layered conductors have a plasmon spectrum that differs fundamentally from 3D metals. In addition to a high-energy "optical" collective mode, the spectrum also contains an important low-frequency part ("electronic" sound; see Fig. 1.7); see Refs. [136–138]. The screening of the Coulomb interaction is incomplete and the *dynamic* nature of the Coulomb interaction becomes important. The contribution of the plasmons in conjunction with the phonon mechanism may lead to high values of T_c.

We consider a layered system consisting of a stack of conducting sheets along the z axis separated by dielectric spacers. Because of the large anisotropy of the conductivity, it is a good approximation to neglect transport between the layers. On the other hand, the Coulomb interaction between charge carriers is effective both within and between the sheets. In order to calculate the superconducting critical temperature T_c, one can use the equations for the superconducting order parameter (cf. Eq. (1.2.6)),

$$\phi_n(\vec{k}) = T \sum_{m=-\infty}^{\infty} \int \frac{d\vec{k}'}{(2\pi)^3} \Gamma(\vec{k}, \vec{k}'; \omega_n - \omega_m) \left. \frac{\phi_m(\vec{k}')}{\omega_m^2(\vec{k}') + \xi_{k'}^2} \right|_{T_c} . \quad (1.5.1)$$

Here, $\phi_n(\vec{k}) = \Delta_n(\vec{k})Z, \omega_n(\vec{k}) = \omega_n Z$, and $\Delta_n \equiv \Delta(\omega_n)$. We shall not write out the expression for Z. The interaction kernel Γ can be written as a sum of electron–phonon and Coulomb interactions. The Coulomb term contains the plasmon excitations and the usual static repulsion.

A detailed analysis [135] based on Eq. (1.5.1) shows that the impact of dynamic screening is different for various layered systems. For example, for the metalintercalated halide nitrides (see, e.g., [139]) the plasmon contribution dominates. As for the cuprates, the plasmon contribution is not so crucial but is noticeable: about 20% of the observed value of T_c is due to acoustic plasmons. The main role is played by phonons, and their impact leads to a high value of T_c.

1.6. Electron–Phonon Interaction and the "Pseudogap" State

A study of the "pseudogap" state of the high-T_c cuprates has attracted much interest. This issue is very interesting and is still controversial. As we know, the superconducting state of usual superconductors is characterized by zero resistance, anomalous diamagnetism, which strongly depends on temperature, by an energy gap, etc. These features are absent above T_c, in the normal state (except for the effect of fluctuations near T_c; see, e.g., [140]). The situation for the cuprates, especially in the underdoped state, is entirely different. According to many experimental results, one can observe, above T_c, along with normal resistance such properties as an energy gap, anomalous diamagnetism, isotopic dependence of the pseudogap temperature T_c^*, a "giant" Josephson effect, etc., that is, many features that are characteristic of a superconducting state.

It is important to realize that, because of doping (carriers are added by substitution or non-stoichiometry), we are dealing with an intrinsically inhomogeneous system. As a result, the compounds display phase separation [141] (see also, e.g., [142]), that is, the coexistence of metallic and insulating phases. According to our approach [143, 144], upon cooling below some characteristic temperature T_c^*, the metallic phase becomes inhomogeneous and represents a mixture of superconducting and normal regions. As temperature decreases toward T_c, the size of the superconducting regions and their number increase. At $T = T_c$ one can observe the percolative transition, that is, the formation of macroscopic superconducting regions. Such a picture was directly observed by Iguchi *et al.* [145] using the STM technique with magnetic imaging.

Recently, Gomes *et al.* [48], using a specially designed variable temperature STM, observed that pairing occurs initially in small regions and can persist at temperatures that greatly exceed the resistive T_c (for $Bi_2Sr_2CaCu_2O_{8+\delta}$, the superconducting nanoregions were observed at $T \approx 160$ K). The observation confirms our predictions. Recent bulk μSR data [146] also support our picture.

One might think that the inhomogeneous nature of the cuprates is an important feature, but it is not directly relevant to the pairing mechanism. However, recent experiments by Gomes *et al.* [48] appear to be important also from this point of view. They measured local values of T_c and the gap; it has been observed that the ratio $2\Delta/T_c$ is rather large [$2\Delta/T_c \approx 8$; here $\Delta \equiv \Delta(0)$ is the energy gap at $T = 0$ K]. Such a large value corresponds to the strong-coupling case and is consistent with the electron–phonon scenario for pairing. Indeed, the ratio $2\Delta/T_c$ is directly related to the strength of the interaction. According to the BCS theory (weak coupling, $\lambda \ll 1$), this ratio is universal and given by $2\Delta/T_c = 3.52$. An increase in λ leads to an increase in this ratio. For example, for Pb $(\lambda \approx 1.5), 2\Delta/T_c \approx 4.3$, and for $Pb_{0.7}Bi_{0.3}$ $(\lambda \approx 2), 2\Delta/T_c \approx 4.85$ (see Ref. [11]). The ratio can be calculated using the general equation [5, 128, 147, 148]

$$2\Delta/T_c = 3.52[1 + 5.3(T_c/\tilde{\Omega})^2 \ln(\tilde{\Omega}/T_c)]. \tag{1.6.1}$$

According to Kresin [149], the ratio $2\Delta/T_c$ for strong coupling lies above the BCS value $(2\Delta/T_c)_{\text{BCS}} = 3.52$ and below the upper limit $(2\Delta/T_c)_{\max} = 13.4$. The measured value of $2\Delta/T_c = 8$ corresponds to strong electron–phonon coupling with values of $\lambda \approx 3 - 3.5$, which are quite large and are sufficient to explain the observed value of T_c. It is interesting to note that this value was observed in a system that contains nanoregions (see Section 1.8).

A large isotope effect on $T_c^*(T_c^* \propto M^{-\alpha}, \alpha \approx -2.2 \pm 0.6)$ has also been observed [150, 151]. This can be explained by the presence of superconducting regions [49] and by the polaronic effect (see Section 1.4), and can be described by a relation similar to Eq. (1.4.3), that is,

$$\alpha = \gamma(n/T_c^*)\partial T_c^*/\partial n. \tag{1.6.2}$$

The experimental observation of the isotope effect on T_c^* also reflects the fact that superconducting pairing persists above the resistive transition. It is interesting to note that the experimentally measured isotope coefficient has a negative sign. This can be explained by Eq. (1.6.2) and by the fact that an increase in doping in the underdoped region leads to a decrease in the value of T_c^* (at optimum doping $T_c^* \cong T_c$); as a result, $\alpha < 0$.

1.7. Proposed Experiment

All experiments described above (Sections 1.3 and 1.6) are interesting and informative and provide strong evidence for the contribution of the electron–phonon interaction to the superconductivity for many of the newly discovered superconductors, especially the cuprates. However, one can propose a different experiment [152, 153] that will allow the unambiguous determination of the coupling boson (excitation) in the cuprate superconductors. This method is based on the generation and detection of the appropriate boson and is analogous to the experiments on the generation of phonons by conventional BCS superconductors.

The method is based on the technique of using Josephson junctions for the generation of phonons [154–157]. One can modify this technique for any boson contributing to the pairing. The generation of excitations caused by pair recombination can be used as a signature of the mechanism of pairing. A non-equilibrium superconducting state is formed by incoming radiation. The creation of excited quasi-particles is followed by a relaxation process. By the end of this process, a noticeable number of quasi-particles are concentrated at or very near the energy gap edge, $\varepsilon \approx \Delta$, where Δ is the pairing gap. The final stage of relaxation is the recombination of Cooper pairs. For conventional superconductors, this stage is accompanied by radiation of phonons.

In a classic experiment [154, 155], the generation and detection of phonons propagating through a sapphire substrate was demonstrated using two Josephson junctions located diametrically on opposite sides of a cylindrical sapphire block. This pioneering work was followed up by several investigations that developed an understanding of the details of the spectroscopy of the phonons generated and detected by similar means. The time and energy distribution of the phonons that were emitted were studied by such experiments. The study was aimed at the generation of almost monochromatic phonons. We now look at such experiments from a different point of view. Indeed, these experiments were possible only because phonons were responsible for pairing in the electrodes of the emitting junction and are thus emitted when quasi-particle excitations relax to the gap edge and recombine to form pairs. In other words, one can observe the recombination of electrons with energies near the gap edge; these electrons can form Cooper pairs, and this process is accompanied by radiation of phonons with $\hbar\omega \approx 2\Delta$.

One can raise the following question: Why are other excitations not radiated, only phonons? The answer is obvious and directly reflects the fact that phonons form the glue for pairing. In fact, radiation of phonons created by recombination is an additional support for the phonon mechanism of pairing in conventional superconductors. If pairing is provided, e.g., by magnetic excitations, the recombination would be accompanied by radiation of magnons.

One can propose a series of experiments analogous to these pioneering efforts; such experiments can provide an unambiguous determination of the appropriate boson responsible for superconductivity in the cuprates. It is crucial that the proposed experiments can distinguish between phonon and non-phonon (e.g., magnon) coupling based on the selection of the propagation medium.

Assume for the moment that superconductivity in the cuprates is mediated by phonons. Then we propose the following experiment. On one side of a high-quality sapphire (or other nearly defect-free single-crystal substrate) one can prepare a Nb or NbN tunnel junction as a detector of phonons. This detector will be most sensitive to phonons that are above the gap energy 2Δ of the electrodes. Phonons with energy lower that the appropriate gap energy will be filtered out since they will not break pairs and will not be detected. On the other side of the substrate, we prepare a cuprate junction or weak link that can be biased into the normal state by a current or infrared pulse. After the current or light has been removed, the quasi-particles generated will relax very rapidly to the gap edge, and as they recombine to form pairs they will emit $2\Delta(T, \mathbf{k})$ phonons. The gap may be anisotropic so that the phonon energy will be dependent on where in $\mathbf{k}$ space the quasi-particles are located. In addition, many cuprates appear to have a gapless superconducting density of states. Although this density of states is peaked near some value of Δ, it is characterized by the presence of states all the way down to $E = 0$. In any event, a large number of phonons will have energies well above the gap of the detector, which is a conventional, low-T_c superconductor. In this case, we expect to see a very well defined signal similar to what was observed for conventional junctions. To calibrate the experiment, we propose that on the very same substrate (prior to the deposition of the cuprate junction) we prepare a Nb or NbN junction as the emitter and perform a replication of the original Eisenmenger and Dayem experiment to estimate the sensitivity. Thus, the observation of a signal from the cuprate, similar in magnitude and temporal behavior to that of the control junction, would be extremely strong evidence that phonons were the primary excitation from the recombination of excited quasi-particles.

The relaxation process of excited quasi-particles consists of several stages. As a result of electron–electron (first stage) and electron–phonon (second stage) collisions, a number of quasi-particles near the value $E = 2\Delta$ will appear. One can show [152] that this process is described by

$$\partial W/\partial t = -\lambda(\Delta/\Omega)^2 W^2. \tag{1.7.1}$$

Here, W is the a number of quasi-particles, λ is the electron–phonon coupling constant, and $\Omega \approx \Omega_D$. Using Eq. (1.7.1), it is easy to determine the function $W(t)$, and then the number of generated phonons ΔN,

$$\begin{aligned} \Delta N(t) &= aW(0)t[1 + at]^{-1}, \\ a &\equiv \lambda(\Delta/\Omega)^2 W(0). \end{aligned} \tag{1.7.2}$$

Here, $W(0) = W(t = 0); t = 0$ corresponds to the beginning of the second stage.

A significantly smaller signal would indicate that phonons were not the primary recombination excitation but might be secondarily produced by the decay of the primary boson. In this case, the energy of the secondary phonons will be much smaller than the gap in the cuprate junction and also smaller than the gap in the NbN, which would mean they would not be detected. Such a small signal would indicate that the pairing boson might be a spin fluctuation or magnon, which then could be confirmed by another series of experiments.

If we now assume that spin fluctuations are the primary pairing excitation, then we would replace the substrate that was a very good phonon propagator with a substrate that would not support the propagation of high-energy phonons but was magnetic and would be an excellent propagator of magnons. Perhaps single-crystal yttrium iron garnet (YIG) with appropriate impurities could be prepared into such a substrate. In fact, cuprate films have already been prepared on such YIG substrates. The same NbN junction would be placed on one side of this substrate and the same two experiments should be performed. The conventional emitter should give a very small signal, whereas the cuprate signal should be much larger, an indicator that magnons are the primary recombination excitation.

Note also that according to many experiments the energy range for phonons and magnons is similar. For example, for YBCO both the phonon and magnon spectra range from $E = 0$ up to $E = 40 - 50$ meV. Therefore, both channels are available for the relaxation process, and the dominance of one of them means that the electrons mainly interact with bosons (phonons or magnons) corresponding to this channel. In addition, since $2\Delta \approx E_{\rm ph} \approx E_{\rm magn}$, the pairing (interaction with virtual excitations) and the relaxation are governed by similar matrix elements.

A sharp signal will be observed if the superconducting oxides have a well-defined energy gap. From this point of view, the Nd-based cuprate and BaKBiO systems could be selected for the initial study. In accordance with Murakami *et al.* [158], the LaSrCuO compound also has a sharp gap and could represent a good candidate for this experiment as well. As for YBCO and other cuprates, they are usually characterized by a gapless spectrum. Nevertheless, the signal generated by the recombination can be detected. In addition, as noted above, for these materials $2\Delta \approx E_{\rm ph}$, and the appearance of phonons with a frequency similar to that for the virtual transitions is a strong indication of a key contribution of the phonon mechanism.

1.8. Superconducting State of Nanoclusters

1.8.1. *Nanoparticles: Size Quantization*

As noted above (Section 1.2), the electron–lattice interaction, in principle, can provide a high value of the critical temperature. This aspect of the interaction is apparent in various superconducting systems. For example, MgB_2, which has a relatively high $T_c \approx 42\,K$, is generally accepted to be a phonon-mediated superconductor.

As we know from our understanding of conventional superconductors, an increase in T_c can be achieved by an increase in the density of states at the Fermi level; this is natural, since the density of states enters as a factor in the expression for the coupling constant (see Eq. (1.2.7)). Historically, the highest value of T_c for conventional superconductors was observed in A-15 compounds. These large values are caused by the presence of a Van Hove singularity in the density of states (DOS), that is, by a sharp peak in the DOS at the Fermi level [159].

It has also been observed [160, 161] that the T_c of Al films ($\sim$ 2.1 K) can be nearly double the value for bulk samples. Even larger increases ($T_c \approx$ 3 K) were observed for granular Al [162]. These increases can be explained by size quantization and corresponding increase in the effective density of states in films and isolated granules; this was explained by Kresin and Tavger [178] for films and by Parmenter [163] for granular structures.

The most distinctive feature of nanoparticles is the discrete nature of their electronic spectra. The superconducting state of nanoparticles has been studied by Tinkham *et al.* [164]; see also [165]. They studied nanoparticles that contained $N \approx 10^4 - 10^5$ delocalized electrons and were placed inside a tunneling barrier.

We focus below on smaller nanoparticles, so-called nanoclusters, with $N \approx 10^2 - 10^3$ delocalized electrons.

1.8.2. *Nanoclusters and the High-T_c State*

High-T_c, potentially up to room temperature, should be observed for specific metallic nanoclusters [49, 166, 167]. This attractive possibility could be realized thanks to a remarkable feature of metallic clusters, namely, the shell structure of their electronic spectra. This phenomenon was discovered by Knight *et al.* [168]. Initially the presence of a shell structure was observed for alkalimetal clusters. Later the presence of energy shells has been detected for many other nanoclusters including Al, Ga, Zn, Cd, and In (see [169]). The importance of the shell structure for superconductivity was discussed by Knight [170] and Friedel [171]. Friedel stressed the possibility of a large increase in T_c. The appearance of a superconducting state requires that $\delta E \lesssim \Delta$ [172], where Δ is the gap parameter and δE is the spacing between discrete electronic levels. One important aspect of the shell structure is that this criterion could be met.

As noted above, metallic clusters contain delocalized electrons whose states organize into shells, similar to those in atoms or nuclei (see, e.g., [173]). In some clusters, shells are completely filled all the way up to the highest occupied shell, e.g., those with $N = N_m = 20, 40, 58, 92, 138, 168, \ldots$. These values are known as "magic" numbers. Such clusters are spherical. The electronic states in such magic clusters are labeled by their orbital momentum l and radial quantum number n. Cooper pairs are formed by electrons with opposite projections of orbital momentum (such a pairing is similar to that in atomic nuclei; see, e.g., [174]). If the orbital momentum l is large, the shell is highly degenerate [$2(2l + 1)$ is large]. This factor drastically increases the effective density of states. In addition, the energy spacing ΔE between neighboring shells varies, and some of them are separated by only a small ΔE. One can show that the combination of high degeneracy and a small energy spacing between the highest occupied shell (HOS) and the lowest unoccupied shell (LUS) leads to the possibility of a large increase in the strength of the superconducting pairing interaction in the corresponding clusters. Qualitatively, this can be understood in the following way. If the HOS is highly degenerate, this means that the shell contains many electrons, which can be viewed as a sharp peak in the density of states at the Fermi level. An increase in the density of states leads to an increase in the value of the electron–phonon coupling constant; this can be seen directly from Eq. (1.2.7). As a result, one can obtain very large values of T_c. This situation is similar to that studied by Labbe *et al.* [159] for bulk materials; the presence of a peak in the density of states results in a noticeable increase in T_c.

The equation for the pairing order parameter $\Delta(\omega_n)$ has the following form [cf. Eq. (1.2.6)]:

$$\Delta(\omega_n)Z = \eta \frac{T}{2V} \sum_{\omega_{n'}} \sum_{s} D(\omega_n - \omega_{n'}) F_s^+(\omega_{n'}),$$

$$D(\omega_n - \omega_{n'}, \tilde{\Omega}) = \tilde{\Omega}^2[(\omega_n - \omega_{n'})^2 + \tilde{\Omega}^2]^{-1} \tag{1.8.1}$$

and

$$F_s^+(w_{n'}) = \Delta(\omega_{n'})[\omega_{n'}^2 + \xi_s^2 + \Delta^2(\omega_{n'})]^{-1}$$

are the vibrational propagator and the pairing function [15], respectively, $\xi_s = E_s - \mu$ is the energy of the sth electronic state referred to the chemical potential μ, V is the cluster volume, $\eta = \langle I^2 \rangle / M\tilde{\Omega}^2$ is the so-called Hopfield parameter [cf. Eq. (1.2.7)], and Z is the renormalization function.

Equation (1.8.1) contains a summation over all discrete electronic states. For magic clusters that have a spherical shape, one can replace the summation over states by summation over the shells $\sum_s \to \sum_j G_j$, where G_j is the shell degeneracy: $G_j = 2(2l_j + 1)$, where l_j is the orbital momentum. Then Eq. (1.8.1) can be written in the form

$$\Delta(\omega_n)Z = \lambda \frac{2E_F}{3N} \sum_{w_{n'}} \sum_{j} G_j \frac{\tilde{\Omega}^2}{\tilde{\Omega}^2 + (\omega_n - \omega_{n'})^2} \frac{\Delta^2(\omega_{n'})}{\omega_{n'}^2 + \xi_j^2} \Big| T_c. \qquad (1.8.2)$$

We used the expression for the bulk coupling constant $\lambda = \nu \langle I^2 \rangle / M\tilde{\Omega}_2$ [Eq. (1.2.7)], where E_F is the Fermi energy. Note that the characteristic vibrational frequency is close to the bulk value because pairing is mediated mainly by the short-wave part of the vibrational spectrum.

If the shell is incomplete, the cluster undergoes a Jahn–Teller deformation, so that its shape becomes ellipsoidal, and the states s are classified by their projection of the orbital momentum $|m| \leqslant l$, and each level contains up to four electrons (for $|m| \geqslant 1$). Note that in the weakcoupling case ($\eta/V \ll 1$ and correspondingly $\pi T_c \ll \tilde{\Omega}$), one should put in Eq. (1.7.1) $Z = 1, D = 1$, recovering the usual BCS scenario.

Equation (1.8.1) looks similar to the equation appearing in the theory of strong-coupling superconductivity, see Eq. (1.2.6), but is different in two key aspects. First, it contains a summation over discrete energy levels E_S, whereas for a bulk superconductor one integrates over a continuous energy spectrum (over ξ). Second, as opposed to a bulk superconductor, we are dealing with a finite Fermi system, so that the number of electrons N is fixed. As a result, the position of the chemical potential μ differs from the Fermi level E_F and is determined by the values of N and T.

It is essential that the value of the critical temperature T_c is determined by parameters that can be measured experimentally. These parameters are as follows: the number of valence electrons N and the energy spacing $\Delta E = E_L - E_H$. The magnitude of T_c for a given nanocluster depends on these parameters and on the values of λ_b, E_F, and $\tilde{\Omega}$, which are already known for each material. Remarkably, for perfectly realistic values of these parameters, a high value of T_c can be obtained. Consider, for example, a cluster with the following parameter values:

$$\Delta E = 65 \text{ meV}, \quad \tilde{\Omega} = 25 \text{ meV}, \quad m^* = m_e,$$
$$k_F = 1.5 \times 10^8 \text{ cm}^{-1}, \quad \lambda_b = 0.4,$$

radius $R = 7.5$ Å and $G_H + G_L = 48$ (e.g., $l_H = 7, l_L = 4$). For this set of values, one obtains $T_c \cong 10^2$ K. The large degeneracies of the HOS and the LUS play an important role. Qualitatively, these degeneracies increase the effective electron–vibrational coupling g_{eff} and, more specifically, the effective density of states. In principle, one can raise T_c up to room temperature.

If we consider specifically a Ga_{56} cluster (the Ga atom has three valence electrons, so that $N = 168$), one can use the values $\tilde{\Omega} \approx 270$ K, $\lambda_b \approx 0.4, m^* \approx 0.6m_e$, and $k_F = 1.7 \times 10^8$ cm^{-1}. The calculation leads to $T_c \approx 145$ K, which greatly exceeds the bulk value ($T_c^b \approx 1.1$ K). It is important to stress that these high values of T_c are caused by the electron–vibrational interaction.

A remaining question is how can one observe the appearance of pairing in an isolated cluster? Pairing leads to a strong temperature dependence of the excitation spectrum. Below T_c and especially at low temperatures close to $T = 0$ K, the excitation energy is strongly modified by the gap parameter and noticeably exceeds that in the region $T > T_c$. For example, the minimum absorption energy for Gd_{83} clusters at $T > T_c$ corresponds to $\hbar\omega \approx 6$ meV, whereas for $T \ll T_c$ its value is much larger: $\hbar\omega \approx 34$ meV. Such a large difference can be observed experimentally and is a manifestation of the superconducting state. It would be interesting to perform such experiments.

Recently, Cao *et al.* [175] used a specially developed technique [176] that allows one to measure the heat capacity of an *isolated* cluster. They observed jumps in heat capacity for selected Al clusters (e.g., for Al_{35}^- ions) at $T \approx 200$ K. The values of T_c as well as the amplitude of the jump and its width are in good agreement with the theory.

An anomalous diamagnetic moment can be also observed. In principle, a tunneling network of such nanoclusters can be built, and a macroscopic superconducting current could be observed.

1.9. Conclusion

In this colloquium, we have described as comprehensively as possible our view regarding the role of the electron–lattice (phonon) interaction in a number of novel superconducting systems, paying special attention to the cuprates. We have indicated how this interaction can give rise to high-temperature superconductivity, and we showed theoretically that even room temperature is possible within this framework. We have presented a variety of experimental observations that are consistent with this view. Furthermore, we have described a set of experiments for the cuprates that can provide an unambiguous answer to the question of the pairing boson. We hope that these experiments will be carried out in the near future.

Theoretically, the superconducting state can occur not only through the exchange by phonons, but also with the help of various bosons (e.g., of magnons). Only some experiments can dissociate between various channels and rule out those that do not provide any noticeable contribution. There has been sufficient experimental evidence for the importance of the electron–lattice (phonon) interaction. We think that the proposed experiments will provide additional crucial evidence for the concepts described above.

Acknowledgments

We are grateful to R. Dynes and H. Morawitz for interesting discussions, and to L. Friedersdorf for help in the preparation of the manuscript. The research of V. Z. K. is supported by DARPA.

References

[1] H. K. Onnes, *Leiden Commun.* 124C. (1911).
[2] L. Landau, and E. Lifshitz, *Statistical Physics* (Pergamon, Oxford, 1969), p. 1.
[3] G. Bednorz, and K. Müller, *Z. Phys. B: Condens. Matter* **64** (1986), 189.
[4] M. Wu, J. Ashburn, C. Torng, G. Peng, F. Szofran, P. Hor, and C. Chu, *Phys. Rev. Lett.* **58** (1987), 908.
[5] V. Kresin, and S. Wolf, *Fundamentals of Superconductivity* (Plenum, New York, 1990).
[6] D. Ginsberg, *Physical Properties of High Temperature Superconductors* (World Scientific, Singapore, 1994).
[7] N. Ashcroft, and N. Mermin, *Solid State Physics* (Holt, New York, 1976).

[8] J. T. L. Devreese, *Polarons* , in *Encyclopedia of Physics*, Vol. 2, edited by R. G. Lerner and G. L. Trigg (Wiley-VCH, Weinheim, 2005), pp. 2004–2027.
[9] J. Bardeen, L. Cooper, and R. Schrieffer, *Phys. Rev.* **108** (1957), 1175.
[10] P. Allen, and R. Dynes, *Phys. Rev. B* **12** (1975), 905.
[11] E. Wolf, *Principles of Electron Tunneling Spectroscopy* (Oxford University Press, New York, 1985).
[12] G. Eliashberg, *Sov. Phys. JETP* **13** (1961), 1000.
[13] G. Eliashberg, *Sov. Phys. JETP* **16** (1963), 78.
[14] A. Migdal, *Sov. Phys. JETP* **37** (1960), 176.
[15] L. Gor'kov, *Sov. Phys. JETP* **7** (1958), 505.
[16] A. Abrikosov, L. Gor'kov, and I. Dzyaloshinski, *Methods of Quantum Field Theory in Statistical Physics* (Dover, New York, 1963).
[17] D. Scalapino, *Superconductivity*, edited by R. Parks (Dekker, New York, 1969), p. 449.
[18] G. Grimvall, *The Electron-Phonon Interaction in Metals* (North-Holland, Amsterdam, 1981).
[19] V. Kresin, H. Morawitz, and S. Wolf, *Mechanisms of Conventional and High Tc Superconductivity* (Oxford University Press, New York, 1993).
[20] N. Bogoliubov, N. Tolmachev, and D. Shirkov. *A New Method in the Theory of Superconductivity* (Consultants Bureau, New York, 1959).
[21] I., Khalatnikov, and A. Abrikosov, *Adv. Phys.* **8** (1959), 45.
[22] P. Morel, and P. Anderson, *Phys. Rev.* **125** (1962), 1263.
[23] W. McMillan, *Phys. Rev.* **167** (1968), 331.
[24] Y. Browman, and Y. Kagan, *Sov. Phys. JETP* **25** (1967), 365.
[25] B. Geilikman, *J. Low Temp. Phys.* **4** (1971), 189.
[26] B. Geilikman, *Sov. Phys. Usp.* **18** (1975), 190.
[27] R. Dynes, *Solid State Commun.* **10** (1972), 615.
[28] V. Kresin, H. Gutfreund, and W. Little, *Solid State Commun.* **51** (1984), 339.
[29] C. Owen, and D. Scalapino, *Physica (Amsterdam)* **55** (1971), 691.
[30] V. Kresin, *Phys. Lett. A* **122** (1987), 434.
[31] S. Tewari, and P. Gumber, *Phys. Rev. B* **41** (1990), 2619.
[32] L., Bourne, A. Zettl, T. Barbee, and M. Cohen, *Phys. Rev. B* **36** (1987), 3990.
[33] W. Little, *Phys. Rev.* **156** (1967), 396.
[34] W. McMillan, and J. Rowell, *Phys. Rev. Lett.* **14** 1965, 108.
[35] W. McMillan, and J. Rowell, *Superconductivity*, Vol. 1, edited by R. Parks (Dekker, New York, 1969), p. 561.
[36] E. Burstein, and S. Lindqvist, *Tunneling Phenomena in Solids* (Plenum, New York, 1969).
[37] R. Dynes, P. Sharifi, and J. Valles, *Lattice Effects in High T_c Superconductors*, edited by Y. Bar-Yam, T. Egami, J. Mustre-Leon, and A. Bishop (World Scientific, Singapore, 1992).
[38] A. Abrikosov, and L. Gor'kov, *Sov. Phys. JETP* **12** (1961), 1243.

[39] V. Kresin, and S. Wolf. *Phys. Rev. B* **51** (1995), 1229.
[40] B. Aminov, M. Hein, G. Müller, H. Piel, Y. Ponomarev, D. Wehler, M. Boeckholt, L. Buschmann, and G. Guntherodt, *Physica C* **235–240** (1994), 1863.
[41] Y. Ponomarev, E. Tsokur, M. Sudakova, S. Tchesnokov, S. Habalin, S. Lorenz, S. Hein, S. Muller, S. Piel, and S. Aminov, *Solid State Commun.* **111** (1999), 513.
[42] B. Renker, F. Compf, E. Gering, N. Nucker, D. Ewert, W. Reichardt, and H. Rietschel, *Z. Phys. B: Condens. Matter* **67** (1987), 15.
[43] B. Renker, F. Compf, D. Ewert, P. Adelmann, H. Schmidt, E. Gering, and H. Hinks, *Z. Phys. B: Condens. Matter* **77** (1989), 65.
[44] Y. Shiina, D. Shimada, A. Mottate, Y. Ohyagi, and N. Tsuda, *J. Phys. Soc. Jpn.* **64** (1995), 2577.
[45] D. Shimada, N. Tsuda, U. Paltzer, and F. de Wette, *Physica C* **298** (1998), 195.
[46] N. Tsuda, D. Shimada, and N. Miyakawa, *New Research on Superconductivity*, edited by B. Martins (Nova Science, New York, 2007), p. 105.
[47] N. Miyakawa, J. Zasadzinski, L. Ozyuzer, T. Kaneko, and K. Gray, *IEEE Trans. Appl. Supercond.* **4** (2002), 47.
[48] K. Gomes, A. Pasupathy, A. Pushp, S. Ono, Y. Ando, and A. Yazdani, *Nature* **447** (2007), 569.
[49] V. Kresin, and Y. Ovchinnikov, *Phys. Rev. B* **74** (2006), 024514.
[50] J. Lee, K. Fujita, K. McElroy, J. A. Slezak, M. Wang, Y. Aiura, H. Bando, M. Ishikado, T. Mazur, J. Zhu, A. Balatsky, H. Eisaki, S. Uchida, and J. C. Davis, *Nature* **442** (2006), 546.
[51] H. Shim, P. Chauduri, G. Lovgenov, and I. Bozovic, *Phys. Rev. Lett.* **101** (2008), 247004.
[52] G. Levy de Castro, C. Berthod, A. Piriou, E. Giannini, and O. Fischer, *Phys. Rev. Lett.* **101** (2008), 267004; we are grateful to O. Fischer for sending this paper prior to its publication.
[53] C. Renner, and O. Fischer, *Phys. Rev. B* **51** (1995), 9208.
[54] W. Little, K. Collins, and M. Holcomb, *J. Supercond.* **12** (1999), 89.
[55] M. Holcomb, J. Coleman, and W. Little, *Rev. Sci. Instrum.* **64** (1993), 1862.
[56] M. Holcomb, J. Coleman, and W. Little, *Phys. Rev. Lett.* **73** (1994), 2360.
[57] M. Holcomb, C. Perry, J. Coleman, and W. Little, *Phys. Rev. B* **53** (1996), 6734.
[58] W. Shaw, and J. Swihart, *Phys. Rev. Lett.* **20** (1968), 1000.
[59] H. Gutfreund, and W. Little, *Highly Conducting Onedimensional Solids*, edited by J. Devresee, R. Evrard, and V. den Doren (Plenum, New York, 1979), p. 305.
[60] V. Ginzburg, *Sov. Phys. JETP* **20** (1965), 1549.
[61] V. Ginzburg, and D. Kirznits, *High Temperature Superconductivity* (Consultants Bureau, New York, 1982).
[62] B. Geilikman, *Sov. Phys. JETP* **48** (1965), 1963.

[63] B. Geilikman, *Sov. Phys. Usp.* **16** (1973), 17.
[64] P. Yu, and M. Cardona, *Fundamentals of Semiconductors* (Springer, Berlin, 1999).
[65] W. Little, M. Holcomb, G. Ghiringhelli, L. Braicovich, C. Dallera, A. Piazzalunga, A. Tagliaferri, and N. Brookes, *Physica C* **460** (2007), 40.
[66] A. Lanzara, P. Bogdanov, X. Zhou, S. Kellar, D. Feng, E. Lu, T. Yoshida, H. Eisaki, A. Fujimori, K. Kishio, J.-I. Shimoyama, J.-I. Noda, S. Uchida, Z. Hussain, and Z. Shen, *Nature* **412** (2001), 510.
[67] N. Gedik, D. Yang, G. Logvenov, I. Bozovic, and A. Zewait, *Science* **316** (2007), 425.
[68] E. Maxwell, *Phys. Rev.* **78** (1950), 477.
[69] C. Reynolds, B. Serin, W. Wright, and L. Nesbitt, *Phys. Rev.* **78** (1950), 487.
[70] B. Geilikman, *Sov. Phys. Solid State* **18** (1976), 54.
[71] V. Kresin, A. Bill, S. Wolf, and Y. Ovchinnikov, *Phys. Rev. B* **56** (1997), 107.
[72] J. Carbotte, M. Greeson, and A. Perez-Gonzales, *Phys. Rev. Lett.* **66** (1991), 1789.
[73] J. Franck, J. Jung, and A. Mohamed, *Phys. Rev. B* **44** (1991), 5318.
[74] J. Franck, *Physical Properties of High Temperature Superconductors*, edited by D. Ginsberg (1994, World Scientific, Singapore), p. 189.
[75] H. Keller, *Structure and Bonding*, Vol. 114, edited by K. Müller and A. Bussmann-Holder (Springer, Heidelberg, 2005), p. 143.
[76] G. Gweon, T. Sasagawa, S. Zhou, J. Graf, H. Takagi, D. Lee, and A. Lanzara, *Nature* **430** (2004), 187.
[77] J. Douglass, H. Iwasawa, Z. Sun, A. Fedorov, M. Ishikado, T. Saiton, H. Eisaki Bando, T. Iwase, A. Ino, M. Arita, K. Shimada, H. Namatame, M. Tanigushi, T. Masui, S. Tajima, K. Fujita, S. Uchida, Y. Aiura, and D. Dessau, *Nature* **446** (2007), E5.
[78] V. Kresin, and G. Zaitsev, *Sov. Phys. JETP* **47** (1979), 983.
[79] M. Reeves, D. Ditmars, S. Wolf, T. Vanderah, and V. Kresin, *Phys. Rev. B* **47** (1993), 6065.
[80] S. Pekar, *Zh. Eksp. Teor. Fiz.* **16** (1946), 341.
[81] S. Pekar, and L. Landau, *Zh. Eksp. Teor. Fiz.* **18** (1948), 419.
[82] K. Hock, H. Nickisch, and H. Thomas, *Helv. Phys. Acta* **56** (1983), 237.
[83] M. Born, and R. Oppenheimer, *Ann. Phys.* **84** (1927), 457.
[84] M. Born, and K. Huang, *Dynamic Theory of Crystal Lattices* (Oxford University Press, New York, 1954).
[85] I. Bersuker, *The Jahn-Teller Effect and Vibronic Interactions in Modern Chemistry* (Plenum, New York, 1984).
[86] V. Kresin, 2008, unpublished.
[87] V. Kresin, and S. Wolf, *Phys. Rev. B* **49** (1994), 3652.
[88] D. Haskel, E. Stern, D. Hinks, D. Mitchell, and J. Jorgenson, *Phys. Rev. B* **56** (1997), R521.

[89] L. Landau, and E. Lifshitz, *Quantum Mechanics* (Pergamon, Oxford, 1977).
[90] K. Müller, *Z. Phys. B: Condens. Matter* **80** (1990), 193.
[91] L. Gor'kov, and V. Kresin, *Phys. Rep.* **400** (2004), 149.
[92] T. O'Malley, *Phys. Rev.* **162** (1967), 98.
[93] C. Dateo, V. Kresin, M. Dupuis, and W. Lester Jr., *J. Chem. Phys.* **86** (1987), 2639.
[94] A. Bussmann-Holder, and H. Keller, *Eur. Phys. J. B* **44** (2005), 487.
[95] D. Zech, H. Keller, K. Conder, E. Kaldis, E. Liarokapis, N. Poulakis, and K. A. Müller, Nature (London) **371** (1994), 681.
[96] H. Keller, *Physica B* **326** (2003), 283.
[97] R. Khasanov, A. Shengelaya, E. Morenzoni, M. Angst, K. Conder, I. Savic, D. Lampakis, E. Liarokapis, A. Tatsi, and H. Keller, *Phys. Rev. B* **68** (2003), 220506.
[98] A. Bill, V. Kresin, and S. Wolf, *Phys. Rev. B* **57** (1998), 10814.
[99] D. Zech, K. Conder, H. Keller, E. Kaldis, and K. Müller, *Physica B* **219** (1996), 136.
[100] Khasanov, R., D. Eshehenko, H. Luetkens, E. Morenzoni, T. Prokcha, A. Suter, N. Garifianov, M. Mali, J. Roos, K. Conder, and H. Keller, *Phys. Rev. Lett.* **92** (2004), 057602.
[101] H. Keller, *IBM J. Res. Dev.* **33** (1989), 314.
[102] R. Khasanov, A. Shengelaya, K. Conder, E. Morenzoni, I. Savic, J. Karpinski, and H. Keller. *Phys. Rev. B* **74** (2006), 064504.
[103] H. Keller, 2008, private communication; we are grateful to H. Keller for discussion.
[104] H. Oyanagi, A. Tsukada, M. Naito, and N. Saini, *Phys. Rev. B* **75** (2007), 024511.
[105] A. Bianconi, N. Saini, A. Lanzara, M. Missori, T. Rossetti, H. Oyanagi, H. Yamaguchi, K. Oka, and T. Ito, *Phys. Rev. Lett.* **76** (1996), 3412.
[106] M. Schafroth, *Phys. Rev.* **100** (1955), 463.
[107] V. Vinetskii, *Sov. Phys. JETP* **13** (1961), 1023.
[108] D. Eagles, *Phys. Rev.* **186** (1969), 456.
[109] S. Alexandrov, and J. Ranninger, *Phys. Rev. B* **23** (1981), 1796.
[110] D. Emin, *Phys. Rev. Lett.* **62** (1989), 1544.
[111] A. Broyles, E. Teller, and B. Wilson, *J. Supercond.* **3** (1990), 161.
[112] R. Micnas, J. Ranninger, and S. Robaszkiewicz, *Rev. Mod. Phys.* **62** (1990), 113.
[113] S. Alexandrov, and A. Andreev, *Europhys. Lett.* **54** (2001), 373.
[114] S. Alexandrov, and N. Mott, *High Temperature Superconductors and Other Superfluids* (Taylor and Francis, London, 1994).
[115] B. Chakraverty, J. Ranninger, and D. Feinberg, *Phys. Rev. Lett.* **81** (1998), 433.
[116] H. Ding, T. Yokota, J. Campusano, T. Takahashi, M. Randeria, M. Norman, T. Mochiki, T. Kadovaki, and J. Giapintzakis, *Nature* **382** (1996), 51.

[117] D. Marshall, D. Dessau, A. Loeser, C. Park, A. Matsuura, J. Eckstein, I. Bozovic, A. Kapitulnik, W. Spicer, and Z. Shen, *Phys. Rev. Lett.* **76** (1996), 4841.
[118] N. Hussey, M. Abdel-Jawad, A. Carrington, A. Mackenzie, and L. Balicas, *Nature* **425** (2003), 814.
[119] K. Müller, G. Zhao, K. Conder, and H. Keller, *J. Phys.: Condens. Matter* **10** (1998), L291.
[120] K. Müller, *J. Phys.: Condens. Matter* **19** (2007), 251002.
[121] P. Anderson, *Phys. Rev. Lett.* **34** (1975), 953.
[122] H. Schuttler, M. Jarrell, and D. Scalapino, *Novel Superconductivity* , edited by S. Wolf and V. Kresin (Plenum, New York, 1987), p. 481.
[123] V. Oganesyan, S. Kivelson, T. Geballe, and B. Moyzhes, *Phys. Rev. B* **65** (2002), 172504.
[124] T. Geballe, *J. Supercond. Novel Magn.* **19** (2006), 261.
[125] A. Leggett, *J. Phys. (Paris), Colloq.* **41** (1980), C7-19.
[126] P. Nozieres, and S. Schmitt-Rink, *J. Low Temp. Phys.* **59** (1985), 195.
[127] P. Nozieres, *Bose-Einstein Condensation*, edited by A. Griffin, D. Snoke, and S. Stringari (Cambridge University Press, Cambridge, 1995), p. 15.
[128] B. Geilikman, and V. Kresin, *Sov. Phys. Solid State* **7** (1966), 2659.
[129] H. Froelich, *J. Phys. C* **1** (1968), 544.
[130] J. Ihm, M. Cohen, and S. Tuan, *Phys. Rev. B* **23** (1981), 3258.
[131] J. Ruvalds, *Phys. Rev. B* **35** (1987), 8869.
[132] J. Ashkenazi, C. G. Kuper, and R. Tyk, *Solid State Commun.* **63** (1987), 1145.
[133] V. Kresin, *Phys. Rev. B* **35** (1987), 8716.
[134] V. Kresin, and H. Morawitz, *Phys. Rev. B* **37** (1988), 7854.
[135] A. Bill, H. Morawitz, and V. Kresin, *Phys. Rev. B* **68** (2003), 144519.
[136] A. Fetter, *Ann. Phys. (N. Y.)* **88** (1974), 1.
[137] V. Kresin, and H. Morawitz, *Phys. Lett. A* **145** (1990), 368.
[138] H. Morawitz, I. Bozovic, V. Kresin, G. Rietveld, and D. van der Marel, *Z. Phys. B: Condens. Matter* **90** (1993), 277.
[139] S. Yamanaka, K. Hotehama, and H. Kawaji, *Nature* **392** (1998), 580.
[140] A. Larkin, and A. Varlamov, *Theory of Fluctuations in Superconductors* (Oxford University Press, Oxford, 2005).
[141] L. Gor'kov, and A. Sokol, *JETP Lett.* **46** (1987), 420.
[142] E. Sigmund, and K. Müller, *Phase Separation in Cuprate Superconductors* (1994, Springer, Berlin).
[143] Yu. Ovchinnikov, S. Wolf, and V. Kresin, *Phys. Rev. B* **60** (1999), 4329.
[144] V. Kresin, Y. Ovchinnikov, and S. Wolf, *Phys. Rep.* **431** (2006), 231.
[145] I. Iguchi, T. Yamaguchi, and A. Sugimoto, *Nature* **412** (2001), 420.
[146] J. Sonier, M. Ilton, V. Pacradouni, C. Kaiser, S. Sabok-Sayr, Y. Ando, S. Komija, W. Hardy, D. Bonn, R. Liang, and W. Atkinson, *Phys. Rev. Lett.* **101** (2008), 117001.
[147] B. Geilikman, V. Kresin, and N. Masharov, *J. Low Temp. Phys.* **18** (1975), 241.

[148] J. Carbotte, *Rev. Mod. Phys.* **62** (1990), 1027.
[149] V. Kresin, *Solid State Commun.* **63** (1987), 725.
[150] A. Lanzara, G. Zhao, N. Saini, A. Bianconi, K. Conder, H. Keller, and K. Mueller, *J. Phys.: Condens. Matter* **11** (1999), L541.
[151] A. Furrer, *Structure and Bonding*, edited by K. Mueller and A. Bussmann-Holder (Springer, Heidelberg, 2005), 114, p. 171.
[152] Y. Ovchinnikov, and V. Kresin, *Phys. Rev. B* **58** (1998), 12416.
[153] Y. Ovchinnikov, S. Wolf, and V. Kresin, *J. Supercond.* **11** (1998), 323.
[154] W. Eisenmenger, and A. Dayem, *Phys. Rev. Lett.* **18** (1967), 125.
[155] W. Eisenmerger, *Tunneling Phenomena in Solids*, edited by E. Burstein and S. Lundqvist (Plenum, New York, 1969), p. 371.
[156] R. Dynes, V. Narayanamurti, and M. Chin, *Phys. Rev. Lett.* **26** (1971), 181.
[157] R. Dynes, and V. Narayanamurti, *Solid State Commun.* **12** (1973), 341.
[158] H. Murakami, S. Ohbuchi, and R. Aoki, *J. Phys. Soc. Jpn.* **63** (1994), 2653.
[159] J. Labbe, S. Barisic, and J. Friedel, *Phys. Rev. Lett.* **19** (1967), 1039.
[160] I. Kuhareva, *Sov. Phys. JETP* **16** (1962), 828.
[161] M. Strongin, O. Kammerer, and A. Paskin, *Phys. Rev. Lett.* **14** (1965), 949.
[162] G. Deutcher, H. Fenichel, M. Gershenson, E. Grunbaum, and Z. Ovadyahu, *J. Low Temp. Phys.* **10** (1973), 231.
[163] H. Parmenter, *Phys. Rev.* **166** (1968), 392.
[164] M. Tinkham, J. Hergenrother, and J. Lu, *Phys. Rev. B* **51** (1995), 12649.
[165] J. von Delft, and D. Ralph, *Phys. Rep.* **345** (2001), 61.
[166] Y. Ovchinnikov, and V. Kresin, *Eur. Phys. J. B* **45** (2005), 5.
[167] Y. Ovchinnikov, and V. Kresin, *Eur. Phys. J. B* **47** (2005), 333.
[168] W. Knight, K. Clemenger, W. de Heer, W. Saunders, M. Chou, and M. Cohen, *Phys. Rev. Lett.* **52** (1984), 2141.
[169] W. de Heer, *Rev. Mod. Phys.* **65** (1993), 611.
[170] W. Knight, *Novel Superconductivity*, edited by S. Wolf and V. Kresin (Plenum, New York, 1987), p. 47.
[171] J. Friedel, *J. Phys. II* **2** (1992), 959.
[172] P. Anderson, *J. Phys. Chem. Solids* **11** (1959), 59.
[173] S. Frauendorf, and C. Guet, *Annu. Rev. Nucl. Part. Sci.* **51** (2001), 219.
[174] P. Ring and P. Schuck, *The Nuclear Many-Body Problem* (Springer, New York, 1980).
[175] B. Cao, C. Neal, A. Starace, Y. Ovchinnikov, V. Kresin, and M. Jarrold, *J. Supercond. Novel Magn.* **21** (2008), 163.
[176] G. Breaux, C. Neal, B. Cao, and M. Jarrold, *Phys. Rev. Lett.* **94** (2005), 173401.
[177] V. Kresin, and W. Lester Jr., *Chem. Phys.* **90** (1984), 335.
[178] V. Kresin, and B. Tavger, *Sov. Phys. JETP* **23** (1966), 1124.
[179] B. Geilikman, *Sov. Phys. Usp.* **8** (1966), 2032.
[180] V. Kresin, and S. Wolf, *Anharmonic Properties of High Tc Cuprates* , edited by D. Mihailovic, G. Ruani, E. Kaldis, and K. A. Mueller (World Scientific, Singapore, 1994), p. 7.

2

Through a Lattice Darkly: Shedding Light on Electron–Phonon Coupling in the High-T_c Cuprates*

D. R. Garcia[†] and A. Lanzara[†,‡]

[†]*Department of Physics, University of California, Berkeley, CA 94720, USA*

[‡]*Material Science Division, Lawrence Berkeley National Laboratory, Berkeley, CA 94720, USA*

With its central role in conventional BCS superconductivity, electron-phonon coupling appears to play a more subtle role in the phase diagram of the high-temperature superconducting cuprates. Their added complexity due to potentially numerous competing phases, including charge, spin, orbital, and lattice ordering, makes teasing out any unique phenomena challenging. In this review, we present our work using angle-resolved photoemission spectroscopy (ARPES) exploring the role of the lattice on the valence band electronic structure of the cuprates. We introduce the ARPES technique and its unique ability to the probe the effect of bosonic renormalization (or "kink") on near-E_F band structure. Our survey begins with the establishment of the ubiquitous nodal cuprate kink leading to how isotope substitution has shed a critical new perspective on the role and strength of electron–phonon coupling. We continue with recently published work connecting the phonon dispersion seen with inelastic X-ray scattering (IXS) to the location of the kink observed by ARPES near the nodal point. Finally, we present very recent and ongoing ARPES work examining how induced strain through chemical pressure provides a potentially promising avenue for understanding the broader role of the lattice to the superconducting phase and larger cuprate phase diagram.

*Reprinted with the permission from D. R. Garcia, A. Lanzara. Original published in *Adv. Cond. Matter. Physics* (2010), 807412.

2.1. Foreword

The challenge of understanding the origin of high-temperature superconductivity in the cuprates stems from the complicated interplay of differing orders and phenomena believed to exist. The goal of this article is to focus on one such phenomenon, the role of the lattice coupling to electronic states. Though historically significant in conventional superconductivity, it has only lately been receiving attention as a potentially important player in the physics of the cuprate phase diagram. Over the course of this paper, we will be addressing the following areas: (1) how angle resolved photoemission spectroscopy (ARPES) can be used to probe and better understand electron self-energy eeffcts; (2) a brief history of the ARPES "kink" seen in the cuprates and how both the energy scale it defines and its ubiquity in these systems open the door to continued debate over low-energy excitations; (3) how the cuprate isotope effect illuminates issues such as the role of phonons, the nature of the coherent and incoherent parts of the electronic dispersion, the limitations of current theory, and the subtle way differing competing orders that may be interrelated within these systems; (4) recent work mapping phonon dispersion and relating it to ARPES data to underscore how phonon mode nesting may relate to the observed kink; and (5) a survey of recent and ongoing work examining the role lattice strain may play in understanding electron–phonon coupling and potentially the larger phase diagram. Thus, our goal in this review is not to argue how much significance the lattice has to the development of high-temperature superconductivity, but rather the ways in which we find it manifesting within these systems.

2.2. Signature of Electron–Phonon Coupling in ARPES

The use of ARPES to study electron–phonon coupling could be seen as the union of two different approaches to the study of phonons. First, and perhaps most obvious, is the mapping of phonon dispersions using inelastic scattering techniques such as inelastic neutron or X-ray scattering. This momentum space perspective is generally the most intuitive manner for understanding phonon modes within the lattice. Still, if we are seeking information about how electronic states interact with phonons, it is, at best, an indirect technique. Historically, tunneling measurements such as those done on conventional superconductors, such as Pb [1], have provided insight into how electron–phonon coupling directly affects electronic states near the Fermi energy E_F. The previously unexpected features seen in the spectra were then able to be explained within a strong coupling form of Migdal–Eliashberg theory [2]. Still, to be able to have the direct information of how phonon modes affect electronic states yet seen within a momentum space perspective requires a different approach, an approach that ARPES is well suited to offer.

2.2.1. *ARPES and $A(\boldsymbol{k}, \omega)$ Analysis*

Over the last decade, ARPES has become a truly unique experimental probe with an ever growing number of publications in the field of correlated electronic systems. Its central ability to directly probe the single particle spectral function, $A(\boldsymbol{k}, \omega)$, makes its experimental insights highly sought after by condensed matter theorists. With angular resolution approaching 0.1°, a steadily improving energy resolution exceeding 1 meV, as well as numerous experimental advances involving highly localized beam spots (nano-ARPES), spin resolution, and laser-based pump-probe experiments, ARPES continues to be and will likely remain on the cutting edge of experimental solid state physics. Because of the central role that it plays in the work described in this review, we will begin with a brief overview of the theory of ARPES, with an emphasis on the analytical techniques which are critical for studying systems such as the high-T_c cuprates.

It is customary to write the ARPES photocurrent intensity, $I(\boldsymbol{k}, \omega)$, as

$$I(\boldsymbol{k}, \omega) = M(\boldsymbol{k}, \omega) f(\omega) A(\boldsymbol{k}, \omega), \tag{2.2.1}$$

where $A(\boldsymbol{k}, \omega)$ is the crucial single particle spectral function, that is, the imaginary part of the single particle Green's function, $G(\boldsymbol{k}, \omega)$ with $\boldsymbol{k}$ referring to the crystal momentum, while ω is energy relative to the chemical potential. Modifying the spectral function, $M(\boldsymbol{k}, \omega)$ is the matrix element associated with the transition from the initial to final electronic state which can be affected by such things as incident photon energy and polarization as well as the Brillouin zone (BZ) of the photoemitted electrons. Finally, $f(\omega)$ is the Fermi–Dirac function indicating that only filled electronic states can be accessed. Because of the temperature scales used for data in this review, we will not distinguish between the chemical potential and the Fermi energy, E_F, which should match at $T = 0$ for conductors. Since ARPES measures the electron removal part of $A(\boldsymbol{k}, \omega)$, we use high and low energy to refer to large and small negative ω value, respectively. (Additionally, "Binding energy" and "Energy" are often used for the same axis in figures, differing by a minus sign.) As a final point, one might find the contribution of the matrix element $M(\boldsymbol{k}, \omega)$ in (2.2.1) a serious issue to an accurate interpretation of $A(\boldsymbol{k}, \omega)$ from $I(\boldsymbol{k}, \omega)$. In practice, the ω dependence is small over an energy range of order 0.1 eV while the $\boldsymbol{k}$ dependence of $M(\boldsymbol{k}, \omega)$, though important to consider, is reasonably understood by the ARPES cuprate community for the range of the data presented here.

Of particular importance to our exploration of bosonic mode coupling is how electron self-energy effects appear in our ARPES analysis. This is nicely done by introducing the electron proper self-energy $\Sigma(\boldsymbol{k}, \omega) = \mathrm{Re}\Sigma(\boldsymbol{k}, \omega) + i\mathrm{Im}\Sigma(\boldsymbol{k}, \omega)$ which contains all the information on electron energy renormalization and lifetime. This leads to Green's and spectral functions given in terms of the electron self energy $\Sigma(\boldsymbol{k}, \omega)$:

$$G(\boldsymbol{k}, \omega) = \frac{1}{\omega - \epsilon(\boldsymbol{k}) - \Sigma(\boldsymbol{k}, \omega)}, \tag{2.2.2}$$

$$A(\boldsymbol{k}, \omega) = -\frac{1}{\pi} \frac{\mathrm{Im}\Sigma(\boldsymbol{k}, \omega)}{(\omega - \epsilon(\boldsymbol{k}) - \mathrm{Re}\Sigma(\boldsymbol{k}, \omega))^2 + \mathrm{Im}\Sigma(\boldsymbol{k}, \omega)^2}, \tag{2.2.3}$$

where $\epsilon(\boldsymbol{k})$ is the single electron band energy, often referred to as the bare band structure. Finally, causality requires that $\mathrm{Re}\Sigma(\boldsymbol{k}, \omega)$ and $\mathrm{Im}\Sigma(\boldsymbol{k}, \omega)$ are connected to each other by the Kramers-Kronig relation.

With advancements in the late 1990s, the unit information of an ARPES experiment consists of a 2D intensity map of binding energy and momentum along a "cut" though momentum space. These 2D maps offer us two natural and complementary methods for analysis. First, one can hold the energy value of the electronic states studied fixed and observe the photoemission intensity as a function of momentum, a momentum distribution curve (MDC). Similarly, one can fix the momentum space position and observe photoemission intensity as a function of energy at that momentum value, an energy distribution curve (EDC). These two methods constitute the core techniques for analysis of the spectral function $A(\boldsymbol{k}, \omega)$ using ARPES.

Within our review, "MDC analysis" refers to the method of fitting Lorentzian distributions to features in the MDCs as is commonly done in the field. This method of analysis has been very successful and can be understood based on some basic conditions, specifically the condition of "local" linearity in both the self energy $\Sigma(\boldsymbol{k},\omega)$ and $\epsilon(\boldsymbol{k})$. In order for each MDC at a given energy ω to be described with a Lorentzian function, both $\Sigma(\boldsymbol{k},\omega)$ and $\epsilon(\boldsymbol{k})$ need to be linear within the narrow energy and momentum range corresponding to the width of the analyzed peak. This condition is expected to generally hold since both $\Sigma(\boldsymbol{k},\omega)$ and $\epsilon(\boldsymbol{k})$ are able to be expanded using simple Taylor expansions in the following way:

$$\begin{aligned} \Sigma(\boldsymbol{k},\omega) &\approx \Sigma(k_p(\omega),\omega) + \Sigma_k(k_p(\omega),\omega)(k - k_p(\omega)), \\ \epsilon(\boldsymbol{k}) &\approx \epsilon(k_p(\omega)) + \nu(k_p(\omega))(k - k_p(\omega)), \end{aligned} \tag{2.2.4}$$

where k_p is the peak position of the MDC at ω and $\Sigma_k(k_p(\omega),\omega) = [\partial\Sigma/\partial k]_{k=k_p(\omega)}$. Plugging in these expressions into (3), we obtain the following equations:

$$A(\boldsymbol{k},\omega) = -\frac{1}{\pi}\frac{\Gamma(\omega)}{(k - k_p(\omega))^2 + \Gamma(\omega)^2}, \tag{2.2.5}$$

$$\mathrm{Re}\Sigma(k_p(\omega),\omega) = \omega - \epsilon(k_p(\omega)), \tag{2.2.6}$$

$$\mathrm{Im}\Sigma(k_p(\omega),\omega) = \Gamma(\omega)[\nu(k_p(\omega)) + \Sigma_k(k_p(\omega),\omega)]. \tag{2.2.7}$$

It is worth noting that we need not make the standard assumption of momentum independence of $\Sigma(\boldsymbol{k},\omega)$ for these results to be valid. This is important since momentum dependence does exist for states away from the nodal cut (the diagonal direction to the Cu–O bonds.) It is the last two equations which provide us with a precise meaning for $k_p(\omega)$ and $\Gamma(\omega)$ as determined in the MDC analysis and, thus, our determination of $\Sigma(\boldsymbol{k},\omega)$ from ARPES. Equation (2.2.6) demonstrates that a reasonable assumption for $\epsilon(\boldsymbol{k})$ is needed to determine $\mathrm{Re}\Sigma(k_p(\omega),\omega)$ while $\mathrm{Im}\Sigma(k_p(\omega),\omega)$ presents the additional challenge of requiring the derivative $\Sigma_k(k_p(\omega),\omega)$. This is further complicated since though $\Sigma(\boldsymbol{k},\omega)$ is a causal function for a fixed $\boldsymbol{k}$ value, $\Sigma(k_p(\omega),\omega)$ is not. Thus, one cannot invoke the Kramers-Kronig relation to relate real and imaginary parts. Nevertheless, so long as these considerations are kept in mind to prevent over-interpretation, qualitatively $\Gamma(\omega)$ and $k_p(\omega)$ do offer access to the causal $\Sigma(\boldsymbol{k},\omega)$ since important structures such as the ARPES kink appear in both self-energies.

Before the instrumental advances which pushed the unit information of ARPES towards a 2D map, 1D data was taken, making EDC analysis the more traditional method. Indeed, there are many advantages of this line of analysis. (1) Fixed momentum helps simplify the matrix element contribution to the photocurrent. (2) Momentum is a good quantum number in a single crystal approximation making the EDC a more physical quantity, opening up spectral weight sum rules, as well as providing a clear physical meaning to the dispersion of EDC peaks. (3) In principle, an EDC analysis should be able to provide us with the causal $\Sigma(\boldsymbol{k},\omega)$ throughout in the entire 2D plane rather than a particular path determined by k_p in the plane as with MDC analysis. However, EDC analysis is uniquely complicated by contributions from the Fermi function cutoff, $f(\omega)$, as well as both elastic and inelastic photoelectron background. This leads to a challenging lineshape to analyze in practice. Still, employing a similar method of Taylor expansion analysis as used earlier, we can expand the self-energy locally near the EDC peak yielding

$$\Sigma(\boldsymbol{k},\omega) \approx \Sigma(\boldsymbol{k},\omega_p(\boldsymbol{k})) + \Sigma_\omega(\boldsymbol{k},\omega_p(\boldsymbol{k}))(\omega - \omega_p(\boldsymbol{k})), \tag{2.2.8}$$

where $\Sigma_\omega(\boldsymbol{k},\omega)$ is the ω-partial derivative of $\Sigma(\boldsymbol{k},\omega)$. Like before, we can insert these expressions into (2.2.3) getting the following relations in the vicinity of the peak

$$A(\boldsymbol{k},\omega) = \frac{Z(\boldsymbol{k})}{\pi}\frac{\Gamma(\boldsymbol{k},\omega)}{(\omega-\omega_p(\boldsymbol{k}))^2+\Gamma(\boldsymbol{k},\omega)^2}, \tag{2.2.9}$$

$$\mathrm{Re}\Sigma(\boldsymbol{k},\omega_p(\boldsymbol{k})) = \omega_p(\boldsymbol{k}) - \epsilon(\boldsymbol{k}), \tag{2.2.10}$$

$$\mathrm{Im}\Sigma(\boldsymbol{k},\omega_p(\boldsymbol{k})) = \frac{\Gamma(\boldsymbol{k},\omega)}{Z(\boldsymbol{k})} - \mathrm{Im}\Sigma_\omega(k,\omega_p(k))(\omega-\omega_p(k)), \tag{2.2.11}$$

$$Z(\boldsymbol{k}) = \frac{1}{1-\mathrm{Re}\Sigma_\omega(\boldsymbol{k},\omega_p(\boldsymbol{k}))}. \tag{2.2.12}$$

When we compare these with our results for MDC analysis, the complementary nature of these two approaches begins to appear. Unlike the Lorenzian lineshape of the MDCs, the EDC lineshape is modified by an asymmetry, which makes EDC analysis less favorable for extracting self energy near E_F than MDC analysis. However, the spectra at large ω is better analyzed with EDCs thanks to the spectral sum-rule requiring $A(\boldsymbol{k},\omega) \to 1/\omega$, leading us to consequently expect $Z(\boldsymbol{k}) \to 1$ and $\Sigma_\omega(\boldsymbol{k},\omega) \to 0$. This means that the portion of the spectral function which we associate with incoherent excitations should begin approaching a Lorenzian lineshape and is better explored with EDCs, although inelastic background contributions at higher energy remain important. In contrast, MDCs at higher energies begin to be affected by momentum dependent matrix element contributions as well as potential deviations of $\epsilon(\boldsymbol{k})$ from a locally linear behavior. Thus, with both tools in our ARPES arsenal, we can undertake a more complete understanding of self-energy effects as they appear in $A(\boldsymbol{k}, \omega)$.

2.2.2. *Visualizing the Kink with ARPES*

Turning our attention to the physics of electron–phonon coupling in the superconducting cuprates, our prior discussion on how self-energy manifests in the ARPES spectral function points us towards the now well-known "kink" feature. As (2.2.6) and (2.2.10) quickly indicate, a sudden increase in the real part of $\Sigma(\boldsymbol{k}, \omega)$ at a particular energy ω would lead to a deviation of the measured peak from the single electron band structure $\epsilon(\boldsymbol{k})$ at this energy scale.

The result is seen in Fig. 2.1 which shows superconducting phase data taken on the well-studied cuprate Bi2212 at its optimal doping ($T_c = 92\,\mathrm{K}$). The two ARPES cuts are taken for states both at (Figs. 2.1(a)–2.1(c)) and off (Figs. 2.1(d)–2.1(f)) the nodal point. The different visualization methods used for each cut are designed to enhance some key characteristics of the ARPES kink phenomenon prior to a more detailed, quantitative approach involving fittings. As labeled in the figure caption, the "MDC map" allows us to track the MDC dispersion and width. This is similarly true for EDCs in the "EDC map". The color scaling is chosen to give the peak maximum and half maximum distinct colors, red and blue, respectively.

From these maps, we can observe the following features. (1) The anisotropic d-wave nature of the superconducting gap is immediately apparent in the MDC "backbending" observed in panel (e) near E_F within the gap energy scale. Evidence of this gap disappears for the nodal cut (panels (a–c)) as expected. (2) There is an abrupt deviation in the electron dispersion around 70 meV below E_F (large black arrow) for both cuts. In both cuts, this corresponds to slower electron dispersion at lower energy while there is faster dispersion at higher energies above the 70 meV energy scale. (3) Focusing particularly on the off-nodal cut, one sees evidence, even in the raw map, of an intensity decrease forming a local minimum at the 70 meV energy scale. This lineshape, further enhanced by the EDC map (panel (f)), is known as a "peak-dip-hump" and is associated with the presence of self-energy effects due to the coupling of electrons with a bosonic mode leading to a redistribution of the spectral weight in the EDC

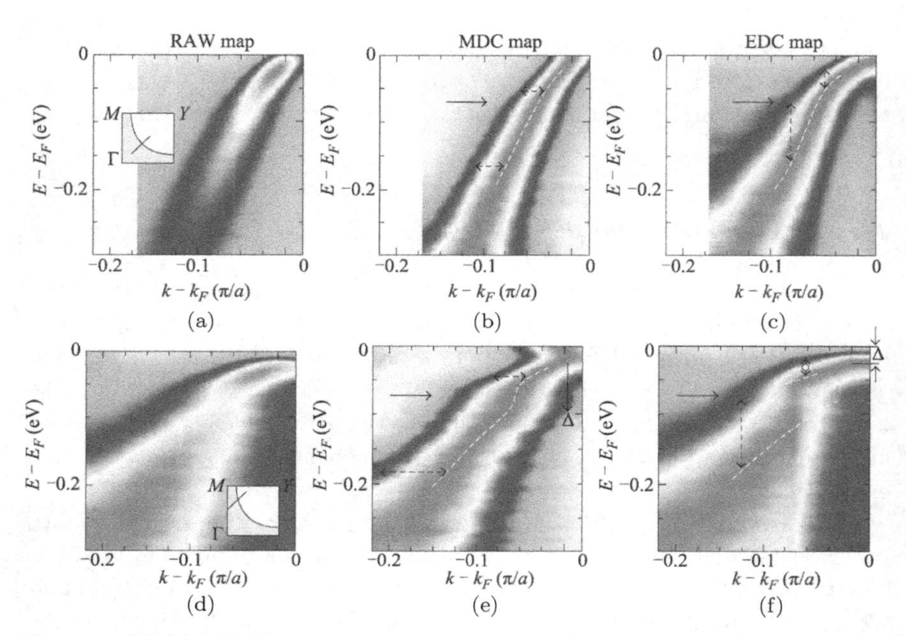

Fig. 2.1. (a)–(c) ARPES data taken at 25 K on optimally doped Bi2212 superconductor ($T_c = 92$ K), for a cut along the ΓY direction through the nodal point in momentum space (indicated in panel (a) inset). (d)–(f) The same sample and orientation, but taken nearer the Brillouin zone (BZ) edge (or anti-nodal point) in momentum space (indicated in panel (d) inset). (a) and (d) Raw ARPES data taken with a color scale where intensity increases from pale yellow to green to blue to white to red. Here, blue (white) corresponds to 1/2 (3/4) of the maximum intensity. (b) and (e) Same data but as an "MDC map", where each MDC has been normalized so that its maximum and minimum intensities are 1 and 0, respectively. (c) and (f) Same data but now each EDC has been appropriately normalized to create an "EDC map". Thick black arrows indicate the energy of the bosonic mode while the Δ is the superconducting gap. Energy resolution used here is ~15 meV.

spectra. (4) Although one would expect quasi-particles in a Landau Fermi liquid paradigm to become sharper (i.e., longer living) as one approaches E_F, it is significant that the kink energy scale also marks a sudden change in coherence. One can make out from the panels, in particular the EDC maps, abrupt changes in the MDC and EDC linewidths as one passes from states above and below the kink energy scale. To bring it all together, we can define the ARPES kink as an energy crossover separating sharp, slowly dispersing, coherent states nearer E_F from broader, quickly dispersing incoherent states at higher energy.

2.2.3. *The Nodal Kink*

With the initial discovery of the ARPES "kink" in the superconducting cuprate Bi2212 [4], we have begun to develop a fuller picture of how low-energy many-body effects manifest in the cuprates. On the heels of this discovery and the resulting debate, a systematic study regarding the origin of this kink discovered the feature's remarkable ubiquity across all cuprate families and dopings accessible by ARPES [3]. Figure 2.2 summarizes these results particularly in double-layered Bi2212 and single-layered Bi2201 and LSCO showing that the kink in the nodal direction exists all across these systems at essentially the same energy, ~70 meV. One can give an estimate of the coupling constant, λ, between the electrons and this bosonic mode by comparing the ratio of the group velocities above and below the kink energy, $(1 + \lambda) = \nu_{\mathrm{High}}/\nu_{\mathrm{Low}}$ [5]. From this analysis, one finds evidence for a trend between doping and the strength of the mode with an enhancement of $\Sigma(\boldsymbol{k}, \omega)$ as one tends to the underdoped side of the superconducting dome (Fig. 2.2(f)).

Additionally significant is the continued existence of the kink below and above the superconducting transition temperature (Figs. 2.2(d) and 2.2(e)), casting doubt on scenarios based on superconducting gap opening and particularly the magnetic mode scenario [6–11]. Comparing the photoemission data with the neutron phonon energy at $q = (\pi, 0)$ (thick red arrow in Fig. 2.2(a)) and its dispersion (shaded area) [12, 13] it was proposed [3] that the nodal kink results from coupling between quasi-particles and this zone boundary in-plane oxygen-stretching longitudinal optical (LO) phonon. Although this is the highest phonon mode contributing to the kink, quasi-particles are also coupled to other low-energy phonon modes [14].

In favor of the electron–phonon coupled system is the drop of the quasi-particle width (Fig. 2.3) below the kink energy and the existence of a well-defined peak-dip-hump in the EDCs, a signature of an energy scale within the problem, persisting up to temperatures much higher than the superconducting critical temperature [15].

2.2.4. *The Near Anti-nodal Kink*

Many-body effects near the anti-nodal region of the BZ (Cu–O bond direction) had been suspected for some time from earlier ARPES studies of the cuprates where evidence of the aforementioned peak-diphump lineshape

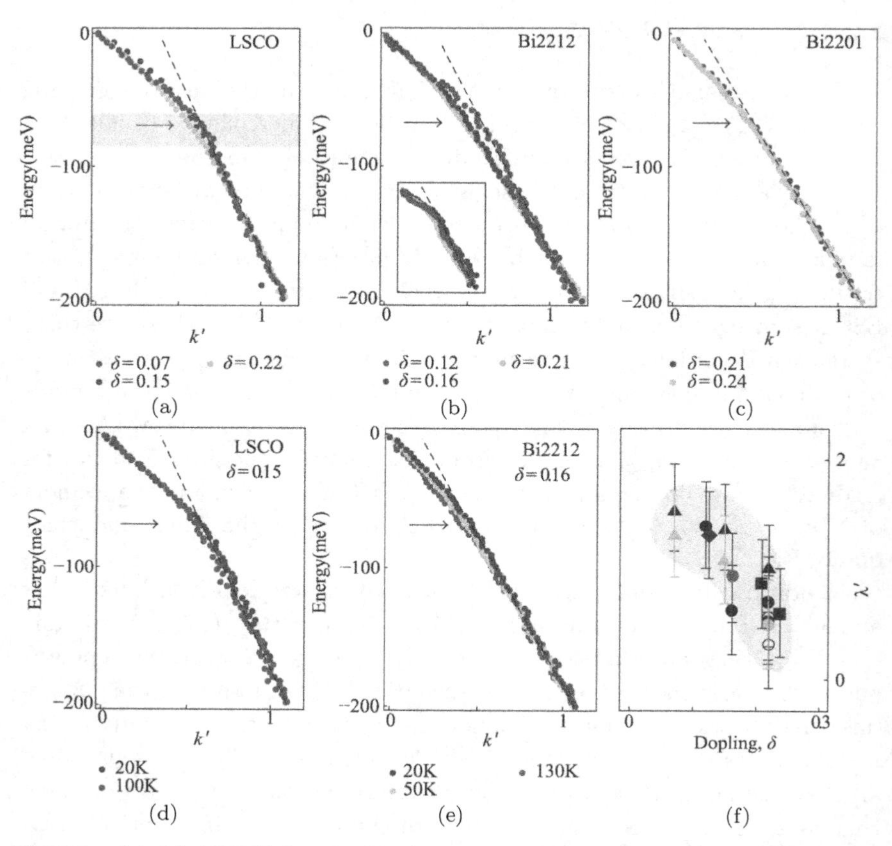

Fig. 2.2. (a)–(e) ARPES data at the nodal point showing the ubiquity of the ARPES dispersion kink as determined from MDC fittings over a variety of cuprates systems, dopings, and temperatures above and below T_c. The kink energy is indicated by the thick arrow and the momentum is rescaled so that k' is 1 at 170 meV binding energy. (f) Estimating the electron coupling constant λ for the different samples as a function of their doping. Filled triangles, diamonds, squares, and circles are LSCO, Nd-LSCO, Bi2201, and Bi2212 in the first BZ, respectively. Open circles are Bi2212 in the second zone. Different shadings represent data from different experiments. The figures are from Ref. [3].

was reported [17, 18]. Although controversy has existed regarding the role of bilayer band splitting on the observed spectra in the double layered Bi2212 compounds, the presence of the peak-dip-hump lineshape was initially interpreted in terms of a magnetic phenomenon observed in YBCO and Bi2212 by inelastic neutron scattering [19–21]. With the resolution of

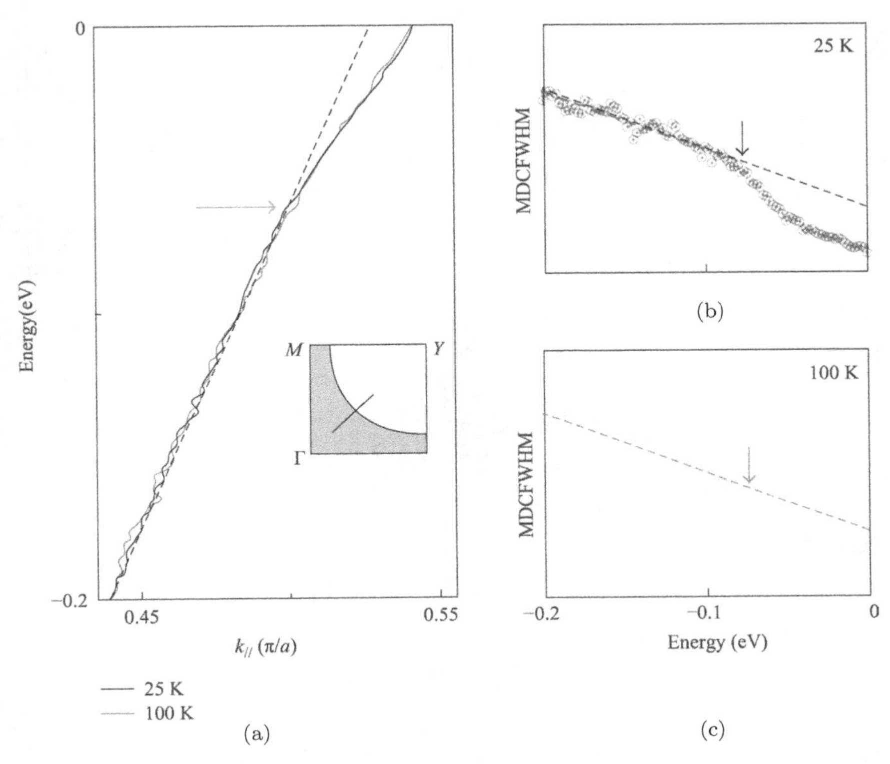

Fig. 2.3. ARPES MDC dispersion data taken optimally doped $Bi_2Sr_2CaCu_2O_{8+\delta}$ ($T_c = 92$ K) as discussed in Ref. [16]. Nodal point data (see inset) comparing (a) dispersion, (b) and (c) MDC full width half max (FWHM) showing little change in the energy of the ARPES kink with T_c. The MDCFWHM is related to the $\mathrm{Im}\Sigma(\boldsymbol{k}, \omega)$.

the bilayer splitting [22–24], a low-energy kink of approximately 40 meV near the anti-nodal region was reported for Bi2212 [25–27]. The disappearance of the kink above T_c and the decrease of its strength moving away from the anti-nodal region has led people to interpret the onset of this kink as coupling to collective magnetic excitations [25–27], despite the absence of these excitations for more heavily doped samples [25].

More recent studies [16, 28] have reported that the near antinode kink also persists above T_c, as seen in Fig. 2.4. However, the energy of this kink shifts towards higher energy, from 40 meV to 70 meV, upon entering the superconducting state. This shift is consistent with the opening of a 30 meV gap below T_c. The persistence of this energy scale above T_c can be clearly

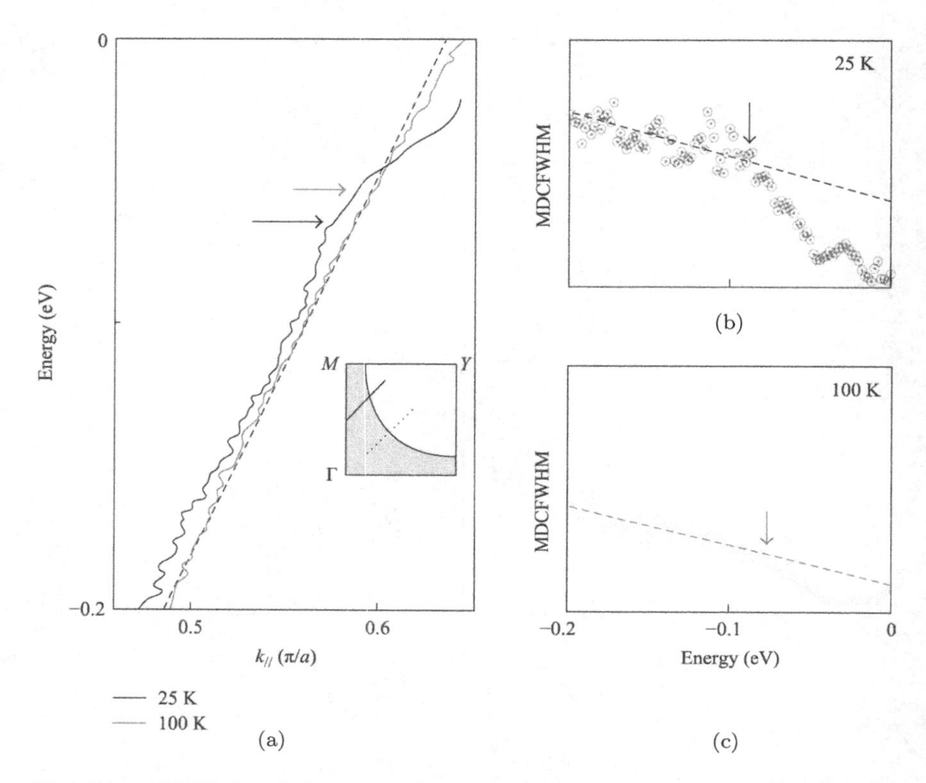

Fig. 2.4. ARPES MDC dispersion data taken optimally doped $Bi_2Sr_2CaCu_2O_{8+\delta}$ (T_c = 92 K) as discussed in [16]. Near anti-nodal point data (see inset) comparing (a) dispersion, (b) and (c) FWHM showing that the ARPES kink energy is shifted to lower energy when passing above T_c.The MDC FWHM is related to the $\mathrm{Im}\Sigma(\boldsymbol{k},\omega)$.

seen both in the dispersion (Fig. 2.4(a)) and in the MDC width (Figs. 2.4(b) and 2.4(c)). This observation has led to a new interpretation of the anti-nodal kink in terms of electron-phonon coupling. In this case, it was proposed that the responsible phonon, with the right energy and momentum, is the B_{1g} mode [28].

Still, it should be noted that spin fluctuations also exist in the normal phase. Indeed they have been used within marginal Fermi liquid (MFL) theory to provide the anomalous self energy [29] and have traditionally been called on to describe ARPES data in the normal phase just above T^*. What is interesting is that although MFL theory can explain the ARPES dispersion Figs. 2.3(a) and 2.4(a), it cannot explain the drop in linewidth seen of Figs. 2.3(c) and 2.4(c), presenting diffculties to the original theory [29]. Nevertheless, what we will find in the next section may suggest a profound connection between the physics of the lattice and spin.

The results presented so far clearly suggest that the electron–lattice coupling could not be so easily neglected in any microscopic theory of cuprate superconductivity. As discussed in Section 2.2.2, the kink and its energy consistently indicate a sudden transition from sharp coherent electronic excitations into broader more incoherent ones. This makes understanding the origin of this phonon mode and its energy scale critical since it has such a substantial effect on low-energy electronic states.

2.3. ARPES Isotope Effectin Bi2212

The role of the isotope effect (IE) in establishing the electronphonon nature of Cooper pairing for traditional BCS superconductors is well known. However, when we consider the IE in the superconducting cuprates, its effect on T_c is substantially less, leading researchers away from the electronphonon paradigm of the BCS superconductors. But with the ubiquitous cuprate kink seen by ARPES, the importance of the lattice returns to the forefront. Additionally from our discussion in Section 2.2, the kink brings up questions about the relationship between the coherent peak (CP) seen at lower energies near E_F and the incoherent peak (IP) seen at higher energies. Indeed, hole doping affects the formation of these peaks dierently [30] with the CP strongly affected while the IP appears minimally changed. Should we be thinking of the CP and IP as different objects or fundamentally connected to each other? In this light, the kink energy scale, and thus the phonon mode responsible for it, becomes increasingly significant as the key crossover between these two domains. It is in light of such questions that we undertook our ARPES study of the IE to better understand the role of the lattice in these issues.

2.3.1. *Prelude-isotope Effect in ARPES Dispersion of H/W*

Using APRES to explore isotope substituted samples presents us with an entirely new and wide open field of study. As of our work, the only other study in the literature explores the surface state on W induced by H chemisorption [31]. Figure 2.5 shows data taken for H on W (indicated in blue) compared to D on W (indicated in red). In spite of the broad

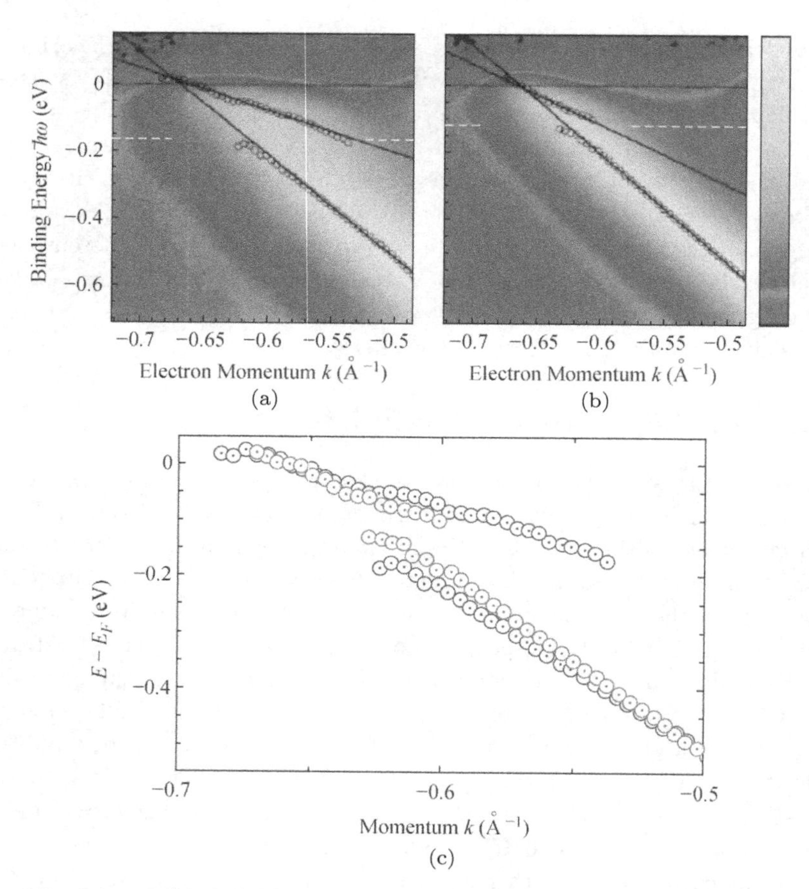

Fig. 2.5. (a) and (b) ARPES dispersions of a surface state for (a) H monolayer on W and (b) D monolayer on W. The dispersions are determined by peak positions from EDC fits and appear as circles while the lines serve as guides to the eye. (c) These two dispersions are compared where H = blue and D = red. The figure is from Ref. [31].

peaks due to the instrumental resolution when compared to data on the cuprates, extracting the EDC peak dispersion clearly shows two types of dispersions akin to that seen with the cuprate kink studies. Specifically, the slower low energy and faster high-energy dispersions were understood in terms of the CP and IP, respectively. These results could be compared to predictions from the strong coupling form of Migdal–Eliashberg (ME) theory which, as discussed earlier, already explained the existence of peak-dip-hump features in tunneling spectroscopy as coupling to phonon modes.

There was fair agreement with ME since the high-energy linewidth (noted from Figs. 2.5(a) and 2.5(b)) and the kink energy position (dotted lines in Figs. 2.5(a) and 2.5(b)) were found to approximately scale as $1/\sqrt{(M)}$ where M is the mass of H or D. Additionally as expected from ME theory, the electron–phonon coupling was ~0.5, and the linewidth at high-energy was $\sim \hbar\omega_p$. Finally, Fig. 2.5(c) illustrates how the dispersions near the kink energy are affected by the isotope change, deviating most substantially near the kink energy while decreasing along both directions in energy, consistent with ME theory.

2.3.2. *Isotope Effect in ARPES Dispersion of Bi2212*

When we compare the ARPES IE seen in the H/W system to that measured on optimally doped $Bi_2Sr_2CaCu_2O_{8+\delta}$, we find surprisingly different behavior. First, let us consider the behavior of the ARPES kink energy as summarized in Fig. 2.6. Figures 2.6(a) and 2.6(b) are MDC maps, as discussed in Fig. 2.1, from data taken at the nodal point in the Γ–Y direction. The two samples examined contain ^{16}O and ^{18}O in their Cu–O planes which, as indicated by the horizontal arrow, already reveals a potential shift in the kink to lower energy with the substituted ^{18}O sample.

Analyzing this more carefully, we turn to the MDC fitted dispersion for both isotopes plus a reexchanged sample, $^{16}O_{\text{Re-exch}}$, whereby we can take a studied ^{18}O sample and re-exchange ^{16}O back into the lattice. This provides us with a unique check on the IE. As indicated in panel (c), we can observe, from the dispersions, a subtle shift in the kink energy between the isotopes of approximately 5 meV. We can additionally quantify the kink by estimating the bare single electron dispersion (using a linear approximation) and extracting the real part of the electron self-energy, $\text{Re}\Sigma(\boldsymbol{k},\omega)$, as described by (6). The location of the peak in $\text{Re}\Sigma(\boldsymbol{k},\omega)$ corresponds to the kink energy and we similarly observe a shift in this peak with isotope change. Thus, the IE does have measureable effect on the nodal ARPES kink energy, as later work would also conrm [32], further establishing its phonon origin.

A second aspect of Fig. 2.6(c) worth noting is the energy range where the IE is most pronounced. The maximum change in the dispersion occurs at higher energies, particularly, beyond the kink energy, and is nearly nonexistent at energies closer to E_F. This is particularly significant when we compare this result to the H/W work where the greatest deviation occurs

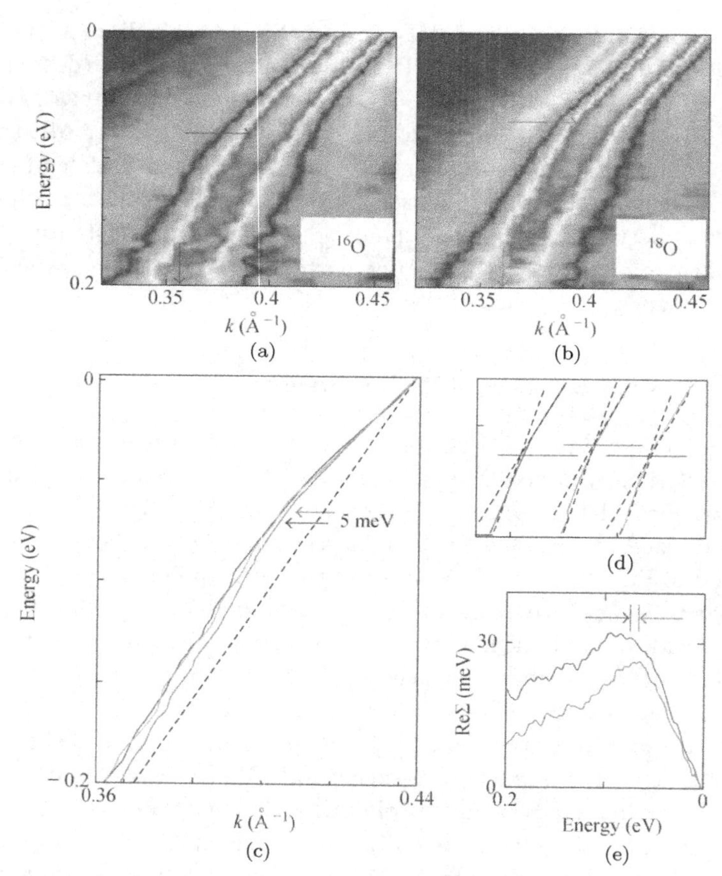

Fig. 2.6. (a) and (b) MDC maps of the nodal point electronic states for cuts along the Γ–Y direction. (a) The ^{16}O sample and (b) the ^{18}O substituted sample with the horizontal arrows indicating the shift in ARPES kink energy with oxygen isotope. (c) The MDC dispersions determined from the ^{16}O, ^{18}O, as well as a resubstituted ^{16}O samples for the cuts in (a) and (b). (d) Cartoon illustration of the kink shift in (c). (e) Real part of the electron self-energy, $\text{Re}\Sigma(\boldsymbol{k}, \omega)$, determined from the MDC dispersion using a linear approximation for the single electron bare band. As before, the ARPES kink position, defined by the peak in $\text{Re}\Sigma(\boldsymbol{k}, \omega)$, is shifted to higher energy as indicated by the arrows.

near the kink energy; Fig. 2.5. We can explore this further by looking at the EDCs taken from this nodal cut, as presented in Fig. 2.7. Figure 2.7(a) shows the dispersion of the EDC peak from where it crosses E_F at k_F as a sharp CP, to higher binding energy where it broadens and becomes the IP. Consistent with the MDC dispersions, we see very little change in the

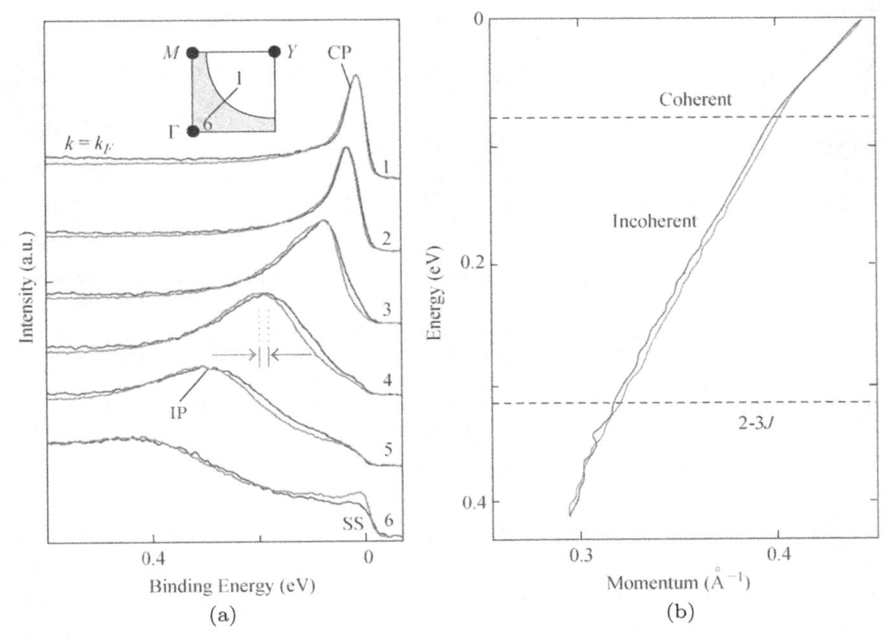

Fig. 2.7. (a) EDCs taken on optimally doped $Bi_2Sr_2CaCu_2O_{8+\delta}$ samples with different oxygen isotopes at $T = 25$ K. The EDCs are from the same cut as in Figure 2.6 and indicated by the inset. The sharp coherent peak (CP) near E_F and broader incoherent peak (IP) at higher energy are identified. The CP has nearly no isotope dependence while the IP has a more substantial IE, most strongly seen in curve 4. The small peak at E_F in curve 6 is the well-known superstructure (SS) replica of the main band. This figure is from [33]. (b) MDC dispersion taken to higher binding energy indicating that the IE, more pronounced for the IP as in (a), disappears again above roughly 2–3 times the anti-ferromagnetic coupling energy, J.

lineshape between the two isotopes for the CP near E_F. However, the IP at higher energy has a lineshape clearly affected by the IE.

In light of this change at higher energy as compared to localization around the kink, a crucial question to ask is at what energy the IE goes away? Figure 2.7(b) attempts to address this issue by following the MDC dispersion to even higher energies. Apparently, the IE seems to disappear around an energy scale of 2 to 3 times the anti-ferromagnetic coupling constant J (where $J = 4t^2/U$ in the $t - J$ model). This could suggest a profound interconnection between the effects of the lattice and spin on the electronic states in the superconducting cuprates. Further work is needed to better understand the connections between these phenomena.

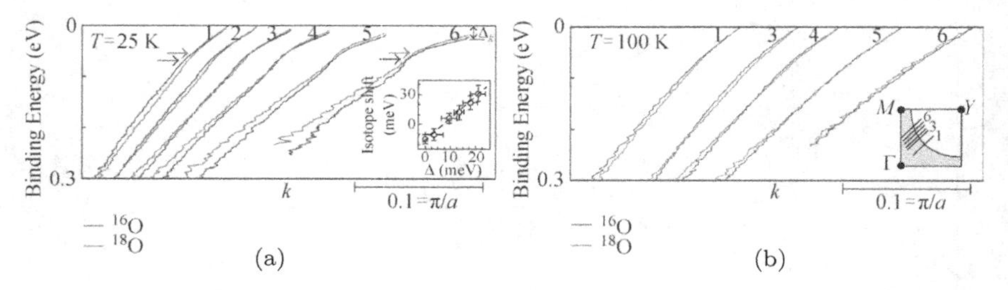

Fig. 2.8. MDC dispersions from cuts parallel to Γ–Y taken off node towards the antinodal point. (a) Data taken in the superconducting phase ($T = 25$ K). The inset shows the isotope energy shift versus the isotope-averaged superconducting gap, Δ. The isotope shift is measured at the momentum value where the isotope-averaged binding energy is 220 meV. The apparent linear correlation indicated by the dashed line is independent of the binding energy used. (b) MDC dispersions from the same cuts measured above $T_c(T = 100$ K). The inset illustrates the location of the cuts. The figures are from Ref. [33].

Up until now, we have focused entirely on the nodal point. Figure 2.8 provides MDC dispersions for Γ–Y slices moving outward from the nodal point towards the antinode, both above and below T_c. There are a few important observations to make from this data. First, the kink energy shows a subtle shift of approximately 5 meV for all momentum cuts. Second, from comparing Fig. 2.8(a) with 2.8(b), it appears that the magnitude of the IE may be, for all curves, diminished above T_c. Third, the IE remains relatively weak near the node while comparatively more pronounced near the antinode leading to a general correspondence between the kink strength, λ, and the IE at high-energy. Plotting this IE shift with respect to the isotope averaged superconducting gap gives a linear relationship seen in the inset of panel (a). Finally, and potentially most surprising, there appears to be a sign change between the two dispersions as we transition from the node towards the antinode, which also appears both above and below T_c.

This sign change is significant since one can examine its location in momentum space. When we plot these crossover points, surprisingly we find that they fall along a line defined in momentum space as $\mathbf{q_{CO}} = 0.21\pi/a$, where a is the lattice constant of the CuO_2 plane. This is illustrated in Fig. 2.9. This particular wavevector (Fig. 2.9(a)) is in excellent agreement with the charge ordering wavevector seen in the far underdoped Bi2212 cuprate at low temperatures as explored by STM [34] (Fig. 2.9(b)), implying that the high-energy part of the electronic structure is strongly coupled to the order parameter, which is in turn strongly coupled to the lattice.

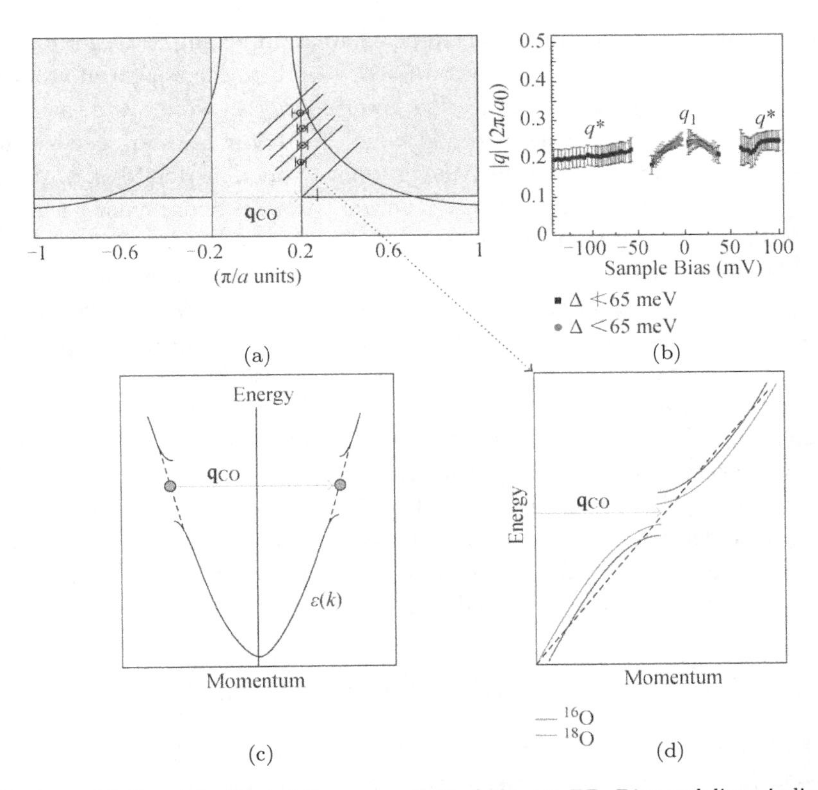

Fig. 2.9. (a) Fermi surface of the upper half of the rst BZ. Diagonal lines indicate cuts used in Figure 2.8 while circles indicate the location in momentum space of the sign change crossover point for each of those cuts. These lie on a line indicated by the wavevector, $\mathbf{q}_{\mathbf{CO}}$ and illustrated by the colored regions. (b) STM data independently determining this wavevector taken from Ref. [34]. (c) Cartoon illustrating how the charge ordering wavevector, $\mathbf{q}_{\mathbf{CO}}$, can open a gap at a binding energy where the electronic states are nestable. (d) Additional cartoon illustrating how the splitting, if slightly different in magnitude between the isotopes, can explain the observed sign change in the bands and its evolution as the dispersions intersect with $\mathbf{q}_{\mathbf{CO}}$.

To understand why at the $\mathbf{q}_{\mathbf{CO}}$ line the isotope effect changes sign, we used a simple charge density wave formation model, to show how an ordering mechanism can affect the quasi-particles dispersion at all energies. In Figs. 2.9(c) and 2.9(d) we present the opening of a gap in the dispersion at the $\mathbf{q}_{\mathbf{CO}}$ vector, due to a charge density wave formation. Based on the report that the pseudogap temperature is strongly isotope dependent and increases for the ^{18}O sample [35–37], we assume that the magnitude of the gap is different between the two isotope samples, for example, larger for the ^{18}O sample (Fig. 2.9(d)). This automatically leads to the appearance of the sign change at $\mathbf{q}_{\mathbf{CO}}$ that migrates to lower energy as we move away from the node, as observed in the data. Although further studies are needed, we believe that the main reason why the pseudogap opening due to an ordering phenomena has never been observed in any ARPES experiment so far, is likely due to the short range nature of such ordering [38, 39].

This result has led us to consider a possible correlation between the charge or spin instability, with $4 \sim 5a$ periodicity, observed in the underdoped regime [34, 40] and the lattice effects relevant at optimal doping. In summary, these results suggest that the lattice degrees of freedom play a key role in the cuprates to "tip the balance" towards a particular ordered state. Simply put, the raw electron–phonon interaction may be small, but it can assist a certain kind of order through a cooperative enhancement of both the assisted order and the electron–lattice interaction [35, 36]. In particular, using a model which considers this electron-coupling boson as the critical source of charge order fluctuations [41], the sign change observed in our dispersion can be well reproduced [39].

2.3.3. *Isotope Effect in ARPES width of Bi2212*

In addition to the shape of the electron dispersion, information about linewidth, $\Gamma(\omega)$, can be extracted and provide critical information on the electronic states. As previously discussed, MDC analysis is not ideally suited to determining the linewidth at high energies, leaving us to examine the linewidth as it is obtained from EDC analysis. Additionally, we have already observed a significant difference between states above and below the kink energy, even prior to our discussion of IE. Thus, Fig. 2.10 divides up the electronic states into those roughly between E_F and the kink energy (0 to 70 meV) and those beyond the kink energy (70 meV to ~250 meV). Taking the average change in width of the EDC peaks shown in Fig. 2.7 between the isotopes for each of these two energy regions we find that IE on the linewidth is very similar to its effect on the dispersion. (1) It is very small for the low-energy coherent electronic states, while much more significant for the higher-energy incoherent states. (2) The magnitude of the IE is small at the node for these higher energy states, but it grows more substantial as one moves towards the antinode. (3) The effect is roughly linear with respect to the isotope averaged superconducting gap, Δ, as seen in Fig. 2.8.

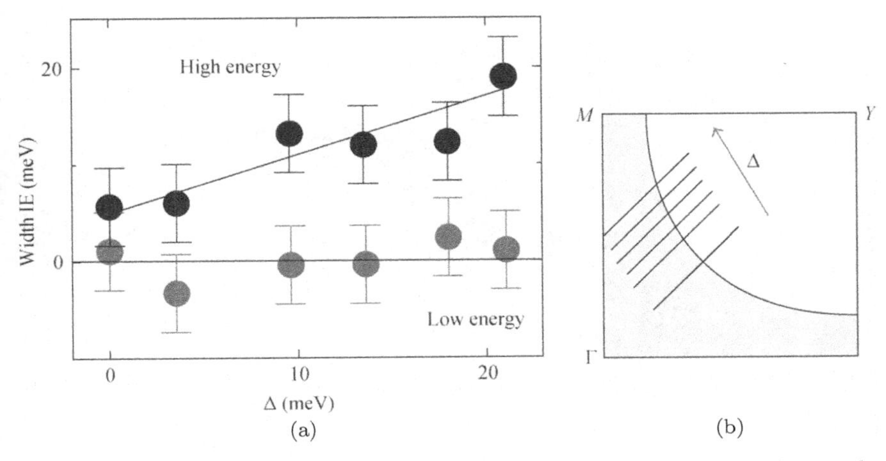

Fig. 2.10. Isotope effect on EDC widths determined from ARPES data taken at the slices indicated in the smaller panel (same as Fig. 2.8) within the superconducting phase. The width change for each cut comes from an average over binding energies below and above the kink energy: low (70 meV to 0) and high (~250 meV to 70 meV) energies. These widths are plotted with respect to the isotope averaged superconducting gap, Δ, of each cut, showing little change at low energies but significant change above the kink energy. The figure is from Ref. [42].

Yet, it does differ from the IE in the dispersion since it lacks the sign change previously discussed. As will be addressed in depth later, the corresponding ME IE linewidth change is much smaller, about 2 meV, making the trend in the high energy linewidth a serious failing of the theory for explaining the cuprate IE.

2.3.4. *Doping Dependence*

So far, the data shown has been on optimally doped Bi2212 with a hole doping determined by our ARPES experiment to be $x = 0.16$. But the question naturally arises: how is the IE on the ARPES data affected by a change in doping? Figure 2.11(a) shows angle integrated photoemission data obtained on two samples at optimal doping ($x = 0.16$) and one at slight over doping ($x = 0.18$). The effect is dramatic that given such a small change of only 2%, the IE, which normally shifts the IP position by ~30 meV, is substantially reduced. This work was initially inspired by a separate study claiming not to see the IE in optimally doped samples of Bi2212 [43]. However, from superstructure analysis and studying the MDC

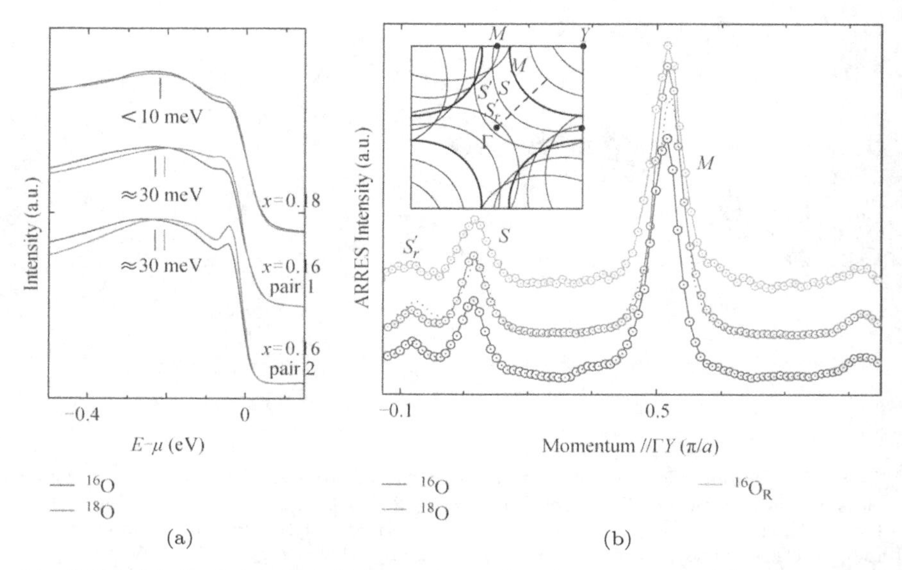

Fig. 2.11. (a) Angle integrated photoemission data from two sets of data at optimal doping ($x = 0.16$) and one set at a slight over-doping ($x = 0.18$). Data has been normalized to the area under curves for energies 0.5 and above. (b) MDCs taken near E_F along the Γ–Y nodal direction as seen from the inset. The peaks correspond to the intersection of the cut with the main Fermi surface band (M) as well as firstorder superstructure reflections (S) and a second-order reflection (S'_r). Overlaid on top of the red curve is a thin grey line representing the ^{16}O curve shifted to model a 0.01 change in doping for comparison. Panel (a) is from Ref. [45].

dispersions, the samples discussed in Ref. [43] were actually slightly over doped. This doping dependence is intriguing but potentially puzzling given such sensitivity. Still, work in other correlated electronic systems, such as the manganites, demonstrates that such a small doping change does cause a qualitative change in the electronic structure [44]. Thus, the IE has a strong sensitivity to optimal doping and, therefore, there exists a rapid change in the electron–lattice interaction near optimal doping.

2.3.5. *Failure of Conventional Explanations*

Although the work of [43] had proposed that the IE was not apparent from ARPES on Bi2212, there are additional explanations that one could invoke to explain the presence of this subtle effect which are important to address. As we will show, all of these conventional explanations turn out to be inadequate. Indeed, the strong temperature dependence by itself rules out all of the following explanations as possible candidates. Still, this position can be made stronger when we consider only the large low-temperature IE in light of the following scenarios.

2.3.5.1. *Doping issue*

As the previous section discussed, establishing the doping of our samples, particularly that their optimal doping has remained unchanged after the isotope substitution, is critical to establishing the veracity of our claims. Doping level was preserved during the sample growth process by annealing the two samples (^{16}O and ^{18}O) in the exactly same environment save the oxygen gas. Yet, ARPES on the cuprates provides us with *in situ* signatures of the doping level and its consistency. Most notably, one can use the Fermi surface size to quite precisely determine doping level, taking advantage of the well-known superstructure reflections of the main hole band structure.

Using the nodal cut and making use particularly of the second-order reflection from the opposite side of the Γ point, Fig. 2.11(b) shows the associated MDCs near E_F for three samples, including the resubstituted sample. All three curves have good agreement with respect to peak positions. Of particular interest is the peak S_r' which is due to the second-order superstructure Fermi surface replica from the opposite side of Γ. This means that any doping change in the sample will affect the distance between M and S_r' twice as fast as the distance between M and Γ, while the distance between S and M remains fixed by the superstructure wavevector. This makes the distance between M and S_r' a sensitive measure of doping change. In fact, plotted on top of the red ^{18}O curve is a grey curve modeling a doping change of 0.01 based on a tight binding fit. From looking at how the ^{18}O peak positions compare to this, it is clear that the uncertainty in doping value is well below 0.01.

However, one might be initially inclined to argue that the doping values between the two isotope samples are different based on the small difference in energy gap that has been previously reported [33] which would explain the observed IE. But this argument loses its plausibility for a few reasons. First, examining the two sets of data we report [33], the difference in energy gap $\Delta_{16} - \Delta_{18}$ actually differed in sign, with an ~ -4 meV change for [33, Fig. 2.1] and an $\sim$5 meV change for [33, Figs. 2.2 and 2.3]. Clearly, an empirically consistent change of gap is not obvious from our data. Second, this becomes more evident over the many measurements ($\sim$20) we have done finding the $\Delta_{16} - \Delta_{18}$ gap difference averaging to zero with a less than

1 meV difference, even while each individual value of Δ_{16} or Δ_{18} uctuated by as much as 5 meV. This was consistent with a typical uncertainty in the gap value from other sources at the time [46]. Third, even if there were a consistently measured gap difference of 5 meV and this were taken to mean that the doping values were different, it still does not explain the observed crossover behavior in Fig. 2.8. More quantitatively, the doping change implied by a difference of 5 meV in gap ($\Delta x = 0.017$) is not suffcient to explain the large IE we have seen. With the associated doping change converted into shift in peak position at high energy, it corresponds to 5 meV for the nodal cut, and only 10 meV for the near-anti-nodal cut. So, these numbers not only have the same sign but also are off in magnitude by a factor of 3 to 4. Thus, doping considerations do not offer a convincing explanation of the observed IE.

2.3.5.2. *Alignment issue*

Another conventional explanation for our observed IE could be sample alignment since even a small misalignment could create an apparent difference in the observed dispersion. A careful examination led us to focus on two particular issues: the exact location of the Γ point and the relative azimuthal orientation of samples with respect to each other.

For the cuts 0–6 discussed here, we have found some evidence of deviations from our alignment based on the Fermi surfaces M, S,and S'_r as described Fig. 2.11(b). These slightly different momentum paths are indicated in Fig. 2.12 to give a sense of magnitude. However, we have found that, as Figs. 2.8(b) and 2.8(c) demonstrate, the expected dierence in the dispersion, based again on our tight binding fit, is none for cut 0 and roughly about 4 times too small for our cut 6 when compared to our measured IE. It is also worth noting that the IE was reproduced for data taken with analyzer slits rotated 45° with respect to this geometry, parallel to the MY direction [33]. This geometry has the advantage of being less sensitive to azimuthal tilting. Thus, alignment issues are not a likely explanation for the observed data.

2.3.5.3. *Static lattice issue*

A final concern that should be considered is that a static lattice effect may be responsible for the large IE since there are no high-quality structural studies to rely on for insight. However, when we consider differences in

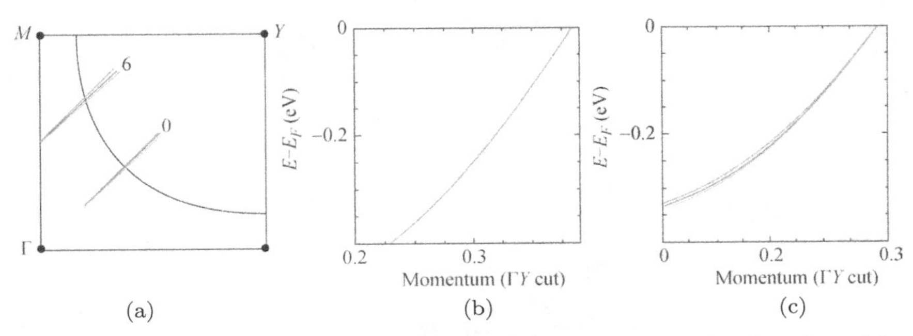

Fig. 2.12. (a) Estimated azimuthal variation for cuts at and far off node indicated by the teal colored cuts relative to the red cut ($\Gamma - Y$). (b) and (c) Tight binding model for the nodal and off nodal cuts, respectively, representing the expected variation in dispersion due to the misalignment.

crystal structure for isotope exchanged LSCO, they are only 0.1%. Given that a static structural effect is known to be more common in the LBCO and LSCO systems than the Bismuth based cuprates, one may reasonably assume that static lattice effects are significantly small in the Bi2212 cuprates and are an unlikely explanation for the observed IE. This is particularly strengthened since even if a static lattice effect were significant, it would need to be a particularly complicated static distortion which would make explaining the IE crossover difficult to accomplish.

2.3.6. *Beyond the Migdal–Eliashberg Picture*

With the IE results more soundly established in light of other potential explanations, we return our attention to the Migdal–Eliashberg (ME) theory and ask whether its applicability is still appropriate for the results seen. One would initially expect ME theory to offer the best theoretical model of the observed ARPES kink in this doping region of the phase diagram. However, in view of the discussion at the start of this section, it is not entirely obvious that ME theory can be used to describe the broader incoherent spectral weight. This is a significant concern because of the prevalent use of ME theory in the context of the ARPES kink in the cuprates despite many experimental [48–56] and theoretical [57–59] works indicating an interaction strength beyond this theory. So, in light of our work, we will distinguish between those aspects which clearly go beyond ME theory as well as those more consistent with the theory.

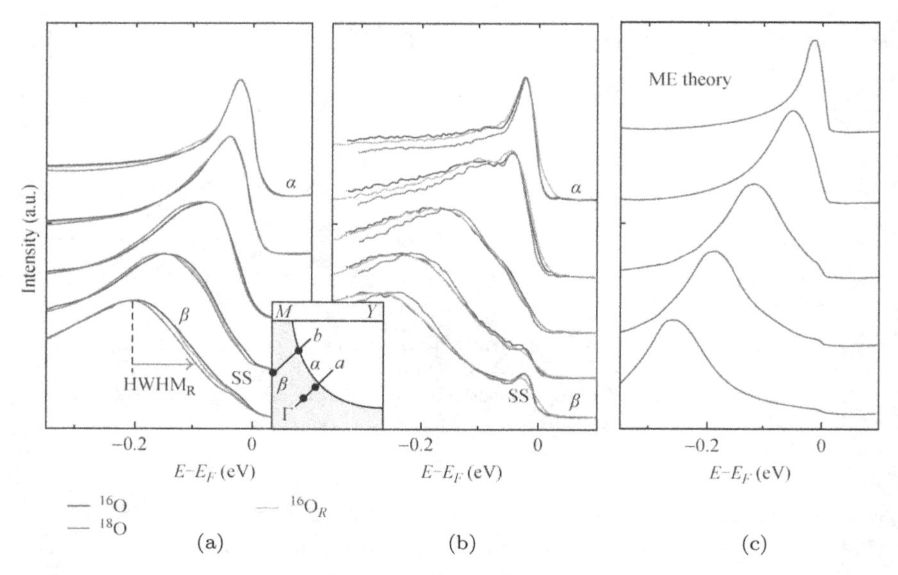

Fig. 2.13. (a) and (b) Comparison of ARPES EDCs for the three samples from Fig. 2.6(c), each normalized to same peak height. The cuts a and b for each panel, respectively, are indicated in the inset spanning the two panels along with the approximate location along the cut in k-space for the EDCs. (c) ME simulations for the expected change to (a) and (b) from ME for ^{16}O and ^{18}O, under-predicting the observed IE as described in text. The figures are from Ref. [42].

To illustrate this former point, Fig. 2.13, begins by showing EDCs for two cuts, a nodal (Fig. 2.13(a)) and a near anti-nodal (Fig. 2.13(b)). Although subtle near the nodal point, there is a significant deviation between the ^{16}O (or the re-substituted ^{16}O$_{R)}$ and the ^{18}O which is not effectively modeled within the expected change from ME theory (Fig. 2.13(c)) and already mentioned previously in regards to the IE on ARPES width. This certainly comes as no surprise since we already knew that the behavior of the dispersion near the kink energy for the cuprate data was not well modeled by ME as compared to the H/W data which better follows its predictions. This is again emphasized in Fig. 2.14, where the IE on the MDC dispersions near the node and antinode (Figs. 2.14(a) and 2.14(b)) is much bigger when compared to the substantially smaller expected change from the ME theory (Fig. 2.14(c)) as well as the aforementioned localization of the change to just near the kink energy. Additionally, as again illustrated in Fig. 2.14, the presence of the momentum dependent sign change in the

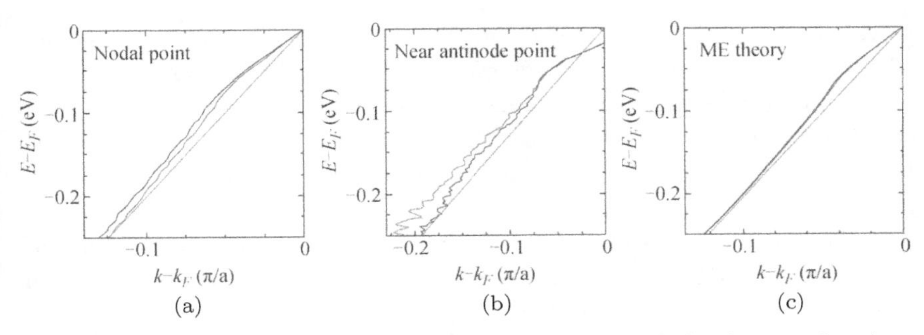

Fig. 2.14. (a) and (b) Comparison of ARPES MDC dispersions for these two locations with the different isotopes. (c) ME simulation for change in dispersion with isotope again under-predicting the magnitude and location of the expected band structure change.

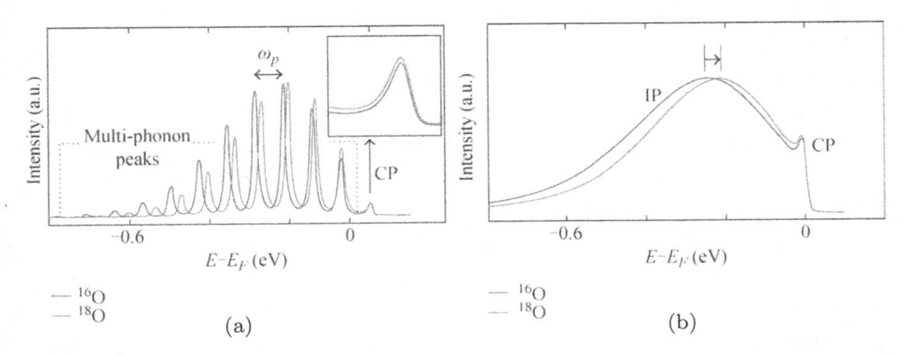

Fig. 2.15. (a) Simulation of EDCs at $k = k_F$ for the off-nodal cut seen in Figure 2.13(b) modeled with a small polaron theory [47]. (b) The expected broadening of this lineshape showing the small IE at low energy (see panel (a) inset) with a more substantial IE shift at higher energy. The figures are from Ref. [42].

dispersion (Figs. 2.14(a) and 2.14(b)) would defy explanation by any simple application of ME theory at least in the absence of an additional ordering phenomenon.

Fundamentally, the failure of the ME theory has its origin in the single phonon loop approximation for the electron self-energy, resulting in a small IE, particularly at higher energy. This suggests to us that we need to increase the electron–phonon coupling in our model. To accomplish this, we employ a Holstein model in the strong-coupling limit as described in Ref. [47]. The resulting comparison of this small polaron theory to the data can be found in Fig. 2.15. These simulate the lineshape at $k = k_F$ for the off-nodal cut Fig. 2.15(b). The multiple peaks seen in Fig. 2.15(a) occur due to the strong multiphonon "shake-up" peaks which appear at harmonics of ω_P. With the expected broadening due to the phonon continuum and strong electron–electron interactions, the result, modeled in Fig. 2.15(b), successfully reproduces the weaker IE for the low-energy CP as well as the larger IE for the broader IP ($\sim$ 30). Additionally, it produces an ARPES linewidth which is more realistic.

These results clearly indicate that the ME theory is insufcient for describing the ARPES data. However, it should be noted that this strong coupling theory used is not quite the right solution either. Within the strong coupling theory, the IP is not expected to have any dispersion while the CP is expected to be nearly non-dispersive as well. Furthermore, it predicts a very small quasi-particle weight ($Z \ll 0.1$) for the CP (panels (c-d)) which is not observed. Both of these issues are better modeled by ME theory; so it is important to note that these shortcomings in the theory are overcome by weakening the interaction strength. This leads us to propose that the proper paradigm for understanding self-energy in the optimally doped cuprates is an intermediate regime. In this regime, there is a significant multiphonon contribution to electron self-energy and both the CP and IP show strong dispersions. Though one may have reasonably expected the IE to also weaken with diminished coupling, some studies [60, 61] show anomalous *enhancement* of the IE at intermediate couplings. Nevertheless, the key message here is that as far as modeling the electron–phonon physics seen in IE, the ME theory is not sufficient and that any subsequent theory must incorporate electron–phonon coupling with a strength beyond the ME paradigm.

2.4. Linking IXS and ARPES through Phonons

With this renewed interest regarding the role of the lattice in the superconducting cuprates as seen by ARPES, a natural direction to explore is to more carefully map the phonon dispersion in addition to that of the electron. A comparison of these dispersions can reveal particularly interesting physics such as when a phonon wavevector matches $2k_F$, where k_F is the Fermi momentum, leading to the well-known Kohn anomaly. Additional, nesting of the Fermi surface in systems with particularly strong electron–lattice coupling can drive the formation of charge density waves. Thus, it becomes important to compare data that directly probes the phonon mode dispersions within the Cu–O plane, such as the one we can get from inelastic neutron scattering (INS) and X-ray scattering (IXS) as discussed in Section 2.2. Then we can compare these results with the observations of the ARPES kink particularly in light of the potential change in the binding energy of the kink as one moves from the nodal point (60–70 meV) towards the anti-nodal point (30–40 meV). Further establishing the phonon nature of the nodal kink as well as shedding light on the lower energy anti-nodal kink is something that a combined ARPES and IXS study could achieve.

2.4.1. *IXS Measurements on La-Bi2201*

We turn our attention to the single-layered $Bi_2Sr_{1.6}La_{0.4}CuO_{6+\delta}$ (La-Bi2201) cuprate which has several advantages for such a study. Like other Bismuth superconducting cuprates, the sample surfaces are high quality for ARPES experiments. These samples have never shown any evidence of magnetic resonance modes, simplifying the comparison between ARPES kink and scattering by removing a potentially additional bosonic mode to couple with the electrons. Moreover, no experimental reports of the optical phonon dispersion exist on these materials to date. The challenge for scattering is the lack of large single crystals, effectively ruling out an INS experiment. Additionally, though IXS can probe the sub-millimeter single crystals available, there is a very low inelastic cross section associated with the bond stretching (BS) mode, a likely candidate for the nodal kink, making observing the mode challenging. Still, both IXS and INS have observed evidence in the past of the Cu–O bond stretching (BS) mode at the metal-insulator phase transition in the superconducting cuprates.

Figure 2.16 encapsulates the IXS experiment with panels (a) and (b) illustrating the relative weakness of the BS phonon peak relative to both the elastic line as well as other modes. Focusing on these higher-energy longitudinal optical modes, we map out their dispersion across the BZ from the center ($\xi = 0$) towards the BZ face. Figure 2.16(a) shows this evolution where the red peak is identified as the BS mode and the results of which are plotted in Figs. 2.16(c) and 2.16(d). We see the two distinct peaks at the zone center but they disperse, becoming indistinguishable around ξ of 0.22–0.25. When $\xi = 0.45$, the two peaks clearly emerge again, leading to two potential scenarios of Figs. 2.16(c) and 2.16(d) depending on the symmetry of the two branches. If they have the same symmetry, they anticross (Fig. 2.16(d)); otherwise they simply cross (Fig. 2.16(c)). Our attempts at distinguishing between these two scenarios using classical shell model

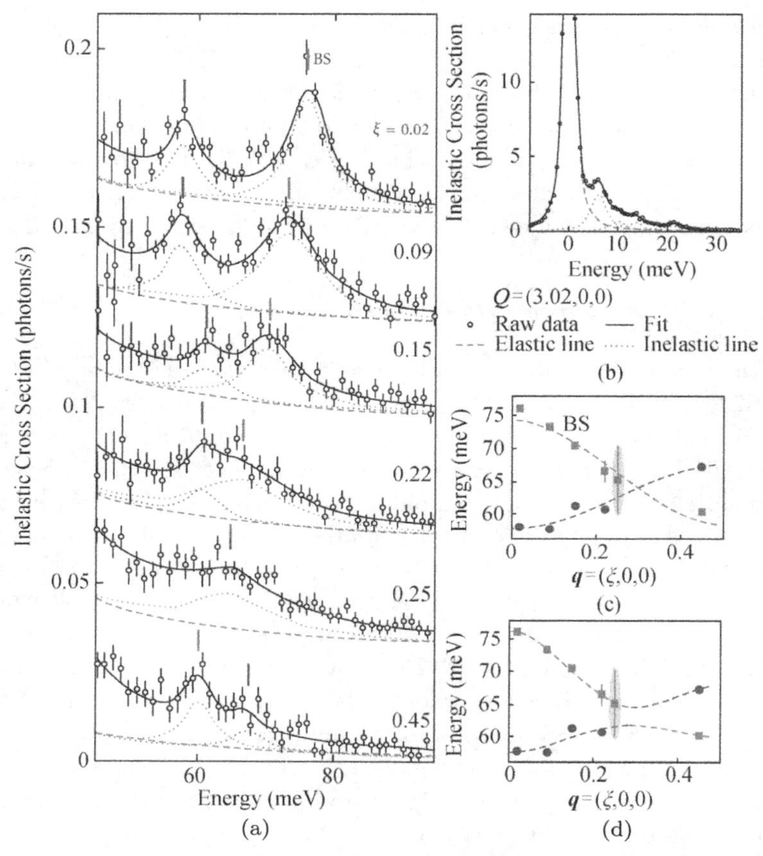

Fig. 2.16. (a) and (b) Raw IXS phonon spectrum taken from optimally doped La-Bi2201 at 10 K. (a) Focusing on the high-energy part and the LO phonon dispersions for $\boldsymbol{Q} = (3+\xi, 0, 0)$ with ξ ranging from the BZ center to the BZ face. (b) Stronger low-energy part of the phonon spectrum. Solid lines indicate fits, dashed lines show the elastic tail, and dotted lines indicate the modes. (c) and (d) The peaks of these dispersions are plotted with cosine-dashed lines for the two potential dispersion scenarios: crossing (c) and anti-crossing (d). The figures taken are from Ref. [62].

calculations could not reproduce the low-and high-energy modes observed in a reliable way. Thus, we have been unable to distinguish between the potential scenarios. Finally, we compared our data with other IXS data in the literature, summarized in Fig. 2.17. We find broad agreement for a softening of the BS mode (panel (a)) between $\xi = 0.2 - 0.3$ as well as a maxima in the full-width half-max (FWHM) of the BS mode peak.

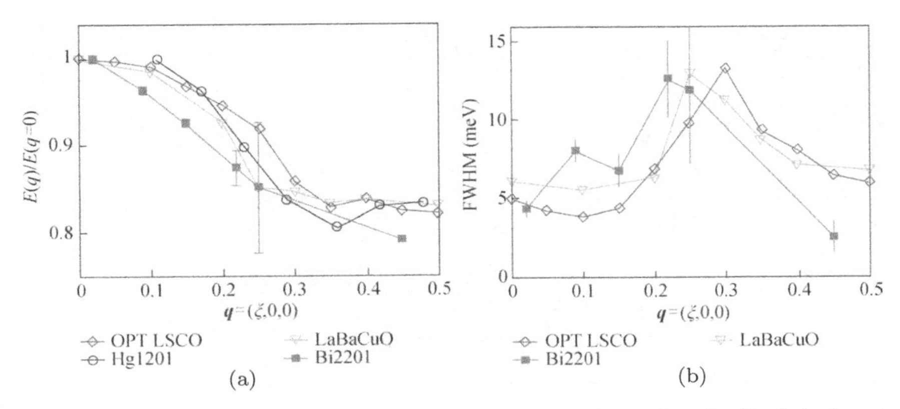

Fig. 2.17. (a) Bond stretching mode softening in a variety of optimally hole-doped cuprates compared to the La-Bi2201 data. (b) Similar comparison of peak FWHM. The figures are from Ref. [62]. Data on LSCO and LaBaCuO come from Ref. [63] and the Hg1201 comes from Ref. [64].

2.4.2. *ARPES Measurements on La-Bi2201*

These results take on a deeper meaning when we compare them with our ARPES studies on La-Bi2201. Figure 2.18 displays the ARPES results in comparison to the IXS data. Figure 2.18(a) shows MDC analysis of the electronic dispersions taken at the nodal point (curve 1) and away from the node (curves 2 and 3) as indicated by the slices along the Fermi surface in Fig. 2.18(b). We see the well-established higher-energy kink at 63 ± 5 meV for the nodal cut. As we move away from the node, this kink abruptly disappears between curves 2 and 3, replaced with only a lower-energy kink of 35 meV. It is significant that this shift occurs at the tips of the so-called "Fermi Arc", region of the Fermi surface which becomes ungapped at temperatures above the superconducting T_c but below the so-called pseudogap temperature, T^*[66]. Beyond the Fermi arc as one moves towards the BZ edge, the gap reopens again for reasons which remain mysterious. It is this transition point in momentum space between the arc and where the gap opens that curves 2 and 3 straddle, though our data was taken in the superconducting phase.

When we compare this with the IXS data as seen in the inset of Fig. 2.18(b), we discover that the 63 meV kink has an energy that corresponds well to the softened BS mode. Even more interesting, as the grey-shaded region in Fig. 2.18(b) illustrates, the region where the 63 meV

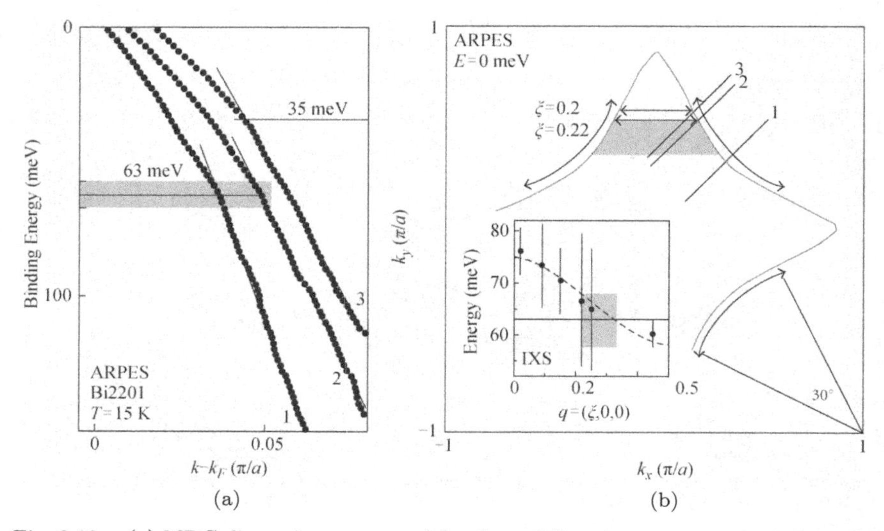

Fig. 2.18. (a) MDC dispersions measured for three different momentum cuts along the G–Y orientation with cut 1 at the nodal point while cuts 2 and 3 are further toward the BZ boundary, near to the edge of the pseudogap phase Fermi arrows [65]. (b) Fermi surface with the cuts from (a) indicated. The solid line indicates a constant energy contour at the kink energy, 63 meV, while the shadow area indicates the region where the nodal kink appears bounded by the indicated nesting wavevectors. The inset shows the IXS dispersion and peak FWHM (seen as error bars) of the BS mode. Note: The apparent Fermi arcs due to the experimental resolution are indicated by the curved arrows. The complete figures can be found in Ref. [62].

kink is observed corresponds to a section of the Fermi surface nestable by wavevectors within the softened part of the phonon mode dispersion from IXS. The sudden disappearance of this kink between curves 2 and 3 corresponds to the stiffening of the BS mode when $\xi < 0.22$. A final insight is that this BS mode is supposed to be nondispersive at about 80 meV along other momentum directions, in particular Ref. [110]. We see no strong feature above 63 meV in our data, helping confirm that the nodal charge carriers are preferentially coupled to the softer Cu–O half-breathing BS mode which disperses along the [100] direction, a result suggested by local spin-density approximation $+U$ results [67]. This work not only provides additional direct evidence for the lattice origin for the 60–70 meV kink seen near the node but also associates it with the softened Cu–O half-breathing BS mode along the [100] direction. Its ability to nest the Fermi surface topology once again underscores the importance of electron–phonon interactions to the physics of the superconducting cuprates.

2.5. Lattice Strain in Bi2201

Our most recent work has continued this exploration into the role of the lattice from yet another perspective. There has been growing independent work from a variety of experimental probes [69–71] positioning the role of the lattice not simply as a source of self-energy effects on the near-E_F low-energy electronic states but potentially as an additional axis within the hole-doped phase diagram. Specifically, it is the effect of lattice strain, both external and internal via chemical pressure, which offers us this new axis to the cuprate phase diagram affecting the super conducting dome. Work using external pressure has indicated critical pressures where the T_c appears to saturate for a range of cuprate hole dopings [69] as well as being coupled to other physical quantities suggesting a significant new critical point along this axis [70]. With chemical pressure, work has suggested that combining doping with strain on the Cu–O layer also reveals that the true quantum critical point is shifted along the strain axis [71]. Additionally, effects related to lattice disorder, particularly in the Sr–O blocks nearest the Cu–O planes, may also have a dramatic effect on the formation of the superconducting phase within the cuprates [72]. Thus, better understanding of this aspect of the role of the lattice is important to our general understanding of electron–lattice physics in these systems.

2.5.1. *Lanthanide Substituted Bi2201*

Due to experimental considerations, the best method for introducing strain into the lattice for an ARPES study is via chemical pressure. Specifically, we can use Lanthanide substituted single-layered $Bi_2Sr_{1.6}Ln_{0.4}CuO_6$ to access this strain in a tunable way [68]. All the samples were grown at optimal doping to simplify the analysis, making the focus solely on the tuning parameter of strain. As Fig. 2.19 illustrates, the substitution of the Lanthanide elements for the Strontium right above the critical Cu–O plane (Fig. 2.19(b)) leads to a monotonic decrease in the measured T_c of the samples. The essential variable to quantify this T_c-competing strain is the atomic radius mismatch, ΔR, as seen on the abscissa of Fig. 2.19(a), which is determined by the difference between the Strontium and the substituted Lanthanide atomic radii, $|R_{\mathrm{Sr}} - R_{\mathrm{Ln}}|$.

We have been able to take data on samples throughout this spectrum of radius mismatch. Although these samples allow us a lattice-based,

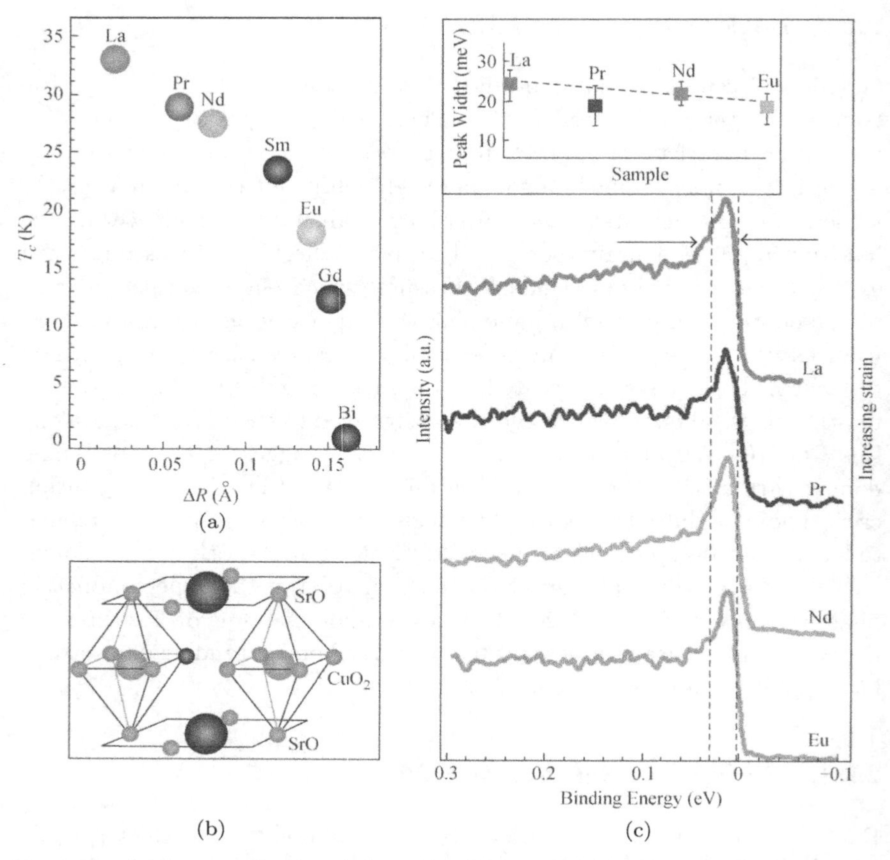

Fig. 2.19. (a) Superconducting T_c for optimally doped $Bi_2Sr_{1.6}Ln_{0.4}Cu_2O_6$ for a series of substituted Lanthanides (Ln) with increasing atomic radius mismatch, $\Delta R = |R_{Sr} - R_{Ln}|$. See Ref. [68]. (b) Cartoon illustrating the location of the substituted Lanthanide right above the Cu–O plane. (c) Nodal point EDCs illustrating the quasi-particle peak for samples with increasing strain, colored in (a), from La to Eu. Inset quanties the half-width half-max of the peaks for these samples.

tunable parameter which competes with superconductivity, one can pose the question if this should be thought of within a lattice strain or lattice disorder paradigm. Figure 2.19(c) provides evidence that, at least for the nodal states, the strain paradigm appears valid. With increasing lattice mismatch, the width of the quasi-particle peak does not increase but, on close examination, may even be decreasing with increased strain. The introduction of lattice disorder should decrease the lifetime of the electronic

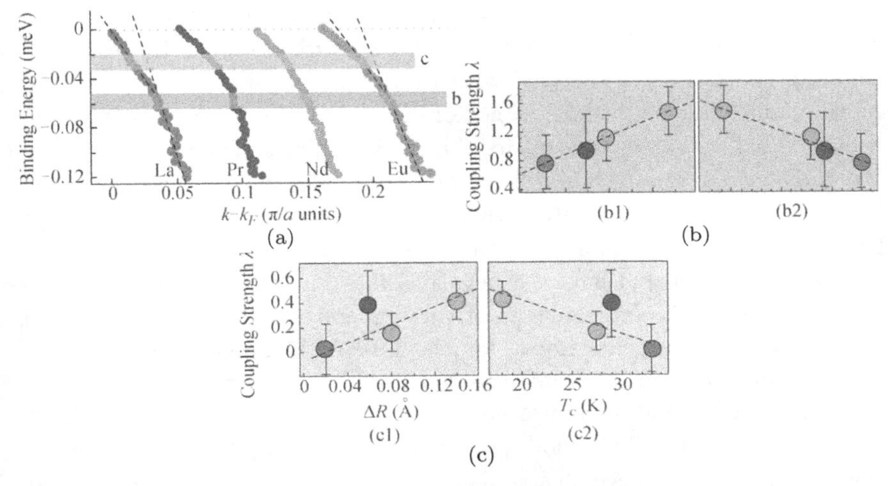

Fig. 2.20. (a) Nodal point MDC dispersions taken from Γ–Y cuts on four different samples of increasing strain (La, Pr, Nd, and Eu). Lines serve as guides to determine deviation from the expected dispersion. Horizontal shaded regions correspond to the two potential ARPES kink energy scales. (b) and (c) Estimating the electron coupling λ on the electronic states from each of the two regions indicated in (a). (b1) and (b2) Higher energy kink λ for each strain as a function of (1) Lattice mismatch Δ, and (2) T_C. (c1) and (c2) Lower energy kink λ analyzed like (b).

states, seen as an increase in the width of the CP. It also has been broadly suggested within the ARPES cuprate community that observing a sharp quasi-particle peak at the nodal point is necessary for confirming that the cleaved surface can provide trustworthy ARPES results.

2.5.2. *ARPES Kink versus Strain*

Throughout this review, we have continually discussed the ARPES kink, particularly near the node, in terms of electron–phonon coupling. So considering the expected effect of the substituted lanthanides on lattice strain and the view that these states at least can be understood within a strain paradigm, we focus our attention again on the MDC dispersions at the nodal point. Figure 2.20 presents our findings for substituted Ln = La, Pr, Nd, Eu [73].

There are two aspects of these dispersions we wish to focus on in the context of this review. First, in agreement with the earlier work on La-Bi2201 discussed in Section 2.4.2, we can observe a kink around 55–60 meV which remains at that energy, for the most part, throughout the strain spectrum. However, what appears to change with strain is the electron-coupling constant, λ associated with the renormalization of these states. In the same manner as was done for the cuprate systems described in Section 2.2.3, we can estimate λ for this mode, which is plotted in panel (b). We find that the strength of this mode appears enhanced by the increasing strain of the lattice mismatch with a generally linear behavior. Equivalently, one can plot λ as a function of sample T_c (Fig. 2.20(b) inset) and, as one would expect, there is a negative, linear relationship between the superconducting T_c and strength of this phonon mode. Secondly, we find that although it appears linear for the La-Bi2201 at energies less than the 60 meV kink, the more strained compounds appear to have additional rounding of the band structure nearer to E_F, most obvious in the highly strained Eu-Bi2201. This mode appears to be around 25–30 meV which could be important since this is closer to the mode energy observed near the anti-nodal point (as discussed in Section 2.2.4) from the peak-dip-hump EDC lineshape. We have also observed this from the kink in the MDC analysis beyond the nestable region of the IXS softened phonon mode in Section 2.4.2.

As with the higher-energy mode, we can attempt to extract the λ from this more elusive mode independent of the 60 meV kink, the result of which is seen in panel (c). Unlike the higher-energy mode which is apparent in all strain, this lowerenergy feature appears to turn on at the node only as strain is introduced, leading to a broadly linear relationship similar to the higher-energy mode. The origin of this feature, remains mysterious, potentially related to an apical oxygen mode particularly in light of the location of the substituted lanthanide above the Cu–O plane.

These results are significant for at least three reasons. First, it has been a prevailing thought that the electronic states at the nodal point are uniquely unaware of the entry into the superconducting phase. With the d-wave symmetry of the gap function, the nodal point states are the only electronic states which still cross through E_F with no gap opening. Additionally, one could argue that the continued appearance of a sharp quasi-particle at the node is merely because these states are protected from the superconducting physics. However, clearly the affect of lattice strain can be seen on these states and the weakening of the superconducting state is present in the electronic dispersion of the nodal quasi-particles. Secondly, one finds still additional evidence that the ~60 meV kink has its origin in the physics of the lattice. Even more intriguing, the appearance and potential enhancement of the lowerenergy kink with lattice strain would tentatively suggest that its origin also is somehow connected to the lattice (such as the aforementioned apical oxygen mode) and not merely a magnetic mode at the nodal point. It could be pointed out that the lanthanides do carry with them magnetic moments whose effect on the Cu–O plane is far from understood. While La has no magnetic moment, Pr, Nd, and Eu all have experimentally determined magnetic moments of around 3.5 μ_B [74]. Whether this may be related to the sudden appearance of this mode at the node for samples beyond La requires further study. Finally, at least for the 60 meV mode, one finds evidence that this phonon mode is somehow connected to the formation of the superconducting phase. From its behavior, it appears to be related to a competing order, associated with the lattice, which may be affecting the formation of the superfluid.

2.6. Summary

Throughout this work, the reoccurring theme has been the growing importance that the lattice and its coupling with electronic states has within the still mysterious phase diagram of the hole doped cuprates. Beginning with the ubiquitous nodal kink and its likely origin in the bond stretching phonon mode, we have explored the effect of the lattice on both the coherent and incoherent parts of the near E_F electronic band structure through the IE. This has confirmed that, near optimal doping, the traditional ME model must give way to a stronger electron–phonon coupling, yet still shy of a polaronic picture. We have traced out the dispersion of the BS phonon mode with IXS and have found its important connection to the ARPES kink near the nodal point. From this, we find that the electronic states within the BZ can be divided up accordingly, with those nearest to the node (often argued to be responsible for the superfluid) uniquely impacted by the mode wavevector's ability to nest them. Finally, the potential for understanding the role of the lattice via the additional axis of pressure/strain has already produced remarkable results for the evolution of the nodal electronic states and their self-energy along this axis as well as pointing towards additional insights. In all, our work has and continues to explore how the cuprate electronic structure is affected by electron–phonon coupling as yet another pillar in our nearly 25 years quests to construct a full understanding of this unconventional superconducting phase.

Acknowledgments

The authors first would like to acknowledge all the members of the group who have greatly contributed to this work: Gey Hong Gweon, Shuyun Zhou, Jeff Graf, and Chris Jozwiak. They also would like to acknowledge many useful scientific discussions with Dung-Hai Lee, Z. Hussain, K. A. Muller, A. Bianconi, T. Sasagawa, and H. Takagi. This work was supported by the Director, Office of Science, Office of Basic Energy Sciences, Materials Sciences and Engineering Division, of the U.S. Department of Energy under Contract no. DE-AC02-05CH11231.

References

[1] J. M. Rowell, P. W. Anderson, and D. E. Thomas, *Phys. Rev. Lett.* **10** (1963), 334.
[2] J. R. Schrieffer, D. J. Scalapino, and J. W. Wilkins, *Phy. Rev. Lett.* **10** (1963), 336.
[3] A. Lanzara, P. V. Bogdanov, X. J. Zhou, *et al., Nature*, **412** (2001), 510.
[4] P. V. Bogdanov, A. Lanzara, S. A. Kellar, *et al., Phys. Rev. Lett.* **85** (2000), 2581.
[5] N. W. Aschcroft and N. D. Mermin, *Solid State Physics.* (Saunders College, Philadelphia, 1976).
[6] L. P. Regnault, P. Bourges, P. Burlet, *et al., Phys. C.* **235** (1994), 59.
[7] H. F. Fong, B. Keimer, D. L. Milius, and I. A. Aksay, *Phys. Rev. Lett.* **78** (1997), 713.
[8] M. Arai, T. Nishijima, Y. Endoh, *et al., Phys. Rev. Lett.* **83** (1999), 608.
[9] P. Dai, H. A. Mook, S. M. Hayden, *et al., Science* **284** (1999), 1344.
[10] E. Demler and S.-C. Zhan, *Phys. Rev. Lett.* **75** (1995), 4126.
[11] P. Dai, M. Yethiraj. H. A, Mook, T. B. Lindemer, and F. Dogan, *Phys. Rev. Lett.* **77** (1996), 5425.
[12] R. J. McQueeney, Y. Petrov, T. Egami, M. Yethiraj. G. Shirane, and Y. Endoh, *Phys. Rev. Lett.* **82** (1999), 628.
[13] Y. Petrov, *et al.*, http://arxiv.org/abs/cond-mat/0003414.
[14] X. J. Zhou, *et al., Handbook of High-Temperature Superconductivity: Theory and Experiment* (springer, Berlin, 2007).
[15] A. Lanzara, P. V. Bogdanov, X. J. Zhou, *et al., Physics and Chemistry of Solids* **67** (2006), 239.
[16] G.-H. Gweon, S. Y. Zhou, and A. Lanzara, *Phys. Chem. Solids* **65** (2004), 1397.
[17] D. S. Dessau, B. O. wells, Z. -X. Shen, *et al., Phys. Rev. Lett.* **66** (1991), 2160.
[18] Y. Hwu, L. Lozzi, M. Marsi, *et al., Phys. Rev. Lett.* **67** (1991), 2573.
[19] J. Rossat-Mignod, L. P. Regnault, C. Vettier, *et al., Phys. C.* **185** (1991), 86.

[20] H. A. Mook, M. Yethiraj, G. Aeppli, T. E. Mason, and T. Armstrong, *Phys. Rev. Lett.* **70** (1993), 3490.
[21] H. F. Fong, P. Bourges, Y. sidis, *et al., Nature* **398** (1999), 588.
[22] D. L. Feng, N. P. Armitage, D. H. Lu, *et al., Phys. Rev. Lett.* **86** (2001) 5550.
[23] Y.-D. Chuang, A. D. Gromko, A. Fedorov, *et al., Phys. Rev. Lett.* **87** (2001), 117003.
[24] P. V. Bogdanov, A. Lanzara, X. J. Zhou, *et al., Phys. Rev. B* **64** (2001), 180505.
[25] A. D. Gromko, A. V. Fedorov, Y.-D. Chuang, *et al., Phys. Rev. B* **68** (2003), 174520.
[26] A. Kaminski, M. Randeria, J. C. Campuzano, *et al., Phys. Rev. Lett.* **86** (2001), 1070.
[27] T. K. Kim, A. A. Kordyuk, S. V. Borisenko, *et al., Phys. Rev. Lett.* **91** (2003), 167002.
[28] T. Cuk, F. Baumberger, D. H. Lu, *et al., Phys. Rev. Lett.* **93** (2004), 117003.
[29] C. M. Varma, P. B. Littlewood, S. Schmitt-Rink, E. Abrahams, and A. E. Ruckenstein, *Phys. Rev. Lett.* **63** (1989), 1996.
[30] K. M. Shen, F. Ronning, D. H. Lu, *et al., Science* **307** (2005), 901.
[31] E. Rotenberg, J. Schaefer, and S. D. Kevan, *Phys. Rev. Lett.* **84** (2000), 2925.
[32] H. Iwasawa, J. F. Douglas, K. Sato, *et al., Phys. Rev. Lett.* **101** (2008), 157005.
[33] G.-H. Gweon, T. Sasagawa, S. Y. Zhou, *et al., Nature* **430** (2004), 187.
[34] K. Mcelroy, D.-H. Lee, J. E. Hoffman, *et al., Phys. Rev. Lett.* **94** (2005), 197005.
[35] S. Andergassen, S. Caprara, C. Di Castro, and M. Grilli, *Phys. Rev. Lett.* **87** (2001), 056401.
[36] A. Lanzara, G. -M. Zhao, N. L. Saini, *et al., Physics Condensed Matter* **11** (1999), L541.
[37] D. Rubio Temprano, J. Mesot, S. Janssen, *et al., Phys. Rev. Lett.* **84** (2000), 1990.
[38] A. Bianconi, N. L. Saini, A. Lanzara, *et al., Phys. Rev. Lett.* **76** (1996), 3412.
[39] G. Seibold and M. Grilli, *Phys. Rev.* **63** (2001), 224505.
[40] J. M. Tranquada, B. J. Sternlieb, J. D. Axe, Y. Nakamura, and S. Uchida, *Nature* **375** (1995), 561.
[41] C. Castellani, C. Di Castro, and M. Grilli, *Phys. Rev. Lett.* **75** (1995), 4650.
[42] G.-H. Gweon, S. Y. Zhou, M. C. Watson, T. Sasagawa, H. Takagi, and A. Lanzara, *Phys. Rev. Lett.* **97** (2006), 227001.
[43] J. F. Douglas, H. Iwasawa, Z. Sun, *et al., Nature* **446** (2007), 7133.
[44] C. Jozwiak, J. Graf, and S. Y. Zhou, *Phy. Rev. B* **80** (2009), 235111.
[45] G.-H. *et al., Gweon Unusual oxygen isotope effects in cuprates — importance of doping.* http://arxiv.rog/abs/0708.1027.
[46] S. V. Borisenko, A. A. Kordyuk, T. K. Kim, *et al., Phys. Rev. B* **66** (2002), 140509.
[47] A. S. Alexandrov and J. Ranninger, *Phys. Rev. B* **45** (1992), 13109.

[48] P. Calvani, M. Capizzi, S. Lupi, P. Maselli, A. Paolone, and P. Roy, *Phys. Rev. B* **53** (1996), 2756.
[49] C. Taliani, R. Zamboni, G. Ruani, F. C. Matacotta, and K. I. Pokhodnya, *Solid State Commun.* **66** (1988), 487.
[50] Y. H. Kim, A. J. Heeger, L. Acedo, G. Stucky, and F. Wudl, *Phys. Rev. B* **36** (1987), 7252.
[51] A. Bianconi and M. Missori, *Phase Separation in Cuprate Superconductors*, edited by E. Sigmund and A. K. Müller (Springer Verlag, Berin-Heidelberg, 1994), p. 272.
[52] I. Bozovic, D, Kirillov, A. Kapitulnik, *et al., Phys. Rev. Lett.* **59** (1987), 2219.
[53] S. J. L. Billinge and T. Egami, *Phys. Rev. B* **47** (1993), 14386.
[54] D. Mihailovic and K. A. Muller, *High-T_c superconductivity 1996: Ten Years after the Discovery*, vol. 343 of NATO Advanced Study Institutes, Series E (Kluwer Academic Publishers, Dordrecht 1997).
[55] T. Imai, C. P. Slichter, K. Yoshimura, and K. Kosuge, *Phys. Rev. Lett.* **70** (1993), 1002.
[56] B. I. Kochelaev, J. Sichelschmidt, B. Elschner, W. Lemor, and A. Loidl, *Phys. Rev. Lett.* **79** (1997), 4274.
[57] A. S. Mishchenko and N. Nagaosa, *Phys. Rev. B* **73** (2006), 092502.
[58] S. Fratini and S. Ciuchi, *Phys. Rev. B* **72** (2005), 235107.
[59] A. S. Alexandrov and J. Ranninger, *Phys. Rev. B* **45** (1992), 13109.
[60] A. S. Mishchenko and N. Nagaosa, *Phys. Rev. B* **73** (2006) 092502.
[61] P. Paci, M. Capone, E. cappelluti, S. Ciuchi, C. Grimaldi, and L. Pietronero, *Phys. Rev. Lett.* **94** (2005), 036406.
[62] J. Graf, M. D'Astuto, C. Jozwiak, *et al., Phys. Rev. Lett.* **100** (2008), 227002.
[63] D. Reznik, L. Pintschovius, M. Fujita, K. Yamada, G. D. Gu, and J. M. Tranquada, *J. Low Temp. Phys.* **147** (2007), 353.
[64] H. Uchiyama, A. Q. R. Baron, S. Tsutsui, *et al., Phys. Rev. Lett.* **92**, (2004), 197005.
[65] T. Kondo, T. Takeuchi, A. K.aminski, S. Tsuda, and S. Shin, *Phys. Rev. Lett.* **98** (2007), 267004.
[66] A. Kanigel, M. R. Norman, M. Randeria, *et al., Nature Phys.* **2** (2006), 447.
[67] P. Zhang, S. G. Louie, and M. L. Cohen, *Phys. Rev. Lett.* **98** (2007), 067005.
[68] H. Eisaki, N. Kaneko, D. L. Feng, *et al., phys. Rev. B* **69** (2004), 064512.
[69] X.-J. Chen, V. V. Struzhkin, R. J. Hemley, H.-K. Mao, and C. Kendziora, *Phys. Rev. B* **70** (2004), 214502.
[70] T. Cuk, V. V. Struzhkin, T. P. Devereaux, *et al., Phys. Rev. Lett.* **100** (2008), 217003.
[71] A. Bianconi, S. Agrestini, G. Bianconi, D. Di Castro, and N. L. Saini, *J. Alloys and Compounds* **317** (2001), 537.
[72] H. Hobou, S. Ishida, K. Fujita, *et al., Phys. Rev. B* **79** (2009), 064507.
[73] D. R. Garcia, *et al., to* be submitted to *Phys. Rev. B.*
[74] R. L. Carlin, *Magnetochemistry* (Springer, New York, 1986).

3
Electron–Phonon Coupling Effects Explored by Inelastic Neutron Scattering*

L. Pintschovius

Forschungszentrum Karlsruhe, Institut für Festkörperphysik, Postfach 3640, 76021 Karlsruhe, Germany

This paper gives a summary of inelastic neutron scattering studies searching for signatures of a strong electron–phonon coupling in high-temperature superconductors. Special emphasis is laid on the anomalous dispersion, the unusually large linewidths and the anomalous temperature dependence observed for plane polarized copper-oxygen bond-stretching vibrations. It will be discussed in how far these results can be understood within conventional density functional theory or require recourse to theories for strongly correlated electrons, in particular the concept of charge stripe order.

3.1. Preliminary Remarks

Inelastic neutron scattering is the most powerful technique for investigations of the phonon properties in solids. In particular, inelastic neutron scattering on single crystals is able to determine the phonon dispersion relations and therefrom to derive the lattice contribution to the specific heat and to infer the interatomic force field. Further, an understanding of the phonon properties is indispensable for an understanding of structural phase transformation. In the case of the high-T_c compounds, it was moreover hoped to

*Reprinted with permission from L. Pintschovius. Original published in *Phys. Status. Solidi. B.* **242** (2005), No. 1, 30–50.

find signatures of a strong electron–phonon coupling which might help to elucidate the still unknown mechanism of high-T_c superconductivity. However, relatively soon after the discovery of the high-T_c superconductors, only a minority of the high-T_c community considered electron–phonon coupling as relevant for the mechanism of high-T_c superconductivity. Therefore, inelastic neutron scattering was primarily used to study the spin fluctuations in the cuprate superconductors rather than the phonons. In recent years, however, a growing number of people working in the field came to the opinion that phonons might have been discarded prematurely. As a consequence, a series of in-depth studies of the phonon properties in various compounds were undertaken providing detailed and often unexpected results. In view of the fact that the inelastic neutron scattering results on the phonon properties in high-T_c superconductors have been reviewed already in 1994 [1] and 1998 [2], the present article will focus on the recent results albeit some of the early results will be presented as well for the sake of completeness.

The paper is organized as follows: In Section 3.2, electron–phonon coupling effects in classical super-conductors will be briefly reviewed to set the stage. Section 3.3 is devoted to the phonon properties of the insulating parent compounds because these compounds are indispensable as a reference. In Section 3.4, the central part of this article, a review will be given on the present knowledge of electron–phonon coupling effects in high-T_c superconductors. Some concluding remarks will be presented in Section 3.7.

3.2. Electron–Phonon Coupling Effects in Classical Superconductors

The search for electron–phonon coupling effects in high-T_c superconductors was guided by the experience gained from investigations on classical superconductors. In general, it was found that the phonon dispersion curves of superconductors exhibit more structure than those of non-superconducting reference compounds. An example is shown in Fig. 3.1 for two compounds of the well-known A15 family. The local minimum seen in the longitudinal acoustic branch in the superconductor Nb_3Sn ($T_c = 18$ K) around $q = 0.3$ reciprocal lattice units (r.l.u.) is a typical example of a phonon anomaly. Often, although not always, the phonons with anomalously low frequencies show also an anomalous temperature dependence, i.e., they soften on cooling rather than to show the usual slight hardening related to anharmonicity.

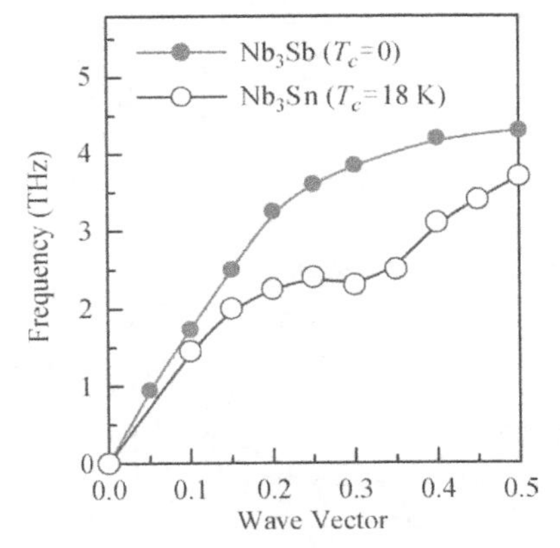

Fig. 3.1. Dispersion of the longitudinal acoustic branch in the (111)-direction in Nb_3Sn [3] and in Nb_3Sb [4].

Such a behaviour is documented in Fig. 3.2 for the phonon branch shown in Fig. 3.1. Theory predicts that the anomalous behaviour should be accompanied by relatively large linewidths. The linewidths are directly related to the electron–phonon coupling constant λ which is proportional to the sum of the linewidths of the individual phonon branches averaged over the whole Brillouin zone. Unfortunately, the broadenings due to electron–phonon coupling are generally quite small, even for strong coupling superconductors like Nb_3Sn: they are of the order of 1% or a few % at maximum. Such small broadenings are very difficult to detect in a neutron experiment due to insufficient resolution. Further, phonon–phonon interaction (anharmonicity) might contribute in as much to the total linewidth as electron–phonon coupling. As a consequence, an experimental determination of the electron–phonon coupling constant λ is unfeasible even in the most favorable cases. This statement holds true even for the recently discovered superconductor MgB_2 with its exceptionally high $T_c = 39$ K. In this compound, a particular phonon mode with E_{2g} symmetry shows an extremely strong coupling and hence a very large linewidth, which can be easily determined by experiment. However, calculations have revealed that the E_{2g} related modes contribute not more than 50% of the total electron–phonon coupling, the rest coming from all the other modes showing rather small linewidths.

Figure 3.2 depicts not only the experimental data but also theoretical curves based on density functional theory within the local density approximation (LDA). The agreement with experiment is quite impressive. Since

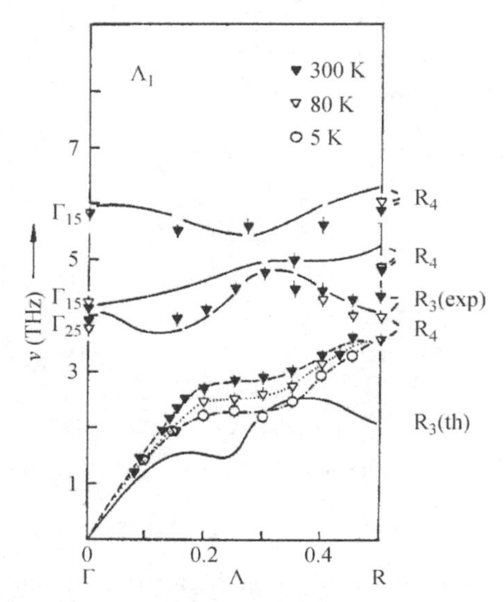

Fig. 3.2. Phonon dispersion curves in Nb_3Sn for the Λ_1 representation at three different temperatures. Full lines have been calculated from density functional theory by Weber [5] prior to the measurements. Dashed, dotted, and dot-dashed lines are guides to the eye only. After [3].

then, theory has made further progress as was demonstrated recently for MgB_2. However, the same theory is not as successful for high-T_c compounds. I will come back to this issue in Section 3.4.

3.3. Phonon Dispersion Relations in the Insulating Parent Compounds

The cuprates have rather complex structures with many atoms in the unit cell and consequently very complex phonon dispersion relations. As a result, it is very difficult if not impossible to discern phonon anomalies just by inspection of the experimentally determined phonon branches. It turned out that the only viable way to identify phonon anomalies in high-T_c cuprates is by comparison to the phonon dispersion of the insulating parent compounds. Therefore, the experimental determination of the phonon dispersion relations in a number of undoped cuprates was indispensable for the later search of electron–phonon coupling effects in the doped compounds. As an example, the phonon dispersion curves of undoped La_2CuO_4 are

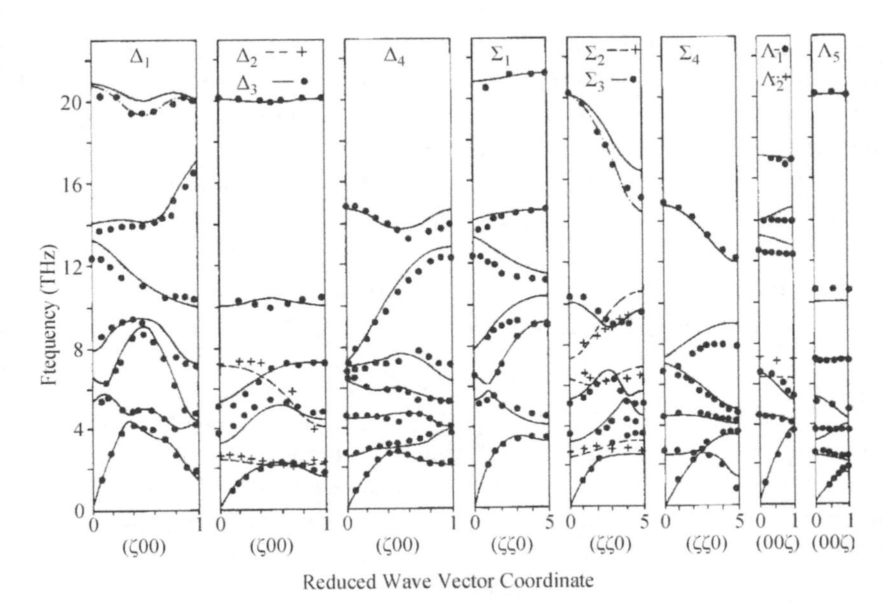

Fig. 3.3. Phonon dispersion curves of undoped La_2CuO_4. The solid lines were calculated from an interaction potential model developed for the structure and the lattice dynamics of several cuprates. The dash-dotted lines are obtained after inclusion of a special quadrupolar force constant. The figure was taken from Ref. [6].

shown in Fig. 3.3. The figure demonstrates that the phonon dispersion in undoped La_2CuO_4 is very well described by phenomenological models developed long time ago for ionic insulators, in particular the shell model. On these grounds, none of the features of the phonon dispersion curves can be considered as anomalous. One might assume that the very steep downward dispersion of the high frequency optical modes of Σ_3 symmetry is somewhat anomalous because (i) such high frequency modes have usually very little dispersion and (ii) a special force constant had to be included into the calculations to arrive at a quantitative description of this phonon branch. However, one has to bear in mind that the simple ionic model without any special force constant leads already to a very decent description of the data. What is more, later investigations on doped $La_{2-x}Sr_xCuO_4$ (LSCO) have shown that the dispersion of this particular branch is slightly reduced upon doping rather than becoming even stronger. Further, these phonons do not acquire large linewidths on doping, in contrast to other phonon modes (see Section 3.4), which also indicates that the electron–phonon coupling is quite small for these modes. I note that the endpoint of the branch under discussionw is the quadrupolar mode which is expected to go soft on approaching a transition to a Jahn–Teller distortion within the basal plane (which would come on top of the well-known elongation along c of the Cu–O octahedra related to the fact that Cu^{2+} is a Jahn–Teller ion). Apparantly, such distortions of the Jahn–Teller type within the basal plane are irrelevant for the physics of the cuprates.

3.4. Phonon Anomalies in High-T_c Superconductors

Figure 3.4 demonstrates that the model used in Fig. 3.3 to describe the phonons in undoped La_2CuO_4 gives a very good description of the phonons in underdoped $La_{1.9}Sr_{0.1}CuO_4$ as well. I note that little change occurs on further doping except for those few phonon modes which are not well described already in underdoped LSCO. This means that the basic character of the interatomic forces remains the same after doping, i.e., they are

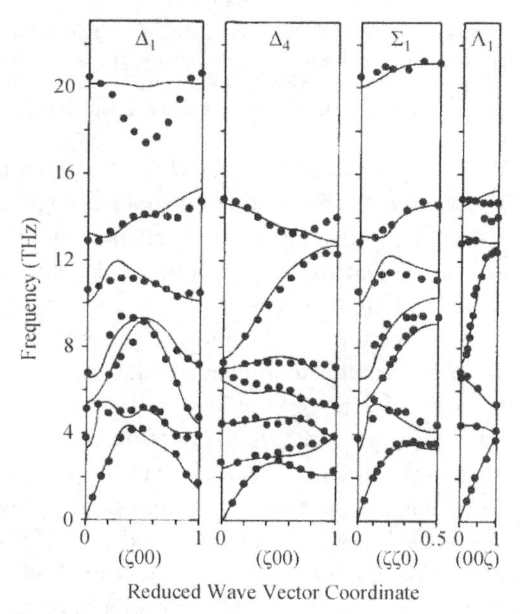

Fig. 3.4. Selected phonon branches of underdoped $La_{1.9}Sr_{0.1}CuO_4$. The lines were calculated from the model as used to describe the dispersion curves of undoped La_2CuO_4 show in Fig. 3.3 implemented with a special term accounting for the screening by free carries. The figure was taken from Ref. [6].

ionic in nature. Of course, the model has to be adapted to account for the insulator-to-metal transition upon doping. To this end, a screening term has been added which suppresses the LOTO splittings observed in the insulating parent compound. However, the inverse screening length in the cuprates is quite small. That is to say, screening affects essentially only long wavelength phonons but not zone boundary modes. This issue will be discussed in somewhat more detail later when dealing with the phonons in electron-doped $Nd_{2-x}Ce_xCuO_4$ (NCCO). I note that the screening effects can be considered as electron–phonon coupling effects only in a very wide sense, but have nothing to do with the coupling leading to superconductivity.

Inspection of Fig. 3.4 shows that the most pronounced deviation between model and experiment occurs in a high-frequency optical branch of Δ_1 symmetry in the (100)-direction. This branch is associated with plane-polarized Cu–O bond-stretching vibrations. The frequency dip observed in this branch closely resembles a phonon anomaly in a classical superconductor. For this reason, these phonon modes have attracted considerable interest and are dealt with in detail in the following subsection.

3.4.1. *Bond-stretching Modes*

The best studied system is $La_{2-x}Sr_xCuO_4$ with measurements ranging from undoped ($x = 0$) to highly overdoped ($x = 0.30$) samples. A summary of the results of the dispersion of the plane polarized Cu–O bond-stretching vibrations is given in Figs. 3.5–3.7. Evidently, the pronounced renormalization of the frequencies of these modes is restricted to those of longitudinal character. Further, the renormalization is considerably stronger in the (100)-direction than in the (110)-direction, at least up to optimal doping. As a consequence, it is the modes propagating along (100) which have attracted most of the attraction, in particular the zone boundary mode for $\boldsymbol{q} = (0.5, 0, 0)$. Its displacement pattern is shown in Fig. 3.7 along with that of the zone boundary mode in the (110)-direction, i.e., for $\boldsymbol{q} = (0.5, 0.5, 0)$. The latter mode is traditionally called "breathing mode". In analogy, the zone boundary mode in the (100)-direction is now generally called "half-breathing mode". I note that in the cuprates, these modes should be in principle called *planar* breathing mode resp. *planar* half-breathing mode because these modes involve primarily in-plane elongations (Fig. 3.8).

Figure 3.5 shows that the frequency renormalization of the half-breathing mode continues on going from optimally doped LSCO to strongly

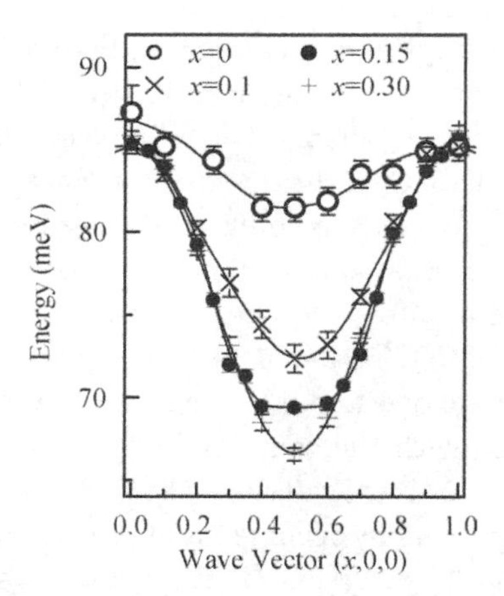

Fig. 3.5. Dispersion of the longitudinal (Δ_1-symmetry) plane polarized Cu–O bond-stretching vibrations in the (100)-direction in $La_{2-x}Sr_xCuO_4$ [2, 7]. Lines are a guide to the eye.

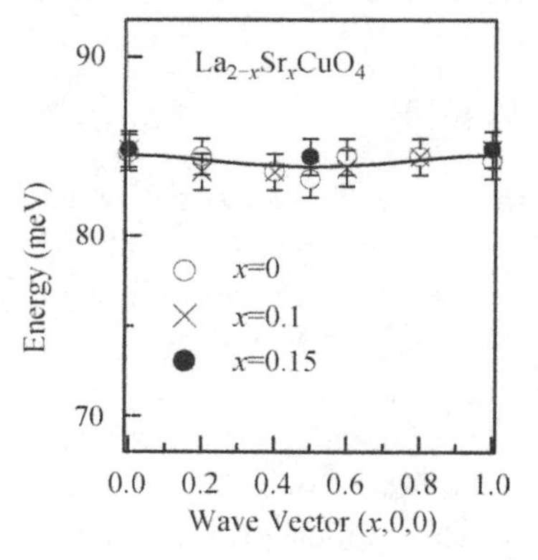

Fig. 3.6. Dispersion of the transverse (Δ_3-symmetry) plane polarized Cu–O bond-stretching vibrations in the(100)-direction in $La_{2-x}Sr_xCuO_4$ [2].

overdoped LSCO, although the additional changes are rather small compared to the effect occuring between undoped and optimally doped LSCO. This result [7] contradicts earlier results obtained on overdoped LSCO by inelastic X-ray scattering [8] claiming that the dispersion in strongly overdoped ($x = 0.29$) LSCO would be close to that in the undoped parent compound. This led to the speculation that the renormalization is strongly correlated with the superconducting transition temperature T_c.

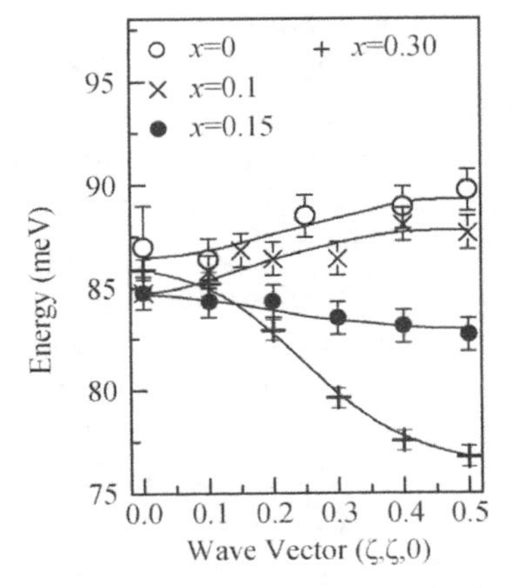

Fig. 3.7. Dispersion of the longitudinal (Σ_1-symmetry) plane polarized Cu–O bond-stretching vibrations in the (110)-direction in $La_{2-x}Sr_xCuO_4$ [2]. Lines are a guide to the eye.

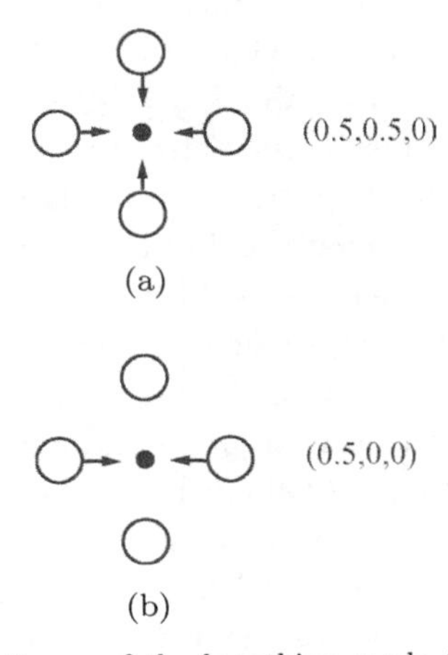

Fig. 3.8. Displacement patterns of the breathing mode (a) and of the half-breathing mode (b), respectively.

However, inspection of the data reported in Ref. [8] shows a two-peak structure for $q < 0.8$ in contrast to what is expected from selection rules for the compound under investigation. This flaw made an assignment of the peaks to the targeted phonon modes ambiguous. I note that no such problem occured in the neutron measurements [7]. Moreover, very recent X-ray measurements on a similar sample fully confirmed the neutron results [9].

Investigations of the doping-induced frequency changes in the electron-doped compound $Nd_{2-x}Ce_xCuO_4$ (NCCO) were for a long time hampered by the lack of suitable single crystals. Inelastic X-ray scattering has the advantage that it can be performed on very tiny crystals and therefore, inelastic X-ray data for doped NCCO [10, 11] were reported prior to inelastic neutron results [12]. According to the X-ray results, NCCO behaves in a somewhat different manner than LSCO: it was claimed [10, 11] that an anomalous softening is observed in the bond-stretching branch in the (100)-direction at wavevector $\boldsymbol{q} = (0.2, 0, 0)$. However, recent neutron experiments [12] revealed unambiguously that the dip at $\boldsymbol{q} = (0.2, 0, 0)$ is simply the consequence of an anti-crossing of two phonon branches with elongations in the Cu–O planes or in the Nd–O planes, respectively. In fact, the interpretation of the doping-induced frequency changes in NCCO is far from being trivial due to an interaction of three phonon branches of the same symmetry at energies above 10 THz. In the undoped compound, the situation is as following: the branch associated with plane-polarized Cu–O bond-stretching vibrations starts at nearly 18 THz and disperses downward to about 15 THz at $\boldsymbol{q} = (1, 0, 0)$. There is an anti-crossing with another branch at about $\boldsymbol{q} = (0.5, 0, 0)$. This branch starts at $\nu = 13$ THz and disperses upward to about 17 THz. This branch is also plane-polarized but is associated with elongations within the Nd–O plane. Finally, there is a c-polarized branch starting from $\nu = 13$ THz and dispersing upward to nearly 14 THz. Inspection of Fig. 3.9 shows that the doping induced changes of this picture are very pronounced. However, the three branches under discussion are affected to a very different degree. I note that the most pronounced changes are observed in the branch associated with Cu–O elongations. As a consequence, the anti-crossing of the two highest branches is shifted from $q = 0.5$ to $q = 0.2$. Further, the strong softening of the Nd–O bond-stretching branch at the zone center leads to an anti-crossing with the c-polarized branch at $\boldsymbol{q}$ about (0.1, 0, 0). Apart from this phenomenon, the c-polarized branch shows very little change upon doping. In order to visualize the doping induced changes more clearly, they have been plotted in Fig. 3.10 according to the character of the two types of bond-stretching vibrations, i.e., Nd–O and Cu–O, respectively. Apparently, very large effects occur for both types of vibrations at the zone center. These changes simply reflect the insulator-to-metal transition on doping by suppressing the LO-TO splittings. The rapid decrease of the screening effects with increasing wavevector

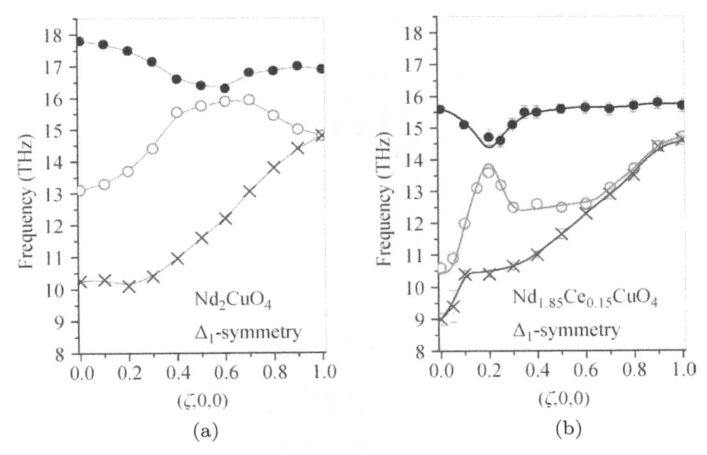

Fig. 3.9. Dispersion of high-energy phonon branches in (a) undoped Nd_2CuO_4 [2] and (b) in optimally doped $Nd_{1.85}Ce_{0.15}CuO_4$ [12] in the (100)-direction. Lines are a guide to the eye.

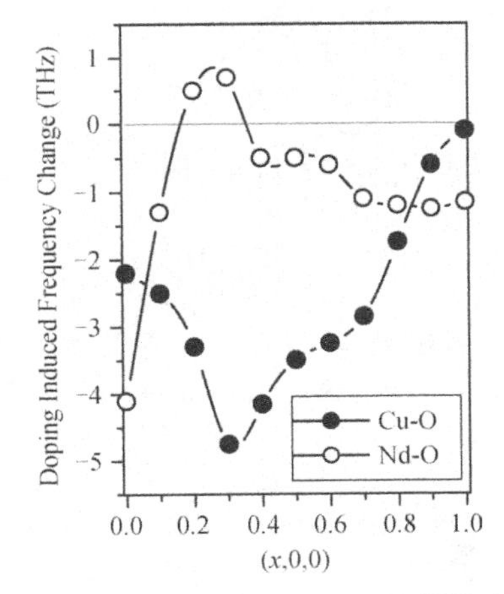

Fig. 3.10. Doping-induced frequency changes in NCCO in plane-polarized bond-stetching vibrations with elongations primarily within the Cu–O planes or within the Nd–O planes, respectively [12].

in the Nd–O bond-stretching branch indicates that the inverse screening length in NCCO is very short or, in other words, that NCCO is a poor metal. On the other hand, the doping-induced frequency changes increase with increasing wavevector for the Cu–O bond-stretching vibrations which clearly shows that a different mechanism is at work for finite wavevectors. By and large, the effects observed for the Cu–O bond-stretching vibrations

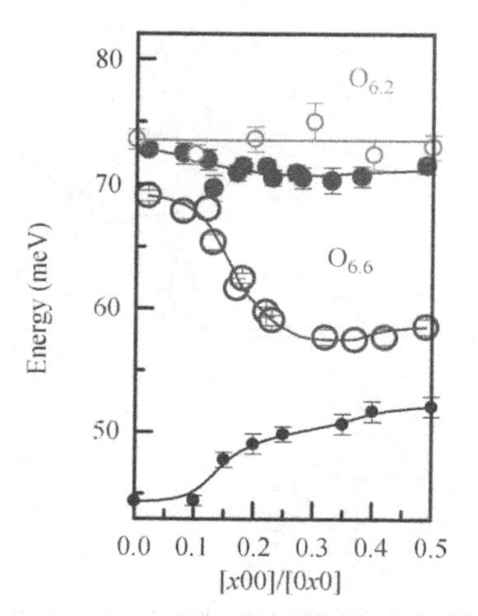

Fig. 3.11. Dispersion of plane-polarized Cu–O bondstretching vibrations in insulating $YBa_2Cu_3O_{6.2}$ ($O_{6.2}$) and in underdoped $O_{6.6}$ at $T = 10$ K [13]. Note that $O_{6.2}$ is tetragonal whereas $O_{6.6}$ is orthorhombic. Full and open symbols refer to the (100)-direction and the (010)-direction, respectively. The phonon branch below 52 meV has bond-bending character. The depicted branches are of Δ_4-symmetry which means that the elongations in the Cu–O bi-layer are out-of-phase. Lines are a guide to the eye.

in optimally doped NCCO are similar to those observed in optimally doped LSCO.

There is a further analogy with optimally doped LSCO: the renormalization of the Cu–O bond-stretching phonons is weaker in the (110)-direction than in the (100)-direction, i.e., 2.2 THz compared to 3.5 THz for the zone boundary modes, respectively [12]. Note that Fig. 3.2 of the X-ray paper [11] suggests a considerably stronger renormalization in the (110)-direction, i.e., 4 THz, which was not confirmed by the neutron results. As is stated in the caption of this figure, the data points close to the zone boundary were of poor quality due to very weak inelastic structure factors. The structure factors are more favorable in the neutron case.

The doping-induced frequency changes in $YBa_2Cu_3O_x$ (YBCO) also resemble those observed in LSCO showing pronounced softenings in the (100) — respectively, the (010) — directions (Figs. 3.11 and 3.12) and sizeable, but less pronounced softenings in the (110)-direction (Fig. 3.13). Note that the crystal structure of $YBa_2Cu_3O_x$ changes with doping from tetragonal to orthorhombic for $x > 6.4$ because the extra oxygen forms Cu–O chains running along the (010)- (or b-)direction. The formation of the chains is accompanied by a slight contraction (expansion) of the a- (b-) axis lattice

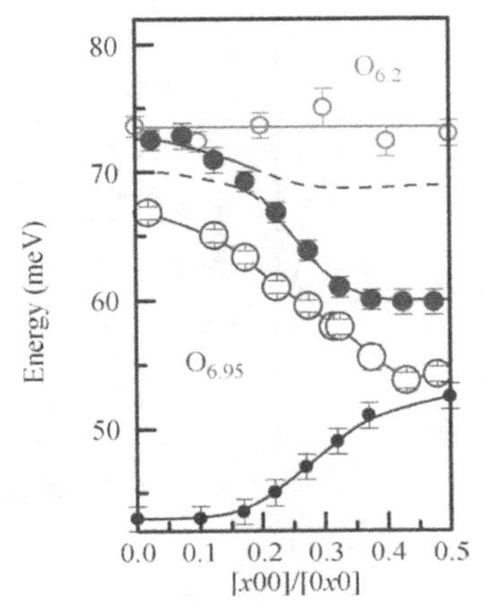

Fig. 3.12. The same as Fig. 3.11 but for optimally doped $O_{6.95}$ [14]. The dashed line depicts a c-polarized branch of Δ_4-symmetry showing an anti-crossing with the bond-stretching branch. The data were taken at $T = 10$ K except for the bond-stretching branch in the (010)-direction for which 200 K data are shown. The low temperature data will be dealt with in Section 3.5.

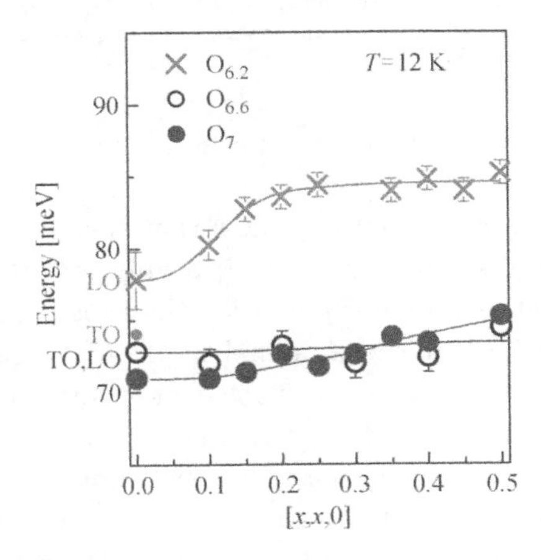

Fig. 3.13. Phonon dispersion of the bond-streching modes in the (110) direction in $YBa_2Cu_3O_x$ [13]. Data for O_7 were taken from Ref. [16]. Lines are a guide to the eye.

constant. As a consequence, the Cu–O bond-stretching force constant is lower in the b-direction than in the a-direction and therefore, the Cu–O bond-stretching vibrations are lower in frequency in the b-direction. However, the pronounced anisotropy of the frequencies of the zone boundary modes in underdopd $O_{6.6}$ cannot be explained by the structural anisotropy

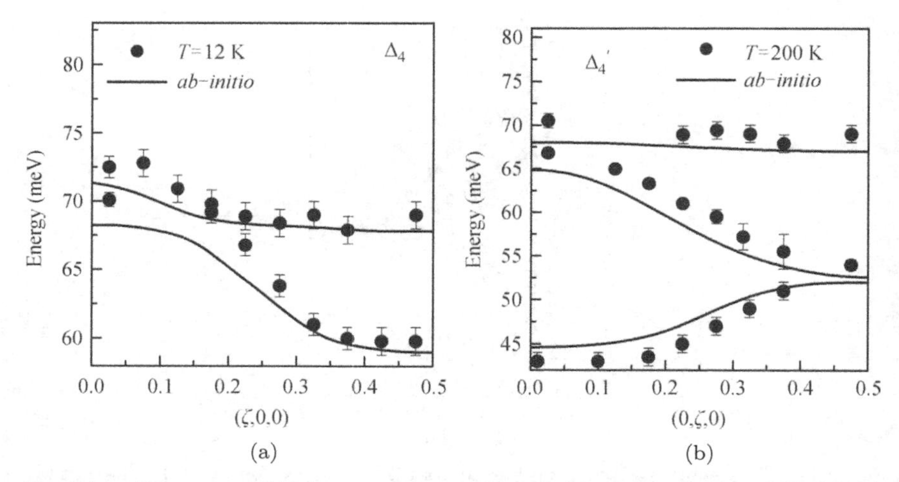

Fig. 3.14. Dispersion of high-energy phonon branches in optimally doped YBCO (a) in the (100)-direction and (b) in the (010)-direction, respectively [15]. These branches are of Δ_4-symmetry which means that the elongations in the CuO_2-bilayer are out-of-phase. The downward dispersing branches are associated with plane-polarized Cu–O bond-stretching vibrations. The flat branches at $E = 68$ meV are c-polarized. The upward dispersing branch has Cu–O bond-bending character. The *ab initio* results were taken from Ref. [17].

alone but is due to a much stronger renormalization in the b-direction. This fact is yet unexplained.

The experimental results clearly show that the doping-induced softening of the Cu–O bond-stretching modes is a universal effect in the high-T_c compounds. Several theoretical explanations have been put forward for this phenomenon. To begin with, the phonon dispersion of high-frequency modes in fully doped YBCO as calculated from density functional theory within the local density approximation (LDA) [17] are shown in Fig. 3.14. Obviously, the theory quantitatively reproduces the anomalous dispersion of the bond-stretching branches. Unfortunately, there is no theoretical result for the insulating parent compound $YBa_2Cu_3O_6$ because LDA does not predict the correct ground state of O_6. Nevertheless, the impressive agreement between the calculated and the experimental frequencies in O_7 gives rise to hopes that the same theory might be used to calculate the electron–phonon coupling strength. Calculation of the Eliashberg function α^2 F led to a small coupling constant $\lambda = 0.27$ and hence a vanishing or very low T_c. Therefore, Bohnen *et al.* [17] conclude that conventional electron–phonon coupling is not important for high-T_c superconductivity.

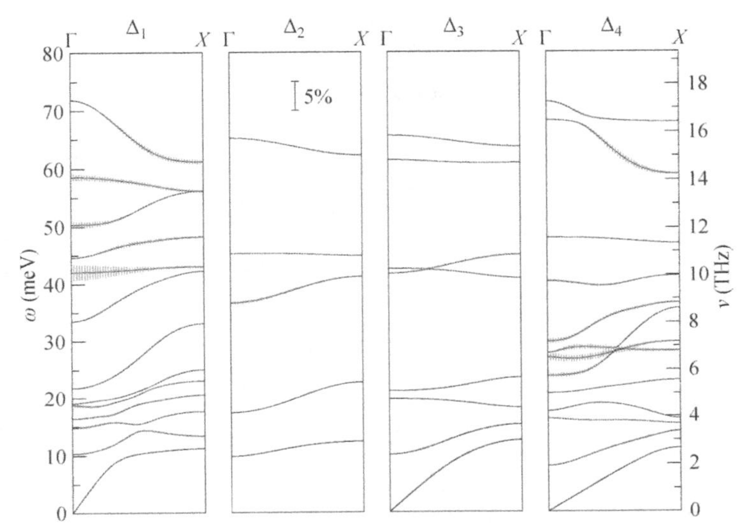

Fig. 3.15. Phonon dispersion curves of $YBa_2Cu_3O_7$ in the (100)-direction as calculated by density functional theory [17]. The verical bars denote the relative linewidth $\Delta\nu/\nu$ (FWHM) related to electron–phonon coupling calculated from the same theory [18]. The scale for the relative linewidth is the same as for the energies.

The theoretical result of a very small λ can be checked experimentally by comparing the calculated phonon linewidths to the observed ones. As stated earlier, the linewidths are directly related to λ via a Brillouin zone average. The calculated linewidths are illustrated in Fig. 3.15 for the (100)-direction. Apparently, the calculated linewidths are generally very small. Somewhat surprisingly, this statement holds even for the Cu–O bond-stretching modes although the theory correctly predicts their anomalous dispersion. Unfortunately, it is technically very difficult to accurately determine the phonon linewidths of the bond-stretching modes in $YBa_2Cu_3O_7$ on the available samples which are either large in volume but twinned or detwinned but quite small. Nevertheless, the available results [19] indicate that the theory is correct in predicting a strong increase of the linewidths from the zone center towards the zone boundary with a local maximum at about (0.3, 0, 0) reciprocal lattice units (r.l.u.). However, the theory certainly underestimates the absolute values by a large factor.

The experimental determination of the linewidths of the bond-stretching modes is easier in LSCO because (i) there is no problem related to twinning and (ii) there is no problem related to anti-crossings with other

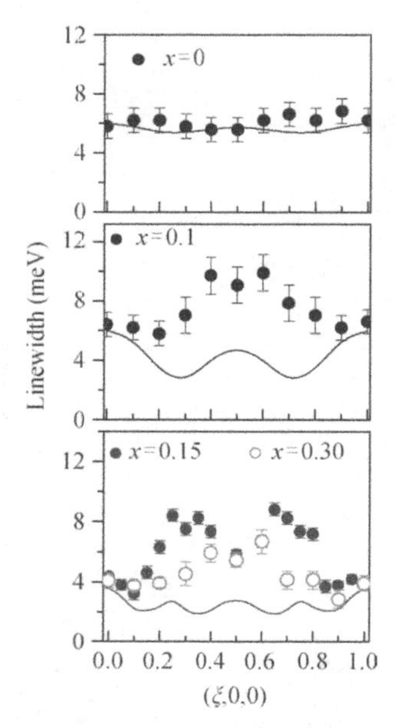

Fig. 3.16. Phonon linewidths observed for the Cu–O bond-stretching modes in $La_{2-x}Sr_xCuO_4$ for undoped [1], underdoped [1], optimally doped [25] and overdoped [7] samples. The lines denote the experimental resolution including focusing effects.

branches. As a consequence, high quality data are available for various doping levels (Fig. 3.16). It turns out that the linewidths are quite small in the undoped parent compound but very large in optimally doped LSCO, i.e., — after correcting for the experimental resolution — up to about 10%, i.e., an order of magnitude larger than was calculated for YBCO. Further, recent data for electron-doped NCCO [12] show a similar behavior for this compound which means that they are typical for high-T_c compounds (Fig. 3.17). Therefore, the prediction of small linewidths for the bond-stretching modes in YBCO appears as a serious shortcoming of the LDA theory.

In this context, I want to emphasize that the linewidths of the bond-stretching modes decrease substantially on going from optimally doped LSCO to overdoped LSCO although the half-breathing mode continues to soften somewhat (Fig. 3.16). This observation indicates that there is no direct relationship between softening and line broadening and that the electron–phonon coupling strength is indeed reduced on overdoping.

The pronounced underestimation of the linewidths of the Cu–O bond-stretching modes by the LDA theory [17] casts doubts on the validity of

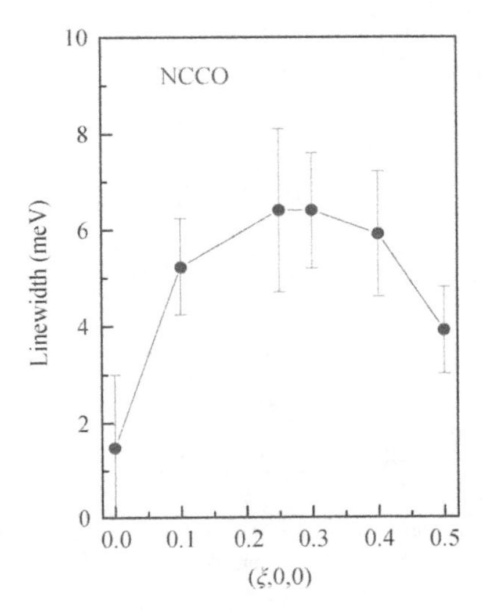

Fig. 3.17. Resolution-corrected phonon linewidths observed for the Cu–O bond-stretching modes in optimally doped $Nd_{1.85}Ce_{0.15}CuO_4$ [7].

the calculated value of λ . Of course, the case of the Cu–O bond-stretching modes might not be representative and therefore, it is difficult if not impossible to estimate by how much λ is underestimated. However, data for other modes which will be dealt with in Section 3.6, also point to a serious underestimation of the phonon linewidths.

There has been another attempt for explaining the phonon anomalies in the cuprates based on the electronic band structure calculated within the LDA, i.e., that reported by Falter and co-workers in a series of papers [20–24]: in this theory, the screening processes producing the softening are described in terms of charge fluctuations on the outer shells of the ions. The results depicted in Fig. 3.18 for LSCO are taken from a recent paper [24] and show a very satisfactory agreement with experiment (see Figs. 3.5 and 3.7). Falter and co-workers conclude from their investigations that electron–phonon interaction is strong in the cuprates and therefore relevant for high-T_c superconductivity, although no quantitative predictions for λ are made.

Other groups have chosen a very different approach to explain the phonon renormalization of the bond-stretching modes upon doping, i.e., using the $t-J$ model and extending it to explicitly include electron–phonon couplings [26, 27]. The correct anisotropy between the [1, 1] and the [1, 0]

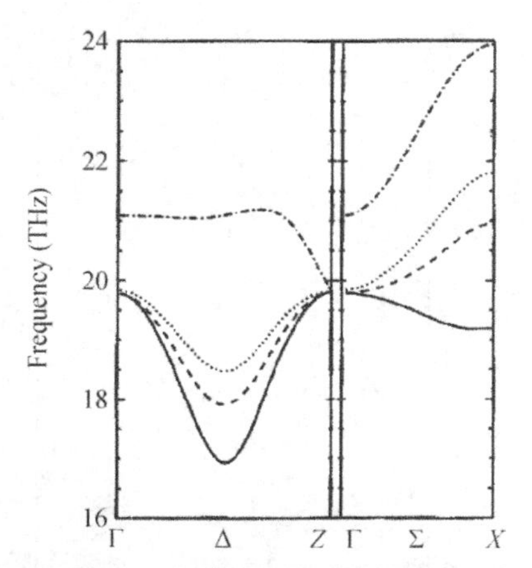

Fig. 3.18. Calculated phonon dispersion of the bond-stretching modes in LSCO for the insulating phase (dash-dotted lines), for underdoped phases (dashed and dotted lines) and for the optimally doped phase (full lines). The figure was taken from Ref. [24].

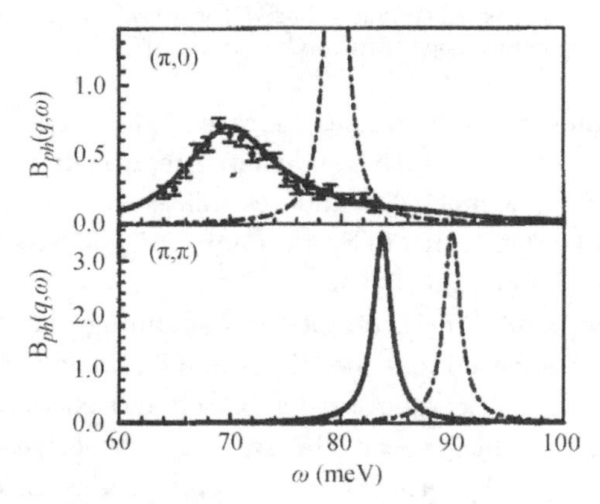

Fig. 3.19. Calulated phonon-spectral function B_{ph} for the half-breathing (top) and for the breathing phonon (bottom) in optimally doped LSCO (solid lines) and the undoped references system (dash-dotted lines). The experimental points were taken from Ref. [28] for the doped compound and from Ref. [1] for the undoped compound.

directions in optimally doped LSCO has been reported for the first time by Khaliullin and Horsch [27] (Fig. 3.19). This figure further demonstrates that the theory predicts a pronounced line broadening for the half-breathing mode. I note that the agreement between theory and experiment is not quite as good as suggested by the figure because high-resolution measurements reported in Ref. [2] have shown a some-what smaller (20%) linewidth of this

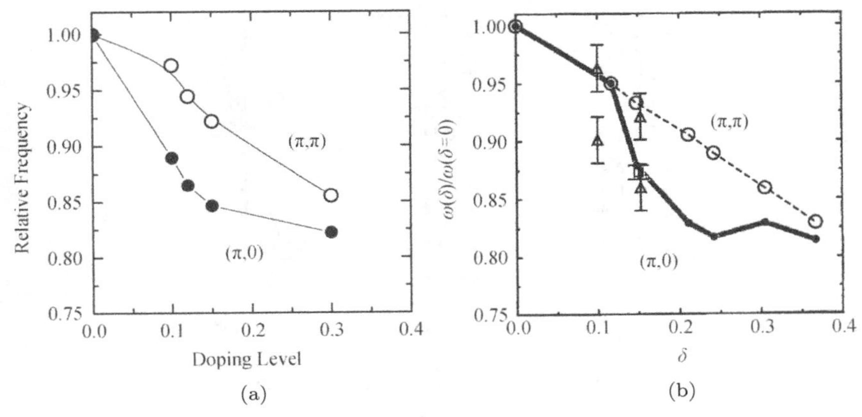

Fig. 3.20. Experimentally observed frequency renormalization in LSCO of the half-breathing mode $(\pi, 0)$ and of the breathing mode (π, π), respectively. This figure was compiled from data shown in Figs. 3.5 and 3.7 as well of unpublished data for $La_{1.48}Nd_{0.4}Sr_{0.12}CuO_4$ [31]. (b) Predicted doping dependence of the frequencies of the half-breathing mode and of the breathing mode, respectively [29]. The triangles denote experimental data known at the time of the report.

phonon. Still, it has to be said that this theory gives a realistic description of both the frequency renormalization and the linewidths in contrast to the LDA theory dicussed above yielding only a correct description of the dispersion. A further success of the theory just mentioned is the very good agreement of the predicted doping dependence of the frequencies of the half-breathing mode and of the breathing mode [29] prior to the experiments on overdoped samples (Fig. 3.20). As is explained in Refs. [27, 30], the rapid drop of the frequency of the half-breathing mode when going from the undoped to the optimally doped compound results from a polaron peak in the electron density fluctuation spectrum $N(\boldsymbol{q}, \omega)$ being in the same energy range as the bond-stretching phonons. Results somewhat similar to those just mentioned have been reported also in a recent paper by Rösch and Gunnarsson [26]. They have evaluated the doping and the $\boldsymbol{q}$ dependence of the O bond-stretching mode of a Cu–O layer through a $t - J$ model with electron–phonon interactions derived from a three-band model.

3.4.2. *c-axis Polarized Modes*

LSCO is so far the only system for which a complete set of phonon dispersion curves has been determined both for doped and for undoped samples.

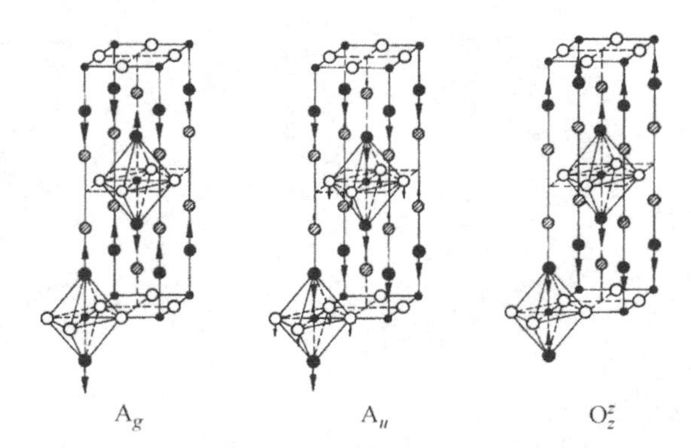

Fig. 3.21. Displacement patterns of c-axis polarized modes in $La_{2-x}Sr_xCuO_4$.

This allowed a systematic search of doping induced phonon effects in addition to those dealt with in the previous chapter. It turned out that a particular c-axis polarized mode exhibits an even stronger frequency renormalization than any plane polarized mode, i.e., the so-called O_z^z-mode. The name was coined by Falter and co-workers [20] signifying that it is a zone boundary mode (the zone boundary is the Z-point) and that the displacements are in the $z(= c)$-direction. The displacement pattern of this mode is shown in Fig. 3.21. In fact, the anomalous behavior of this mode was predicted by Falter *et al.* [20] prior to the experimental observations. As can be seen in Fig. 3.22, the frequency of this mode in optimally doped LSCO is lower than found in undoped LCO by as much as 5.5 THz. At the same time, the mode acquires a massive line broadening of about 3 THz which supports the assumption of a strong electron–phonon coupling. The strong coupling is explained within the theory of Falter *et al.* [20] by charge fluctuations induced by apical oxygen vibrations within the Cu–O plane. The displacement pattern of the O_z^z-mode is particularly favorable for this type of coupling because all the oxygen atoms above and below the Cu–O move in-phase, in contrast to the case of related apical oxygen modes of A_g and of A_u type (see Fig. 3.21).

Falter *et al.* argued that the strong coupling of the O_z^z-mode is presumably relevant for high-T_c superconductivity [21, 22]. Somewhat surprisingly, recent measurements performed on a heavily overdoped (non-superconducting) sample [7] revealed that the O_z^z-mode remains soft and moreover, it also retains its very large linewidth (Fig. 3.23). That is to say, the behavior of the O_z^z-mode is somewhat different from that of the

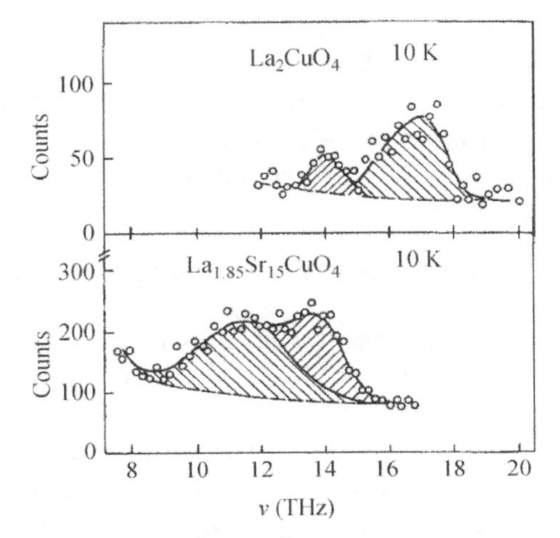

Fig. 3.22. Energy scans taken at $Q = (0, 0, 15)$ on undoped and on optimally doped LSCO [2]. The differently hatched areas correspond to the contributions from the O_z^z-mode at $\nu = 17$ THz and 11.5 THz in the undoped and in the doped sample, respectively, and from another Z-point mode at $\nu = 14$ THz of A_g character. Note that the rather broad linewidth of the O_z^z-peak in the undoped sample is completely attributable to the Q-resolution of the instrument, whereas the O_z^z-peak for optimally doped LSCO is much broader than the instrumental resolution $\Delta\nu = 1$ THz (FWHM).

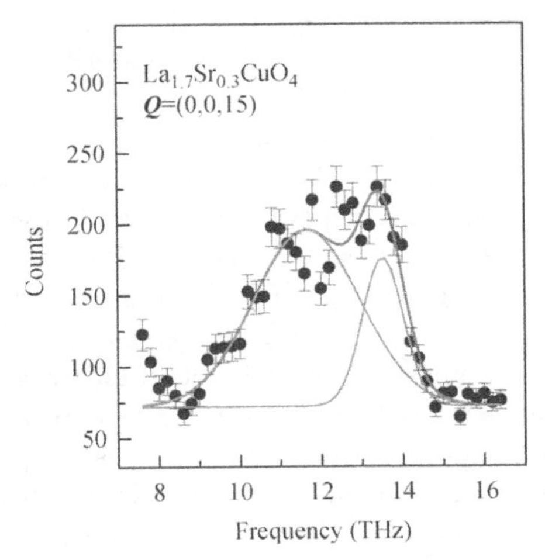

Fig. 3.23. Energy scan taken at $\boldsymbol{Q} = (0, 0, 15)$ on a heavily overdoped sample [7]. The intensity distribution was fitted with two Gaussians corresponding to the contributions of the O_z^z-mode and the A_g type mode.

plane polarized bond-stretching modes which remain soft as well but do show a reduction of the linewidth on overdoping.

In principle, similar effects as those seen for the O_z^z-mode in LSCO could be expected in other systems as well. However, no systematic search was undertaken so far in any other compound.

3.5. Coupling of Phonons to Charge Stripe Order

The discovery of static stripe order in $La_{1.48}Nd_{0.4}Sr_{0.12}CuO_4$ by Tranquada *et al.* [32] boosted the popularity of the stripe picture enormously. It was advocated that stripe order is a universal phenomenon in the cuprates. However, static stripe order is obviously detrimental to high-T_c superconductivity [33]. Therefore, it was argued that stripe order is dynamic in nature in the superconducting members of the cuprate family. The assumption that dynamic stripe order is a general feature in the cuprates, was based on the observation of incommensurate low energy spin fluctuations. Such observations were made very early in LSCO [34] which actually triggered the development of the stripe phase concept by theory [35, 36]. Later, incommensurate spin fluctuations were also observed in YBCO [37, 38] but it remained controversial whether or not these incommensurate spin fluctuations are consistent with the stripe picture. Very recent results on the spin fluctuation spectrum in the static stripe phase $La_{1.88}Ba_{0.12}CuO_4$ reported by Tranquada *et al.* [39] have shown that the high energy spin fluctuations of stripe ordered compounds are quite different from what was naively expected. That is to say, they cannot be understood as spin waves emanating from the magnetic superlattice peaks. Rather, they are quite similar to those observed in optimally doped YBCO [43] which makes some arguments put forward against the stripe picture for YBCO obsolete.

Whatever conclusions are drawn from the spin fluctuations, the stripe picture can be uphold only if evidence is found for dynamic charge stripe order as well. Unfortunately, there is no direct way to observe such dynamic charge order. Inelastic neutron scattering appears as an appealing technique because indirect evidence for dynamic charge order might be obtained from an interaction with phonons. The basic idea is that dynamic charge stripes will strongly couple to phonons with an appropriate displacement pattern leading to a softening as well as to a line broadening. Naively, one might expect effects similar to those observed long time ago in 1D conductors as precursors to charge density wave order. In fact, charge stripe order is

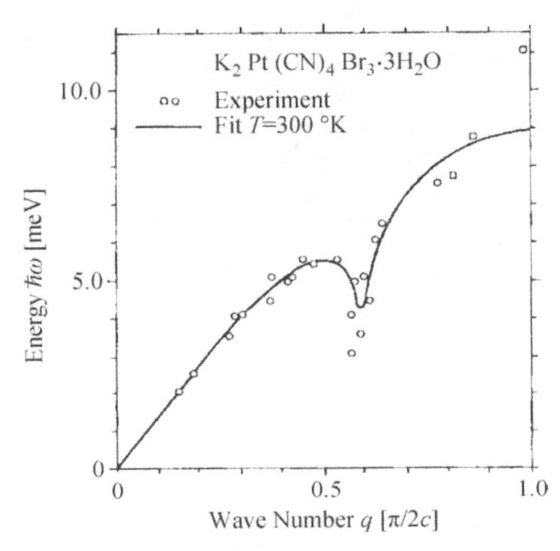

Fig. 3.24. Dispersion of longitudinal acoustic phonon in the 1D conductor KCP at room temperature, i.e., above the transition to static charge-density-wave order. The solid line was calculated from a simple free-electron model. The figure was adapted from Ref. [40].

nothing but a special case of charge density wave order. An example of such a precursor effect is the observation of a phonon anomaly in the longitudinal acoustic branch of the 1D conductor KCP [40] depicted in Fig. 3.24.

In the cuprates, the phonons which are expected to show the strongest coupling are not the acoustic ones but the plane-polarized Cu–O bond-stretching phonons. This expectation is based on the assumption that any charge inhomogeneity will induce a modulation of the Cu–O bond length. The theoretical paper by Kaneshita *et al.* [41] gives an idea about what might be expected from a coupling of phonons to charge stripe order, i.e., basically a strong broadening and also some softening of the phonons in a narrow range of $\boldsymbol{q}$ around the ordering wavevector. Unfortunately, there was no clear-cut experimental evidence to support the predicted effects. I note that there was an attempt to gain some insight into the problem by performing phonon measurements on a related system, i.e., $La_{1.69}Sr_{0.31}NiO_4$ by Tranquada *et al.* [42]. This compound seems to be well suited as a reference system because (i) it is isostructural to LSCO and (ii) it shows rather strong magnetic and nuclear superlattice peaks indicative of static charge stripe order. It was found that the Ni–O bond-stretching phonons do behave in a very anomalous way but on the other hand, there was no obvious relationship of the phonon anomalies to the wavevector of the charge stripe order–which remains to be understood. Compelling evidence

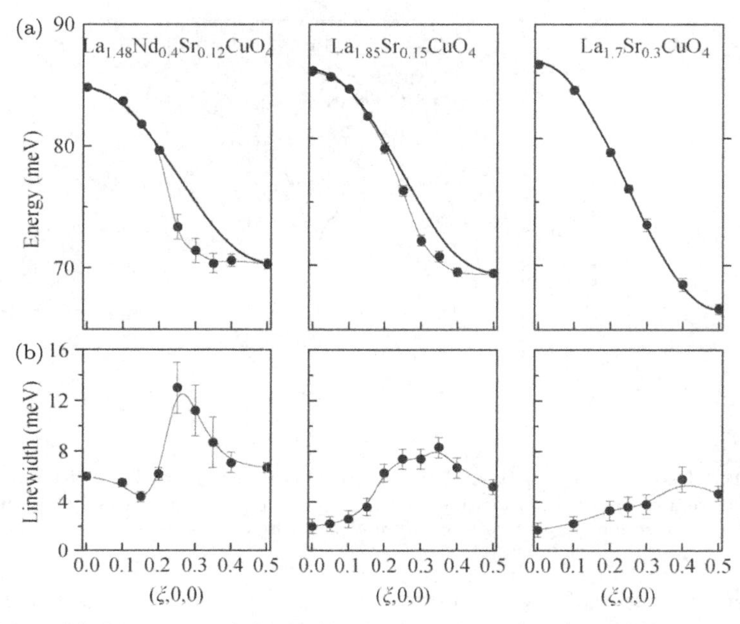

Fig. 3.25. (a) Dispersion of the longitudinal plane-polarized Cu–O bond-stretching vibrations in the static stripe phase $La_{1.48}Nd_{0.4}Sr_{0.12}CuO_4$ (left) [31] as well as in optimally doped LSCO (middle) [25] and in overdoped LSCO (right) [7]. The thick lines are a cosine-function and the thin lines are a guide to the eye. (b) Resolution-corrected phonon linewidths of the phonons shown at the top. Lines are a guide to the eye.

for an intimate relationship between phonon anomalies and charge stripe order was obtained only very recently by measurements on the static stripe phase $La_{1.48}Nd_{0.4}Sr_{0.12}CuO_4$ [31]. Some preliminary results are shown in Fig. 3.25.

Obviously, the dispersion becomes extremely steep at the charge ordering wavevector ($\boldsymbol{q} = (0.24, 0, 0)$) and therefore, the data points for $q > 0.25$ fall considerably below the heavy line depicting the general trend. This means that a special mechanism comes into play in this q range leading to an extra softening–in line with the expectations just mentioned. At the same time, the phonons acquire a very large linewidth, again in complete agreement with the expectations. Apparently, these effect die out on the high q side more slowly than they set in on the low q side. Preliminary temperature dependent measurements revealed a hardening of the anomalous phonons and a reduction of the linewidths on raising the temperature, again as is expected for an electronic instability. Further, I note that the

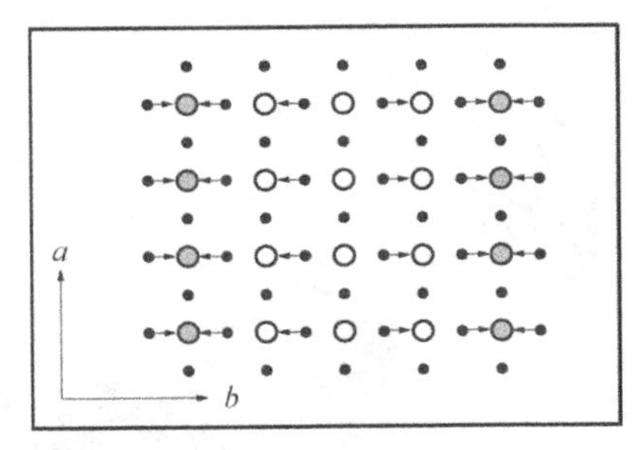

Fig. 3.26. Displacement pattern of the anomalous phonons with wavevector (0, 0.25, 0). Only the displacements in the CuO_2 planes are shown (they are the dominant ones). Some circles are filled with grey to indicate that the displacement pattern favors dynamic charge accumulation on every fourth row of coppers atoms.

displacement pattern of the anomalous phonons is just that which favors dynamic accumulation of charge on every fourth row of atoms consistent with the static charge stripe pattern (Fig. 3.26).

The above results may now serve as a guide to look for fingerprints of dynamic charge stripes in other compounds. Inspection of Fig. 3.25 (middle) for optimally doped LSCO reveals a clear resemblance to those of the static stripe phase although the stripe-order-related effects are somewhat less pronounced. This supports the interpretation given in Ref. [25] that the very steep dispersion and the large linewidths at halfway to the zone boundary are indicative of a coupling to dynamic charge stripes. On the other hand, no such fingerprints are seen in strongly overdoped LSCO. This observation is in line with the widely hold idea that the cuprates behave largely as normal metals in the overdoped regime.

Some results presented in the previous section for electron-doped NCCO also show some similarity to the peculiarities of the static stripe phase, i.e., a local maximum of the doping-induced renormalization (Fig. 3.10) and of the phonon linewidth (Fig. 3.17) around $\boldsymbol{q} = (0.3, 0, 0)$. As for the system YBCO, the very steep dispersion in underdoped $O_{6.6}$ at about $\boldsymbol{q} = (0, 0.25, 0)$ is again reminiscent of the effects seen in the static stripe phase (Fig. 3.11). On the other hand, the dispersion in optimally doped YBCO shown in Fig. 3.12 is less steep and can be fully explained by conventional density functional theory within the LDA (Fig. 3.14). However, it is important to note that this applies only to high-temperature data (the data shown in Fig. 3.12 for the b-direction were taken at 200 K on purpose). As has been explained in Ref. [14], a pronounced downward shift

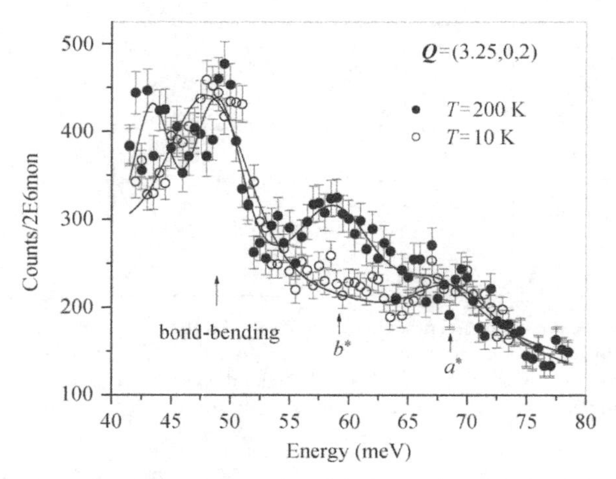

Fig. 3.27. Energy scans taken on a twinned sample of optimally doped YBCO at a nominal wave vector of $\boldsymbol{Q} = (3.25, 0, 2)$. The peaks observed at $E = 68$ meV and 59 meV are assigned to the plane-polarized Cu–O bond-stretching vibrations along the a- and the b-direction, respectively, whereas the 49 meV peak is related to Cu–O bond-bending vibrations. On cooling, the 59 meV peak is depleted which is partly compensated by a gain in intensity around 45 meV. Some of the missing intensity is presumably shifted down to energies below 40 meV.

of spectral weight was observed for a wavevector of about $\boldsymbol{q} = (0, 0.27, 0)$ on cooling to 10 K (Fig. 3.27). This downward shift sets in very sharply at $\boldsymbol{q} = (0, 0.25, 0)$, whereas it appears to be more gradual on the high q side of the anomaly, in striking similarity to the behavior of the static stripe phase. Thus, it appears that optimally doped YBCO is close to a charge-density-wave instability as well. In summary, effects indicative of dynamic charge stripe order are seen in a variety of cuprates and thus seem to be a generic feature.

At present, the theoretical understanding of these effects remains very poor. One might think that the charge-density-wave instability is due to Fermi surface nesting. However, density functional theory for YBCO [17] did not find any indication for an anomaly of the kind observed. A careful analysis of the predicted phonon dispersion did reveal a local frequency minimum in a bond-stretching branch linked to Fermi surface nesting. However, this branch involves primarily elongations within the Cu–O chains. The local frequency minimum can be considered as a Kohn anomaly linked to chain oxygen related electronic states. On the other hand, no such effect was predicted for phonon modes of Δ_4-symmetry for which the chain oxygen atoms are at rest. Therefore, Fermi surface nesting is a somewhat unlikely candidate for the phonon anomaly in YBCO.

Stripe phase order has been predicted by theories for highly correlated electron systems long time ago [35, 36]. The phonon results discussed above lend certainly some support to these ideas. However, the theories remain so far qualitative. For instance, it remains to be understood why the wavevector of the phonon anomaly in YBCO is the same — within experimental accuracy — as in $La_{1.48}Nd_{0.4}Sr_{0.12}CuO_4$. Further, it is completely unclear why the anomaly in YBCO has been observed only in the *b*-direction and not in the *a*-direction.

3.6. Superconductivity-Induced Phonon Effects

Superconductivity-induced phonon self-energy effects were discovered already very early, i.e., 1987, by Raman scattering [44]. Subsequently, these effects attracted considerable attention giving rise to numerous optical studies. Similar, although somewhat weaker, effects were observed also by infrared (IR) spectroscopy. The results of these studies allowed one for the first time to deduce a quantitative estimate of the electron–phonon coupling constant λ. The very first theoretical evaluation of λ resulted in a surprisingly large value of $\lambda = 2.9$ [46]. Later, it was found that the simplifying assumptions made in Ref. [46] are somewhat unrealistic. Subsequent studies of similar kind resulted in much smaller values of λ (see, e.g., [45]). A major drawback of these investigations was the fact that they had to rely on information on zone center modes only whereas the electron–phonon coupling constant λ is related to the Brillouin zone average of the phonon self-energy effects. Therefore, inelastic neutron scattering was a very attractive technique for extending the determination of superconductivity-induced phonon self-energy effects to wavevectors across the whole Brillouin zone. The first study of this kind was published in 1993 by Pyka *et al.* [47] after sufficiently large single crystals of optimally doped YBCO had become available.

The largest superconductivity-induced frequency shifts were observed for a mode with B_{1g}-symmetry with frequency $\nu = 10.2$ THz (42 meV or 340 cm^{-1}), similar to what had been found by optical spectroscopy. On cooling through the superconducting transition, the frequency of this mode was found to decrease rather abruptly by about 1% (Fig. 3.28). Subsequent measurements [47] on a fully oxygenated sample revealed an even larger effect (2%, Fig. 3.29), again confirming optical studies. The major

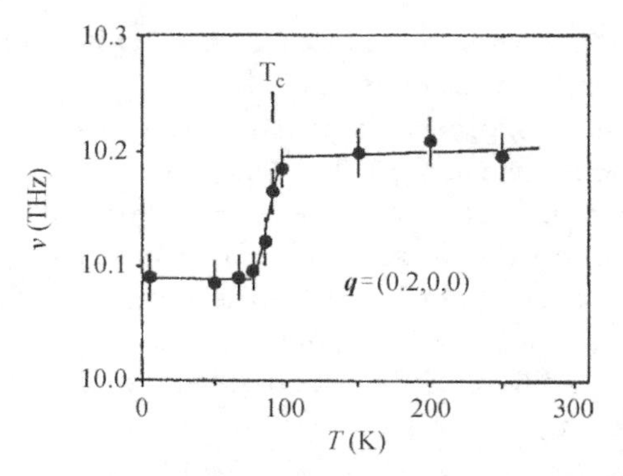

Fig. 3.28. Frequency of a phonon with wavevector $\boldsymbol{q} = (0.2, 0, 0)$ versus temperature in optimally doped YBCO ($YBa_2Cu_3O_{6.92}$) [47]. The line is a guide to the eye.

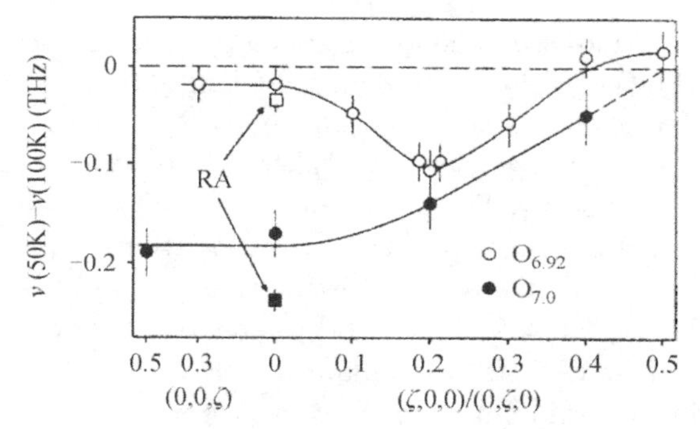

Fig. 3.29. Superconductivity-induced frequency shifts observed in $YBa_2Cu_3O_x$ for phonons related to the B_{1g} mode versus momentum transfer [47]. The lines are a guide to the eye. The Raman results [44] refer to $x = 7.0$ (full square) and $x = 6.87$ (open square), respectively.

contribution of neutron scattering was to show that such effects were not restricted to zone center modes but were observable in large parts of the Brillouin zone (Figs. 3.29 and 3.31). A later study [48] using a sample of larger volume was able to determine not only the frequency shifts but also the changes in linewidth occuring on entering the superconducting state. Again, such effects were found to be of similar size throughout the Brillouin zone (Fig. 3.31).

Qualitatively similar though quantitatively smaller effects were observed for a number of other phonon modes [47]. Their displacement patterns are shown in Fig. 3.30. A later study [52] found quite strong phonon self-energy effects for a c-axis polarized B_{1u} mode with out-of-phase motions

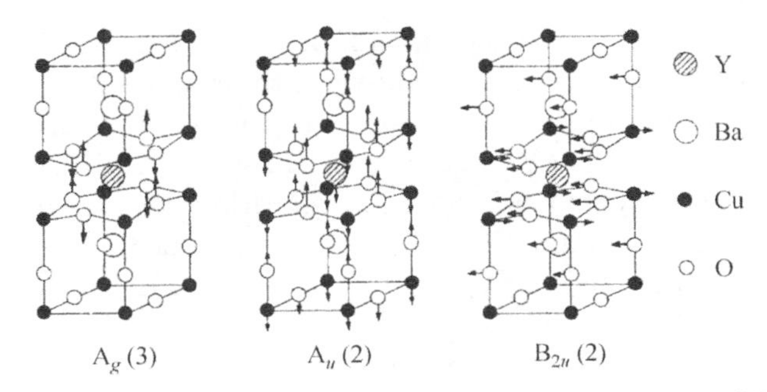

Fig. 3.30. Displacement patterns of three zone center phonons in YBCO. The mode labeled as A_g (3) is often (and correctly) labeled as B_{1g}.

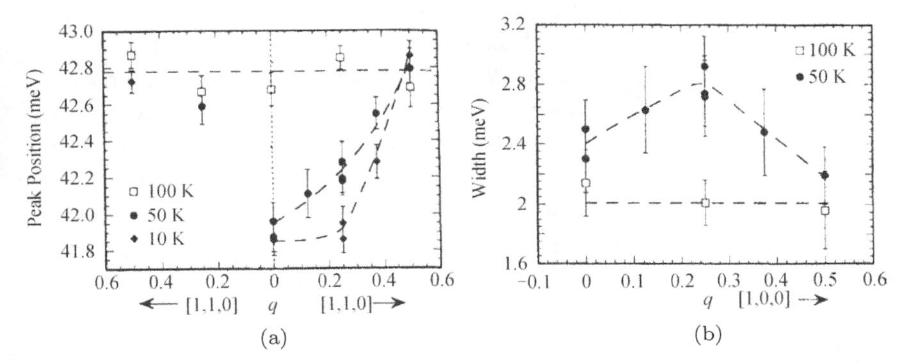

Fig. 3.31. Wavevector dependence of the peak position (a) and of the linewidth (b) of phonons related to the B_{1g} mode at different temperatures [48]. Lines are a guide to the eye.

of the in-plane oxygen in fully doped YBCO (Fig. 3.32). Somewhat smaller effects of the same kind were found in under-doped YBCO as well. No systematic search for further effects of this kind was undertaken so far, presumably because of lack of a valid theory for assessing such results. Whereas the theory of Zeyher and Zwicknagl [46] was very popular in the beginning, it turned out later that several experimental results could not be consistently explained within this theory. For instance, it was concluded that the value of the superconducting gap is precisely $2\Delta = 316$ cm^{-1} [54] or has to be between 440 cm^{-1} and 500 cm^{-1} [55]. In both cases, the conclusions had been based on the frequency dependence of superconductivity-induced phonon self-energy effects and were interpreted within the framework of the Zeyher–Zwicknagl theory. One of the obvious shortcomings of this theory is the fact that it was developed for s-wave superconductivity whereas the cuprates are d-wave superconductors. This flaw was later remedied by Jiang and Carbotte [51] who extended the theory to the case of a d-wave superconductor. Their results are, however, quite similar to those obtained before by Zeyher and Zwicknagl which means that the inconsistencies just mentioned are not removed. Zeyher tried to extend the theory also to phonons with finite wavevector (the initial theory was set up for zone center modes only) but with very limited succes [50]. It seems that the underlying assumptions about the Fermi surface were too crude.

As has been discussed in the previous chapter, there has been recent progress in the calculation of the electron–phonon coupling strength based on a realistic band structure by Bohnen *et al.* [17]. Such a theory is probably also a good starting point for predicting superconductivity-induced phonon self-energy effects. It has, however, yet to be worked out. In a few cases, contact can be made with this theory with out further calculations: line narrowings observed on entering the superconducting state can be taken as an upper bound for the electron–phonon coupling related linewidth in the normal state and can thus be compared directly to the calculated results depicted in Fig. 3.15. For instance, the line narrowing documented in Fig. 3.32(b) corresponds — after deconvoluting the experimental

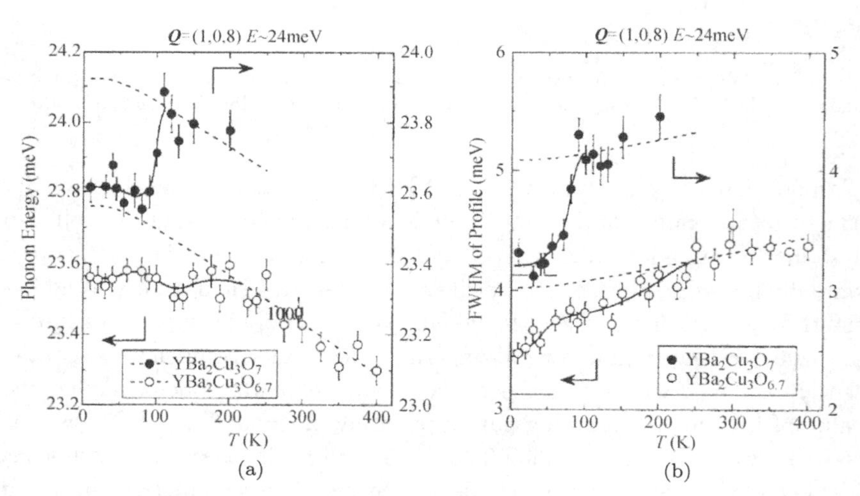

Fig. 3.32. Energies (a) and linewidths (b) of the B_{1u} mode versus temperature in underdoped and in fully doped YBCO. The dotted lines show the behavior expected from phonon–phonon coupling. The solid lines are guides to the eye. The horizontal bars in the right hand panel denote the instrumental resolution (3.13 meV). After Ref. [52].

resolution — to an intrinsic linewidth of about 7%. This value is an order of magnitude larger than calculated. Similarly, the Raman active mode at $120\,\mathrm{cm}^{-1}$ shows a narrowing of about 3% [55], again very much larger than calculated. Considering that the linewidths of the bond-stretching phonons are also seriously underestimated by the LDA theory (see Section 3.3), the above findings cast doubts on the validity of the theoretical result concerning the small value of the electron–phonon coupling constant λ.

3.7. Concluding Remarks

I hope to have shown that inelastic neutron scattering has provided a wealth of information on electron–phonon coupling effects in high-T_c superconductors. The softening of the half-breathing mode, which has been known for quite some time, has recently attracted renewed interest. In particular, the half-breathing mode is sometimes considered to be at the origin of the "kink" in the electronic dispersion seen in ARPES experiments [56]. This issue remains certainly a matter of debate but it has helped to bring phonons back to the attention of the high-T_c community. As outlined above, the softening of the half-breathing mode — and of the breathing mode — is quantitatively accounted for by theories based on the classical band structure approach as well as by theories developed for highly correlated electrons. Somewhat surprisingly, none of these theories predict that the strongest experimentally observed electron–phonon coupling effects occur at half way to the zone boundary, i.e., not at the zone boundary for the half-breathing mode. The experimental evidence points to a close relationship between these effects and the tendency towards charge stripe order. In spite of the fact that stripe order has often been advocated as a universal phenomenon in the cuprates, there is no theoretical framework for a detailed understanding of the phonon effects related to charge-stripe instabilites. That is to say, experiment is again much ahead of theory in the field of high-T_c's.

When advocating that inelasting neutron scattering has provided evidence of electron–phonon coupling effects, I do not want to claim that electron–phonon coupling is strong enough to be responsible for the mechanism of high-T_c superconductivity. As discussed in Section 3.2, such a claim can never made on the basis of the quite limited information obtainable by inelastic neutron scattering. On the other hand, I have pointed out that the available data seriously question the small value of the electron–phonon coupling constant λ calculated from density functional theory [17]: despite the very good success of this theory for reproducing the phonon dispersion curves it appears to underestimate the phonon linewidths. On these grounds, electron–phonon coupling may well be of more relevance to high-T_c superconductivity than is often thought.

Acknowledgments

I am grateful to W. Reichardt and Ch. Meingast for a critical reading of the manuscript.

References

[1] L. Pintschovius and W. Reichardt, *Physical Properties of High Temperature Superconductors IV* (World Scientific, Singapore, 1989).
[2] L. Pintschovius and N. W. Reichardt, *Neutron Scattering in Layered Copper-Oxide Superconductors*, Physics and Chemistry of Materials of Low-Dimensional Structures, Vol. 20 (Kluwer Academic Publishers, Dordrecht, 1998).
[3] L. Pintschovius, H. Takei, and N. Toyota, *Phys. Rev. Lett.* **54** (1985), 1260.
[4] L. Pintschovius, H.-G. Smith, N. Wakabayashi, W. Reichardt, W. Weber, G. W. Webb, and Z. Fisk, *Phys. Rev. B* **28** (1983), 5866.
[5] W. Weber, *Electronic Structure of Complex Systems*, NATO Advanced Study Institute, Vol. 113 (Plenum, New York, 1983).
[6] S. L. Chaplot, W. Reichardt, L. Pintschovius, and N. Pyka, *Phys. Rev. B* **52** (1995), 7230.
[7] L. Pintschovius, K. Yamada, and D. Reznik, unpublished results.
[8] T. Fukuda, *et al.*, cond-mat/0306190.
[9] K. Yamada, private communication.
[10] M. d'Astuto, P. K. Mang, P. Giura, A. Shukla, P. Ghigna, A. Mirone, M. Braden, M. Greven, M. Krisch, and F. Sette, *Phys. Rev. Lett.* **88** (2002), 167002.
[11] M. d'Astuto, P. K. Mang, P. Giura, A. Shukla, P. Ghigna, A. Mirone, M. Braden, M. Greven, M. Krisch, and F. Sette, cond-mat/0210700.
[12] M. Braden, L. Pintschovius, and K. Yamada, unpublished results.
[13] L. Pintschovius, *J. Low Temp. Phys.* **131** (2003), 401.
[14] L. Pintschovius, Y. Endoh, D. Reznik, H. Hiraka, J. M. Tranquada, W. Reichardt, H. Uchiyama, T. Masui, and S. Tajima, cond-mat/0308357.
[15] L. Pintschovius, Y. Endoh, D. Reznik, H. Hiraka, J. M. Tranquada, W. Reichardt, P. Bourges, Y. Sidis, H. Uchiyama, T. Masui, and S. Tajima, *Physica C*, in press.
[16] W. Reichardt, *J. Low Temp. Phys.* **105** (1996), 807.

[17] K. P. Bohnen, R. Heid, and K. Krauss, *Europhys. Lett.* **64** (2003), 104.
[18] R. Heid, private communication.
[19] L. Pintschovius, *et al.*, unpublished results.
[20] C. Falter, M. Klenner, and W. Ludwig, *Phys. Rev. B* **47** (1993), 5390.
[21] C. Falter, M. Klenner, and Q. Chen, *Phys. Rev. B* **48** (1993), 16690.
[22] C. Falter, M. Klenner, and G. A. Hoffmann, *Phys. Rev. B* **52** (1995), 3702.
[23] C. Falter, M. Klenner, and G. A. Hoffmann, *Phys. Rev. B* **55** (1997), 3308.
[24] C. Falter and G. A. Hoffmann, *Phys. Rev. B* **61** (2000), 14537.
[25] L. Pintschovius and M. Braden, *Phys. Rev. B* **60** (1999), R15039.
[26] O. Rösch and O. Gunnarsson, cond-mat/0308035.
[27] G. Khaliullin and P. Horsch, *Physica C* **282–287** (1997), 1751.
[28] R. J. McQueeney, T. Egami, G. Shirane, and Y. Endoh, *Phys. Rev. B* **54** (1996), R9683.
[29] P. Horsch and G. Khaliullin, talk given at the *Int. Conf. on in: Quantum Coherence in Strongly Correlated Fermion Systems*, 22–26. 7. 1996, Pisa, Italy (unpublished).
[30] P. Horsch and G. Khaliullin, *Highlights in Condensed Matter Physics, AIP Conference Proceedings* 695, (AIP, New York, 2003), cond-mat/0312561.
[31] L. Pintschovius, M. Sato, M. Itoh, and D. Reznik, unpublished results.
[32] J. M. Tranquada, *et al.*, *Nature* **375** (1995), 561.
[33] J. M. Tranquada, *et al.*, *Phys. Rev. Lett.* **78** (1997), 338.
[34] T. R. Thurston, *et al.*, *Phys. Rev. B* **40** (1989), 4585.
[35] K. Machida, *Physica C* **158** (1989), 192.
[36] J. Zaanen and O. Gunnarsson, *Phys. Rev. B* **40** (1989), 7391.
[37] G. Shirane, *Physica B* **213–214** (1995), 37.
[38] P. Dai, H. A. Mook, and F. Dogan, *Phys. Rev. Lett.* **80** (1998), 1738.
[39] J. M. Tranquada, *et al.*, cond-mat/0401621.
[40] B. Renker, H. Rietschel, L. Pintschovius, W. Glaeser, P. Bruesch, K. Kuse, and M. J. Rice, *Phys. Rev. Lett.* **30** (1973), 1144.
[41] K. Kaneshita, M. Ichioka, and K. Machida, *Phys. Rev. Lett.* **88** (2002), 115501.
[42] J. M. Tranquada, N. Nakajima, M. Braden, L. Pintschovius, and R. J. McQueeney, *Phys. Rev. Lett.* **88** (2002), 07505.
[43] D. Reznik, *et al.*, cond-mat/0307591.
[44] M. C. Krantz, H. J. Rosen, R. M. MacFarlane, and V. Y. Lee, *Phys. Rev. B* **38** (1988), 4992.
[45] A. P. Litvinchuk, C. Thomsen, and M. Cardona, *Solid State Commun.* **80** (1991), 257.
[46] R. Zeyher and G. Zwicknagel, *Solid State Commun.* **66** (1988), 617.
[47] N. Pyka, W. Reichardt, L. Pintschovius, G. Engel, J. Rossat-Mignod, and J. H. Henry, *Phys. Rev. Lett.* **70** (1993), 1457.
[48] D. Reznik, B. Keimer, F. Dogan, and I. A. Aksay, *Phys. Rev. Lett.* **75** (1995), 2396.
[49] R. Zeyher and G. Zwicknagel, *Solid State Commun.* **66** (1988), 617.
[50] R. Zeyher, *Phys. Rev. B* **44** (1991), 9596.

[51] C. Jiang and J. P. Carbotte, *Phys. Rev. B* **50** (1994), 9449.
[52] H. Harashina, K. Kodama, S. Shamoto, M. Sato, K. Kakurai, and M. Nishi, *Physica C* **263** (1996), 257.
[53] R. M. MacFarlane, M. C. Krantz, H. J. Rosen, and V. Y. Lee, *Physica C* **162–164** (1989), 1091.
[54] B. Friedl, C. Thomsen, and M. Cardona, *Phys. Rev. Lett.* **65** (1990), 915.
[55] K. F. McCarty, H. B. Radouski, Z. J. Liu, and R. N. Shelton, *Phys. Rev. B* **43** (1991), 13751.
[56] A. Lanzara, *et al.*, *Nature* **412** (2001), 510.

4
Phonon Anomalies and Dynamic Stripes*

D. Reznik

Department of Physics, University of Colorado,
Boulder, CO 80304, USA

Stripe order where electrons self-organize into alternating periodic charge-rich and magnetically-ordered charge-poor parallel lines was proposed as a way of optimizing the kinetic energy of holes in a doped Mott insulator. Static stripes detected as extra peaks in diffraction patterns, appear in a number of oxide perovskites as well as some other systems. The more controversial dynamic stripes, which are not detectable by diffraction, may be universally present in copper oxide superconductors. Thus it is important to learn how to detect dynamic stripes as well as to understand their influence on electronic properties. This review article focuses on lattice vibrations (phonons) that might show signatures of the charge component of dynamic stripes. The first part of the article describes recent progress in learning about how the phonon signatures of different types of electronic charge fluctuations including stripes can be distinguished from purely structural instabilities and from each other. Then I will focus on the evidence for dynamic stripes in the phonon spectra of copper oxide superconductors.

4.1. General Considerations

Electrons in a doped Mott insulator sometimes form stripe order, consisting of alternating charge-rich lines and charge-poor anti-ferromagnetic regions. It was predicted theoretically in 1989 [1, 2] and discovered in a

*Reprint with the permission from D. Reznik. Original published in *Physica C Supercond.*, Vol. 481, 75–95 (2012).

nickel oxide perovskite in 1994 by neutron diffraction [3]. Static stripe order was subsequently found in some copper oxide perovskites [4]. Diffraction measurements of copper oxide high-temperature superconductors do not show magnetic and charge Bragg peaks characteristic of static stripes. However, it was proposed that dynamic stripes, which are much harder to observe than the static ones [5], play an important role in the copper oxide superconductors [6–9].

Periodic charge density modulation, which forms as a result of stripe formation, induces a periodic distortion of the crystal lattice with the same propagation vector. Its magnitude is proportional to the electron–phonon coupling strength. Such distortions can be observed by neutron diffraction as well as scanning tunneling microscopy [5]. Charge density fluctuations associated with dynamic stripes may soften and/or broaden certain phonons signaling an incipient lattice instability. However, structural distortions as well as soft phonons may appear not only because of stripe formation, but also for other reasons. Thus, in order to use phonons as a tool to study dynamic stripes, it is crucial to learn how to distinguish between different origins of soft phonons.

Structural phase transitions can result from purely structural instabilities or from atomic positions reacting to charge density modulation associated with electronic instabilities (such as the tendency to form stripes). In both cases, the phase transition occurs due to a small energy difference between the ground-state structure of low symmetry and a high symmetry structure favored at high-temperatures. The high-temperature phase is often characterized by a soft phonon mode whose polarization is the same as the "frozen" lattice distortion of the low-T phase. The softening of this mode is most pronounced at the structural transition temperature. In the low temperature phase, the number of phonon branches increases due to the larger unit cell. This phenomenon is called branch folding.

Diffraction is a direct way of detecting structural phase transitions. Static stripes show up in neutron diffraction spectra as extra lattice/magnetic peaks corresponding to charge/spin stripes, respectively. Many solids do not undergo structural phase transitions, but exhibit incipient electronic and/or lattice instabilities, which are undetectable by diffraction. In these cases, softening and broadening of phonons (which is often temperature-dependent) provides a unique window on competing phases, which often play an important role in many phenomena including stripe formation.

My goal here is to review recent progress in studies of phonon in correlated electron systems with the focus on phonon anomalies that may be associated with the formation of charge stripes. This article is organized as follows. First, I will discuss soft phonons in different systems and demonstrate that, although phonon behavior is very similar near different types of lattice instabilities, it may be possible distinguish between them. Then, I will focus on the features of soft phonons that may be specific to dynamic stripes.

This paper partially overlaps and complements two other recent reviews [10, 11].

4.2. Different Types of Structural and Electronic Instabilities

Crystal structure as well as frequencies of atomic vibrations are ultimately determined by interactions between all atomic nuclei and all electrons in the crystal, so in this sense all lattice instabilities are driven by electron–phonon coupling. However, one can distinguish between the instabilities caused by electron–phonon coupling directly (which I call structural instabilities), and the instabilities where the electronic state itself undergoes a phase transition, which then pulls the lattice along (which I call electronic instabilities). Stripe formation is an example of the latter.

Phonon measurements can provide insights into mechanisms underlying second order displacive structural transitions that involve soft phonon behavior. In these transitions atomic displacements in a crystal change bond lengths and/or angles, without severing the primary bonds. Such transitions are associated with soft phonons whose eigenvectors are close to the character of the atomic displacements that take place during the transition. The contribution to the Hamiltonian from the soft phonon with the generalized coordinate $X, H(X)$, can always be approximated by adding a double-well potential term to the regular harmonic potential. It is useful to approximate the bottom of the double well potential as

$$V_{\text{double-well}}(X) = aX^2 + bx^4, \quad \text{where } b > 0.$$

If $a < 0$, the two sides of the well with the minima at $X = \pm\sqrt{-a/2b}$ represent the lattice distortion in one or another symmetry-equivalent direction. At temperatures higher than the barrier between the potential wells, $(k_B T > a^2/4b)$, the average structure is undistorted, i.e., $\langle X \rangle = 0$. But the anharmonic potential makes the phonon soft and broad. This broadening and softening becomes enhanced on cooling towards the transition temperature, as the amplitude of the vibration decreases and the phonon feels the anharmonic bottom of the potential more and more. Below T_c, the lattice settles at the minimum of one of the wells with $\langle X \rangle = +\sqrt{-a/2b}$ or $\langle X \rangle = -\sqrt{-a/2b}$ and the phonon hardens and narrows.

If $a > 0$, the phase transition will not occur, but soft phonons induced by anharmonicity and/or electron–phonon coupling may be observed.

Usually (but not always) the phonon spectral function is that of a damped harmonic oscillator. In the limit of weak damping it is close to a Lorentzian centered at the phonon frequency whose linewidth is proportional to the inverse phonon lifetime.

This behavior is common to all second order real or incipient displacive transitions regardless of their origin (see Sections 4.2.1 and 4.2.2). It is possible to distinguish between different mechanisms behind them (e.g. structural vs. electronic) only by detailed observations, calculations, and analysis focusing on subtle features of the phonon spectra. Without detailed understanding of soft phonons in general, little can be learned about dynamic charge stripes from phonon measurements on the copper oxides.

Section 4.3 provides highlights of substantial recent advances in understanding of how structural and electronic instabilities affect phonons in real systems. This information is then used in Sections 4.4 and 4.5 to discuss phonon anomalies observed in copper oxides and their possible relationship to dynamic stripes.

4.2.1. *Structural Instabilities*

Structural instabilities can occur when an anharmonic double-well potential for the anomalous phonon is induced by a mismatch of bond lengths, which favors a small distortion away from a high-symmetry structure. We will discuss an example of such an instability: the HTT–LTO transition in $La_{2-x}Sr_xCuO_4$ (Section 4.3.1).

In metals, strong coupling of conduction electrons to phonons can also induce a lattice deformation, which is usually discussed in terms of enhanced coupling of a particular phonon to electron–hole excitations of conduction electrons across the Fermi surface. These phonon anomalies can be better understood in terms of phonon renormalization by conduction electrons (Sections 4.3.3–4.3.6).

4.2.2. *Electronic Instabilities*

Phonon anomalies associated with electronic instabilities are induced by screening of the phonon by charge density fluctuations associated with the instability. Electronic instabilities can be of two types. One occurs when the Fermi surface nesting induces a charge density wave (CDW) or a spin density wave (SDW) instability. (Sections 4.3.2, 4.3.6 and 4.3.7) The second one originates from electronic correlations such as the stripe-ordering instability.

4.3. Phonon Anomalies in Model Systems

4.3.1. *An Example of a Purely Structural Instability without the Participation of Conduction Electrons: HTT–LTO Transition in* $La_{2-x}Sr_xCuO_4$

A good example of a purely structural instability is the transition from the high-temperature tetragonal (HTT) to the low temperature orthorhombic (LTO) phase in $La_{2-x}Sr_xCuO_4$. It occurs due to the mismatch of La–O and Cu–O bond lengths. This mismatch introduces strain into the tetragonal structure, which is stable only at high-temperatures. In the LTO phase, a small rotation of the CuO_6 octahedra around the 110 direction (defined based on the HTT unit cell) relieves the strain. Braden *et al.* [12] investigated the HTT–LTO transition in $La_{1.87}Sr_{0.13}CuO_4$ in detail. In particular, they found soft phonon behavior: the phonon that corresponds to the rotations of the octahedra has a pronounced minimum in its dispersion at the wavevector where the additional Bragg peak appears in the LTO phase (see Fig. 4.1). This dispersion dip is the largest at the phase transition temperature, T_{LTO}.

At temperatures higher than the barrier between the wells of the double-well potential discussed in Section 4.2 ($k_BT > a^2/4b$), the average structure is tetragonal with zero octahedral tilt, but the anharmonic potential already softens and broadens the rotational mode of the CuO_6 octahedra. This broadening and softening becomes enhanced on cooling towards the transition temperature, as the amplitude of the vibration decreases and the phonon increasingly feels the anharmonic bottom of the potential.

One can exclude electron–phonon coupling to conduction electrons from the mechanism of the transition, because the same instability occurs in the

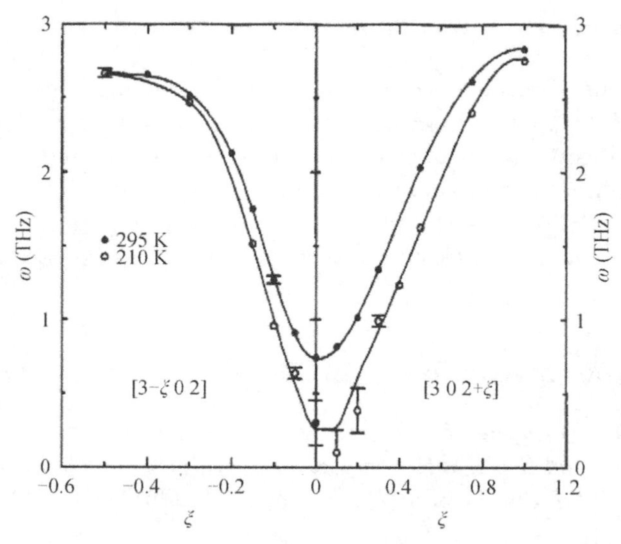

Fig. 4.1. The dispersion of the soft mode in $La_{1.67}Sr_{0.13}CuO_4$ near the structural phase transition (210 K) and at room temperature (from Ref. [12]).

undoped La_2CuO_4, which is an insulator. Furthermore, the metal-insulator transition that occurs with Sr doping has no observable impact on T_{LTO}.

4.3.2. *Kohn Anomalies*

Phonons behave very similarly when the structural distortion results from an electronic instability although the underlying physics is quite different.

In metals with Fermi surfaces, phonons may couple to singularities in the electronic density of states, which appear at specific wavevectors. Here the phonon renormalization associated with the real or incipient phase transition is discussed in terms of electron–phonon coupling. The coupling is between the phonon and the low-energy two-particle electronic response. Singularities in the electronic response can produce sharp features in phonon dispersions called Kohn anomalies [13]. These typically become stronger with reduced temperature, due to the sharpening of the Fermi surface. A classic example of this behavior is 1D conductors such as $K_2Pt(CN)_4Br_{0.30}{\cdot}3D_2O$ (KCP). Due to the one-dimensionality of the electronic states, these systems are characterized by the Fermi surface nesting at $2\boldsymbol{k}_F$ ($\hbar\boldsymbol{k}_F$ is Fermi momentum). This nesting greatly enhances

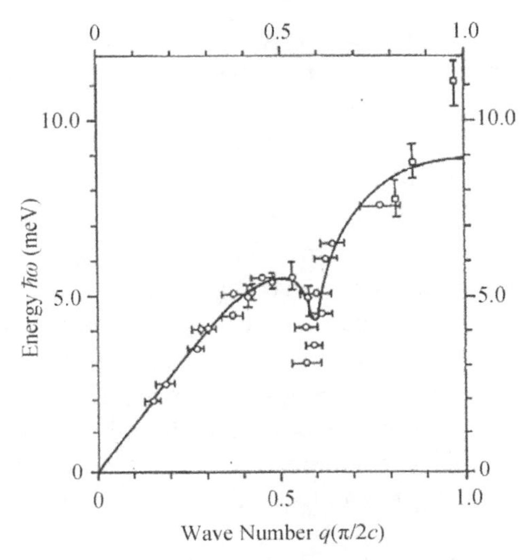

Fig. 4.2. LA phonon branch of $K_2Pt(CN)_4Br_{0.30}$ $3D_2O$ in the [1] direction at room temperature. Solid line represents the result of a calculation based on the simple free-electron model as discussed in Ref. [14].

the number of possible electronic transitions at $2\boldsymbol{k}_F$ compared to other wavevectors, which results in softer and broader phonons. For this reason acoustic phonons in KCP show pronounced dips at $\boldsymbol{q} = 2\boldsymbol{k}_F$($\boldsymbol{Q}$ is the total wavevecor, $\boldsymbol{q}$ is the reduced wavevector) [14] (see Fig. 4.2).

The amount of this phonon softening and broadening depends on the details of the interaction and varies greatly between different systems with Kohn anomalies. Often the broadening is much smaller than the experimental resolution, thus only the softening appears in the experiment.

4.3.3. *Ab Initio Calculations and the Role of the q-dependence of the Electron–phonon Matrix Element*

Electron–phonon scattering as well as anharmonicity influence the phonon frequency and reduce the lifetime (real/imaginary parts of the phonon self-energy).

Isolating the role of electrons near the Fermi surface requires knowing phonon frequencies and linewidths in the absence of coupling of phonons to electrons near the Fermi surface. Determining these accurately is typically a challenge. The simplest way to model phonons is by balls-and-springs models with the atomic nuclei serving as balls and the Coulomb forces screened by the electrons as springs. Shell models are more sophisticated modifications of this approach. Including only short-range interactions gives smooth phonon dispersions. Such models can be fit to the experimental dispersions that do not contain any sharp dips. Then deviations from these dispersions may indicate an incipient electronic or lattice instability. However, this phenomenological approach by itself is unable to distinguish between different types of instabilities discussed above.

In fact even defining the real part of the phonon self-energy, is not at all straightforward, because the unrenormalized dispersions are not easy to determine. Thus very precise measurements of phonon frequencies and linewidths do not, in a general case, provide direct information on electron–phonon coupling. However, combining careful measurements with detailed calculations and general arguments, can often elucidate the mechanism of the phonon anomalies.

It was noticed early on that Fermi surface nesting alone cannot adequately explain phonon softening resembling Kohn anomalies in many cases. For example, dispersions of certain phonons in NbC and TaC dip strongly at wavevectors where the FS nesting is relatively weak (see Fig. 4.3) [15, 16]. These could be modeled by including either very long-range repulsive interactions or by adding an extra shell with attractive interactions [16]. Sinha and Harmon [17] introduced $\boldsymbol{q}$-dependent electron-phonon coupling to explain these effects.

They included $\boldsymbol{q}$-dependent dielectric screening into their model and obtained good agreement with experiment for certain values of adjustable parameters.

Further theoretical development led to *ab initio* calculations based on the density functional theory (DFT) in the local density approximation (LDA) or in the generalized gradient approximation (GGA). These treat the electronic correlations at the mean field level and can very accurately predict phonon dispersions and line-widths in systems with structural instabilities as well as the electronic instabilities related to Fermi surface nesting [18]. However, it is very difficult to distinguish between these two scenarios and isolate contributions of specific electronic states to the phonon self-energy using this approach. Since it is necessary to go beyond the approximations built into LDA and GGA to obtain stripes, any phonon anomaly that is reproduced in these approximations cannot originate from interactions between phonons and stripes.

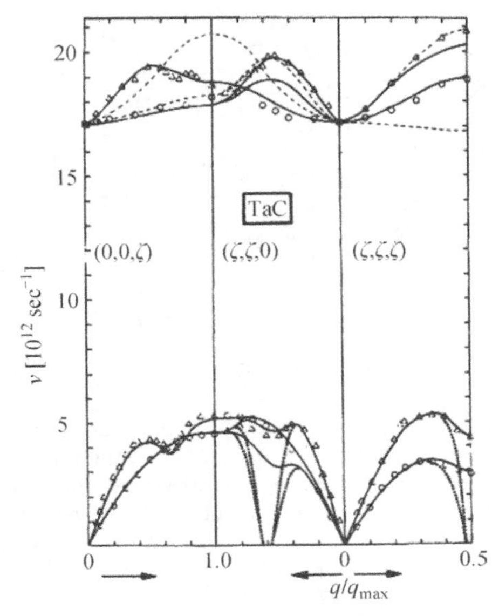

Fig. 4.3. Phonon dispersions of TaC (data points) [15] and calculations based on the double-shell model of Weber [16]. Note that the dispersion dip at $\boldsymbol{Q} = (0, 0, 0.6)$ originates mostly from the $\boldsymbol{q}$-dependence of electron–phonon coupling strength.

Heid *et al.* made an attempt to overcome this problem in Ru [19]. First, they performed both the *ab initio* DFT/LDA calculations and detailed measurements on a high quality single crystal. The phonon dispersions in Ru had a pronounced softening near the M-point, which was well reproduced by theory. The calculated softening became much weaker when electron–phonon coupling was made $\boldsymbol{q}$-independent (see Fig. 4.4). This result indicated that both the Fermi surface nesting and the $\boldsymbol{q}$-dependence of electron–phonon matrix element were important (the latter more than the former). The phonon dispersion dip disappeared entirely upon the exclusion of conduction electrons from the calculation. In addition, the entire dispersion hardened substantially, which may be an artifact of the procedure used to leave out the conduction electrons. This method showed that the phonon dispersion dips in Ru originate from coupling to conduction electrons, but the bare dispersions obtained by excluding them were somewhat arbitrary. This study illustrates the difficulty in separating the bare dispersion from the real part of the phonon self-energy. It is not important for reproducing or predicting phonon anomalies, but is important for making connections with other experiments and for identifying the origin of the phonon dips.

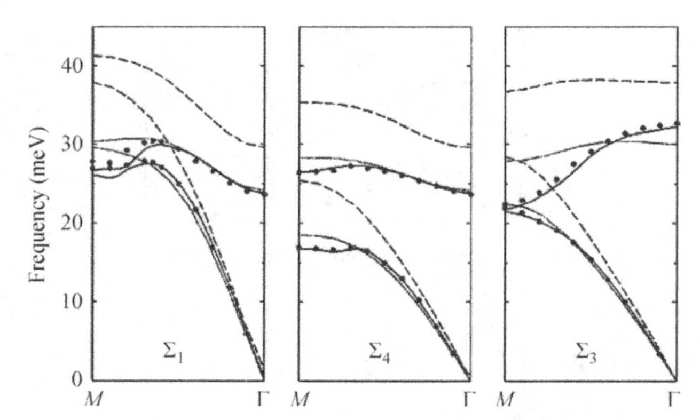

Fig. 4.4. Phonon dispersions in Ru (from Ref. [19]). The data points correspond to experimental results. The solid lines represent results of the full DFT/LDA calculation. Dotted lines were obtained by replacing $\boldsymbol{q}$-dependent electron–phonon matrix element with the average value (which was $\boldsymbol{q}$-independent). The dashed lines represent the DFT calculation without including the conduction electrons. Disappearance of the phonon dispersion dip near the M-point upon exclusion of conduction electrons demonstrates that the dip results from electron–phonon coupling.

4.3.4. *A Reinterpretation of the Origin of the CDW Formation in NbSe₂ Based on Phonon Measurements*

$NbSe_2$ has been considered as a classic quasi-2D CDW system with an incommensurate CDW induced by Fermi surface nesting. However, based on theoretical arguments, it has been suggested that the formation of an incommensurate superstructure in this material results from the $\boldsymbol{q}$-dependence of electron–phonon coupling, not Fermi surface nesting, i.e., that it is not a true CDW but is a structural instability [20–22]. However, these ideas have not been tested by neutron scattering measurements of phonons, because the available samples were too small for a comprehensive temperature-dependent study.

Recently Weber *et al.* performed detailed measurements of phonon dispersions using inelastic X-ray scattering (IXS), which does not require large samples [23]. Figure 4.5 shows the phonon dispersion in $NbSe_2$ as a function of temperature with the phase transition occurring at 33 K. The $\boldsymbol{q}$-dependence of the phonon softening is in marked contrast to the sharp, cusplike dips that normally characterize Kohn anomalies at $2k_F$ due to the Fermi surface nesting. In 2H–$NbSe_2$ the phonon renormalization extends over 0.36 Å^{-1}, or over half the Brillouin zone, and the critically damped

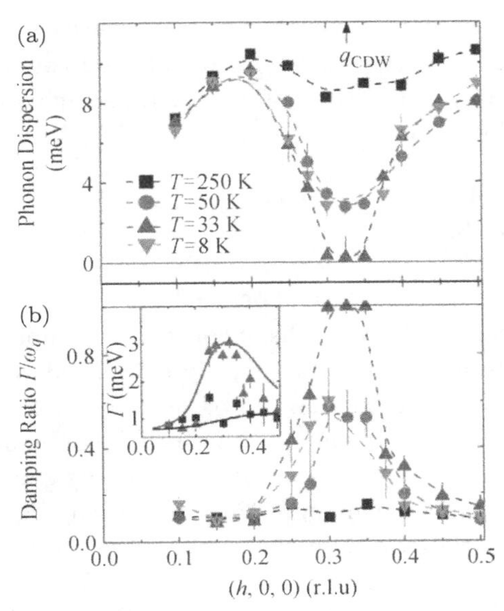

Fig. 4.5. Experimentally obtained dispersion and damping ratio of the soft-phonon branch in 2H-NbSe$_2$ at four temperatures 8 K < T < 250 K. Plotted are (a) the frequency of the damped harmonic oscillator $\omega_q = \sqrt{\tilde{\omega}_q^2 - \Gamma^2}$ and (b) the damping ratio Γ/ω_q. Lines are guides to the eye. Note that phonons at $h = 0.325$; 0.35 and $T = 8$ K were not detectable due to strong elastic intensities. The inset in (b) shows the experimentally observed damping of the damped harmonic oscillator (symbols) and scaled DFT calculations with $\sigma = 0.1$ eV (blue) and 1 eV (black) (from Ref. [23]). (For interpretation of the references to color in this figure legend, the reader is referred to the web version of this paper.)

region (where the phonon frequency goes to zero) extends over 0.09 Å^{-1}, which is much broader than the experimental $\boldsymbol{q}$-resolution. This behavior clearly rules out a singularity in the electronic response in 2H–NbSe$_2$.

By detailed comparison with the DFT/LDA calculations, Weber *et al.* showed that the structural instability results from the $\boldsymbol{q}$-dependence of the electron–phonon matrix element. They smeared the calculated Fermi surface by different amounts and found a strong effect of smearing on the phonon dispersion, which suggests that the interaction with conduction electrons drives the instability.

The observed phonon dispersion result is in contrast with the phonon anomaly reported in a quasi-1D metal ZrTe$_3$ [24], where the dispersion dip occurs over a much narrower region of reciprocal space, which is similar in size to KCP.

Weber *et al.* concluded that the CDW wavevector is determined by the wavevector dependence of the electron–phonon coupling, as proposed in Refs. [20–22], not by the position of the broad maximum of the two-particle electronic response measured by ARPES [25].

4.3.5. *Phonon Anomalies in Conventional Superconductors*

There are two interesting electron–phonon effects observed in measurements of phonon dispersions and linewidths in conventional superconductors. One is the normal state anomalies whose study can elucidate, which phonons contribute the most to the mechanism of superconductivity. The other is the effect of the superconducting gap 2Δ on the phonon spectra below T_c, which can be used as a probe of the superconducting gap.

In MgB_2 the high superconducting transition temperature, T_c, is explained by strong electron–phonon coupling of E_g modes around 80 meV near the zone center. A strong dip in the dispersion of these phonons observed in experiments and reproduced by LDA calculations is a clear signature of this coupling [26, 27].

The *ab initio* calculations based on DFT/LDA can predict all physical properties of materials that depend on electronic band structure, phonon dispersions, and electron–phonon coupling without adjustable parameters. In particular, they can be used to calculate T_c based on Migdal–Eliashberg theory [28]. Several *ab initio* calculations of phonon dispersions as well as T_cs were performed for the transition metal carbides and nitrides in order to explain relatively high transition temperatures in some and not in others [29–31]. They correctly reproduced phonon dispersions including the anomalies discussed in Section 4.3.3 [29, 31, 32], and established a correlation between the phonon anomalies and T_c. They also suggested that the phonon anomalies are associated with the Fermi surface nesting [31], but more detailed calculations along the lines of Ref. [19] need to be performed to separate the role of nesting from the enhancement of the electron-phonon matrix element. One interesting possibility that may need to be explored, is that additional screening near the nesting wavevector may enhance the electron–phonon matrix elements, thus the Kohn anomalies at the nesting wavevectors may be stronger than expected from enhanced electronic response due to nesting alone.

In a conventional superconductor with $T_c = 15$ K, YNi_2B_2C, the transverse acoustic phonons near $\boldsymbol{q} = (0.5, 0, 0)$ soften and broaden on

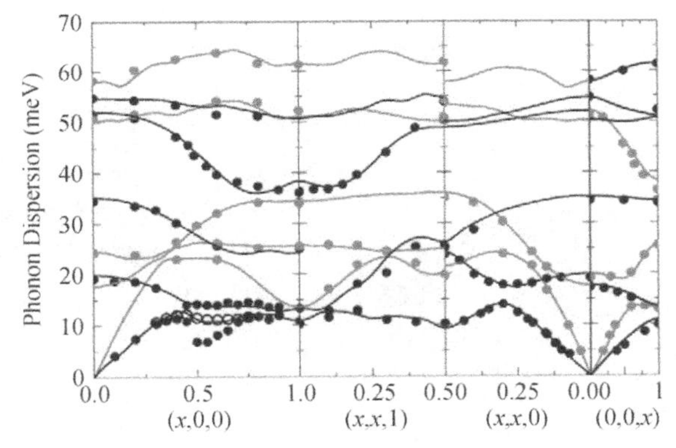

Fig. 4.6. Calculated (lines) and observed (filled dots) phonon frequencies in YNi_2B_2C at 20 K. Open dots along [100] were measured at 300 K. Branches shown in red/ black refer to phonons of predominantly longitudinal/transverse polarization, respectively. The horizontal axes denote different crystallographic directions in reciprocal lattice units (r.l.u.). The theoretical results were scaled up by a factor of 1.03 (from Ref. [34]).

cooling [33] as expected from electron–phonon coupling. Reichardt *et al.* calculated phonon frequencies and linewidths using LDA [34]. These calculations reproduced this phonon anomaly and correctly predicted an additional wavevector ($\boldsymbol{q} = 0.5, 0.5, 0$) where acoustic phonons couple strongly to conduction electrons (see Fig. 4.6) [35]. The energies of both soft phonons are comparable to the superconducting gap energy. In this case the opening of the superconducting gap has a strong influence on the phonon spectral function.

These phonons were so broad, that their normal state spectra extended below the low temperature superconducting gap 2Δ. In this case, when the 2Δ opens in the electronic spectrum in the superconducting state, the damped harmonic oscillator approximation of the phonon breaks down. Allen *et al.* [36] developed a theory precisely for this case, which predicted that phonon line-shapes in the superconducting state should contain either a step or a sharp peak very close to 2Δ depending on the values of the phonon energy, electron–phonon coupling and 2Δ. Normal state lineshape fixes the first two parameters, which leaves only one adjustable parameter, i.e., the superconducting gap. Detailed measurements of Weber *et al.* [37] confirmed this theory (Fig. 4.7). They also suggested how to use phonon measurements to determine the magnitude of 2Δ and to probe gap anisotropy.

The above analysis allows a reliable determination of the origin of phonon renormalization in YNi_2B_2C. Since the superconducting gap affects

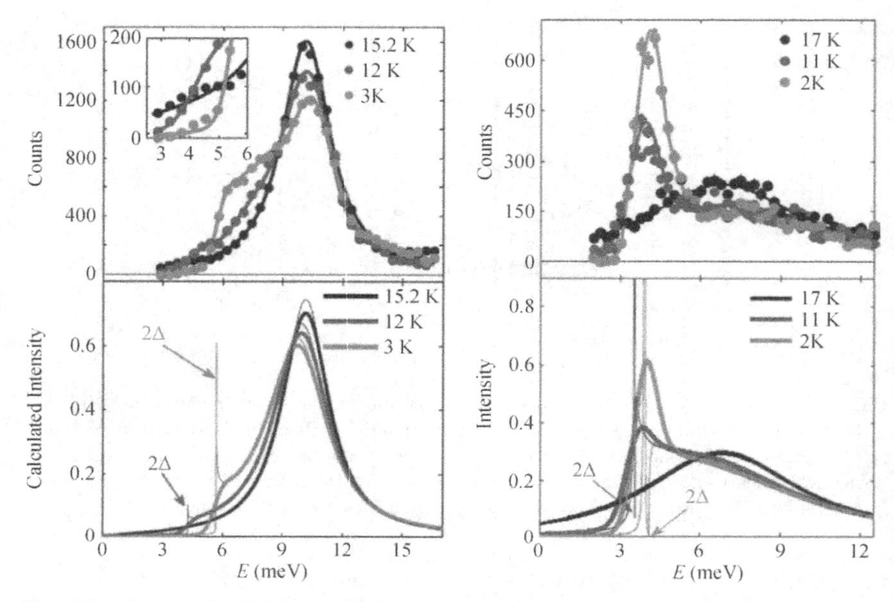

Fig. 4.7. Top panels: Evolution of the neutron-scattering profile measured on YNi_2B_2C at $\boldsymbol{Q} = (0.5, 0.5, 7)$ (left) and $\boldsymbol{Q} = (0.5, 0, 8)$ (right) above and below $T_c = 15$ K. Bottom panels: Thin lines represent calculated phonon lineshapes based on the theory of Allen *et al.* [36], using parameters extracted from the lineshape observed in the normal state. The thick lines are obtained after the convolution of the calculated lineshape with the experimental resolution (from Ref. [37]).

the phonon lineshapes as predicted by Allen *et al.* the phonon renormalization originates from the interaction between conduction electrons and phonons. Breadth of the phonon anomalies in $\boldsymbol{q}$-space argues against the nesting scenario: the $\boldsymbol{q}$-width of the anomalous region is closer to the one in $NbSe_2$ than to the 1D metals. Furthermore, the value of the superconducting gap extracted from phonon lineshapes was different for different wavevectors, which proves that soft phonons with different $\boldsymbol{q}$ connect different parts of the Fermi surface. Thus, the phonon anomaly originates primarily from the $\boldsymbol{q}$-dependence of electron–phonon coupling rather than from the nesting of the Fermi surface. LDA calculations predict phonon softening at the correct wavevectors (with the caveat that the softening is more pronounced at $\boldsymbol{q} = (0.5 - 0.7, 0, 8)$ at low T), so it is not necessary to invoke electronic correlations to explain experimental results.

Similar results were obtained on subsequent phonon measurements of Nb where phonon lineshapes were in almost perfect agreement with conventional theory of Allen *et al.* discussed above [38], although evidence for an effect of correlations on phonon linewidths beyond this theory in Nb and Pb was reported in Ref. [39] based on ultra-high resolution neutron scattering measurements.

4.3.6. *Phonon Anomalies with and without the Fermi Surface Nesting in Chromium*

A different situation appears in Chromium where the Fermi surface nesting is responsible for an incommensurate spin density wave (SDW) at a nesting wavevector $\boldsymbol{q}_{\rm sdw} = (0.94, 0, 0)$ and, as a secondary effect, of the charge density wave (CDW) at $\boldsymbol{q}_{\rm cdw} = (0.11, 0, 0)$ [40]. Shaw and Muhlestein [41] measured phonon dispersions in Chromium by INS. They reported soft phonons around $\boldsymbol{q} = (0.9, 0, 0)$, which is near $\boldsymbol{q}_{\rm sdw}$, as well as near $\boldsymbol{q} = (0.45, 0.45, 0)$ where some nesting has also been calculated. However, the neutron data were not accurate enough to establish their exact wavevectors.

Recently Lamago *et al.* [42] performed a more precise and comprehensive set of measurements by IXS, which showed that the two anomalies actually appear at $\boldsymbol{q}_{\rm sdw}$ and $\boldsymbol{q} = (0.5, 0.5, 0)$, respectively. A surprising result was that a transverse phonon branch softened throughout the zone boundary between $\boldsymbol{q} = (0.5, 0.5, 0)$ and (1,0,0), i.e., for $\boldsymbol{q} = (0.5 + h, 0.5 - h, 0)$, where $0 < h < 0.5$. LDA-based calculations performed as a part of this investigation, successfully reproduced the observed phonon softening (Fig. 4.8). However, the electronic response function that couples to the phonons obtained from the same calculations showed no clear features corresponding to the phonon dips along the zone boundary. Thus, these phonon dips come exclusively from the $\boldsymbol{q}$-dependence of the electron–phonon coupling strength.

This result was further corroborated by the effect of the numerical smearing of the Fermi surface on the calculated phonon dispersions. Such a smearing makes it possible to isolate coupling to electronic excitations near the Fermi surface. If it has a strong effect on the calculated phonon dispersions, electrons near the Fermi surface contribute significantly to the phonon self-energy. Otherwise, the coupling to these electrons is small.

In Cr, this smearing suppressed only the calculated effect at $\boldsymbol{q} = (0.94, 0, 0)$. It had no effect at $\boldsymbol{q} = (0.5 + h, 0.5 - h, 0)$ for any h including $h = 0$ (Figs. 4.8(a) and 4.8(b)). Thus, the phonon dispersion dip at $\boldsymbol{q} = (0.5 + h, 0.5 - h, 0)$ comes exclusively from an interaction with electrons

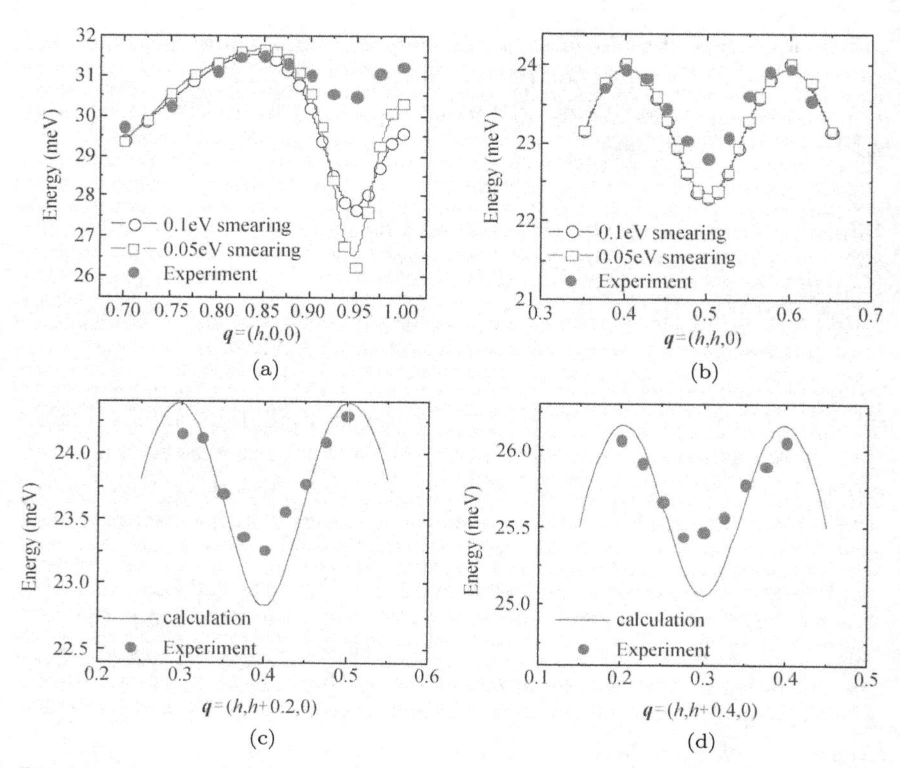

Fig. 4.8. A comparison between measured (red) and calculated (blue, black) phonon dispersions and the LDA calculation in Cr. Only the data for the lowest energy transverse phonons near the zone boundary line connecting $\boldsymbol{q} = (1, 0, 0)$ and $\boldsymbol{q} = (0.5, 0.5, 0)$ are shown. The smearing has a strong effect in (a) but not in (b) (from Ref. [42]). (For interpretation of the references to color in this figure legend, the reader is referred to the web version of this paper.)

far from the Fermi surface. This is true even for $h = 0$, which is near a nesting feature previously thought [41] to be responsible for phonon softening. This nesting feature disappears even at small h, but the phonon renormalization does not become smaller in either the calculation or the experiment. In contrast with the earlier work [41], which did not include this type of analysis or measurements for $h > 0$, Lamago et al. [42] concluded that this nesting feature makes a negligible contribution to the phonon self-energy.

Thus, different types of phonon anomalies can appear in the same material.

4.3.7. *Phonon Anomalies Related to Stripes in $La_{1.69}Sr_{0.31}NiO_4$*

Charge modulation due to stripe formation is not directly related to Fermi surface nesting, but results from electronic correlations (see Section 4.1). Intense research effort has been devoted to trying to understand the physics of stripes because dynamic stripes may play an important role in copper oxide superconductors. Most work focused on the magnetic component of the stripes, and charge stripes still remain relatively unexplored. In particular, the interaction between charge stripes and phonons is still poorly understood.

According to a model proposed by Kaneshita *et al.* [43], steeply dispersing Goldstone modes arising from charge stripe order should mix with phonons at the charge stripe ordering vectors and the resulting anti-crossing would be observed as phonon line splitting or phonon line softening and/or broadening if the experimental resolution is insufficient (see Fig. 4.9).

It is widely believed that charge stripes in the nickelates and cuprates should couple most strongly to Ni–O or Cu–O bond-stretching phonons, hence these phonon branches have been investigated most thoroughly. It is also believed that the L-component (along the direction perpendicular to the Ni–O planes) can be ignored except at low energies and the same effects should be observed at all. Tranquada *et al.* searched for the effect of stripes on Ni–O bond stretching phonons in $La_{1.69}Sr_{0.31}NiO_4$ [44]. The stripe-related anomaly was expected at $\boldsymbol{q} = (0.31, 0.31, L)$, but instead of an anomaly at this wavevector they found that the phonon branch splits on approach towards the zone boundary $\boldsymbol{q} = (0.5, 0.5, 0)$ in the 110 direction (in the tetragonal notation) (see Fig. 4.10). This effect was explained in

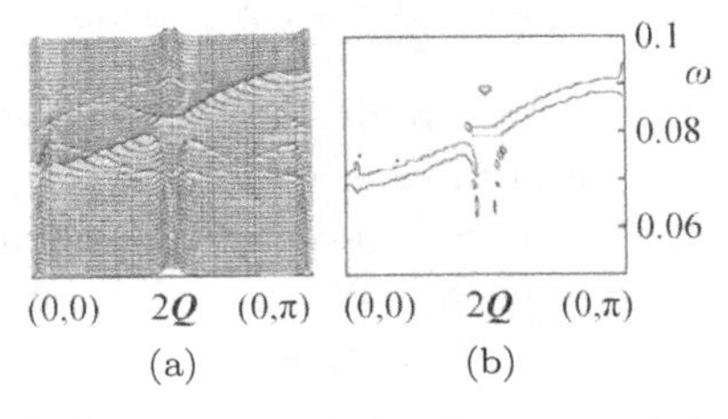

Fig. 4.9. Renormalized phonon spectral functions assuming an interaction between phonons and collective stripe modes: (a) is a logarithmic plot to emphasize the small intensity of the shadow bands and (b) is a contour plot shown for high intensity. The simple sinusoidal form is assumed for the unperturbed phonon [43].

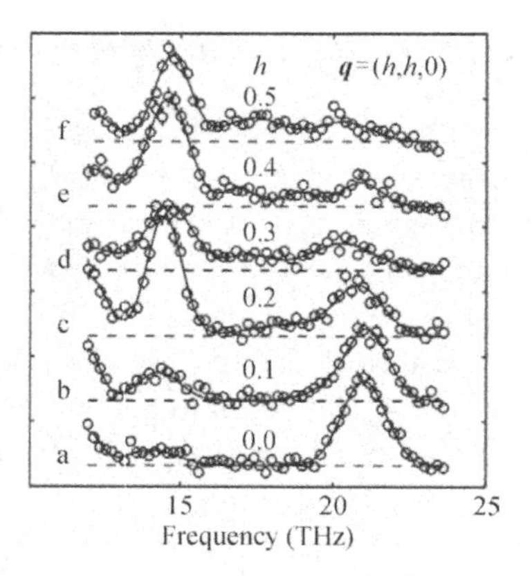

Fig. 4.10. Neutron scattering measurements of phonons in $La_{1.69}Sr_{0.31}NiO_4$ at $T \sim 10$ K for $\boldsymbol{q} = (h, h, 0)$; dashed lines indicate the constant background. Measurements were performed in various equivalent zones, with varying L component, in order to avoid spurious peaks due to accidental Bragg scattering by the sample (from Ref. [44]).

terms of a toy model containing inhomogeneous distribution of force constants as would occur in the case of static stripes with a large amplitude of charge inhomogeneity.

A number of important open questions about the interactions between stripes and phonons in the nickelates remain. In particular, acoustic phonon branches have not been investigated. Reference [44] presented data above the stripe ordering temperature only at the zone center and the zone boundary. Thus it is unknown what happens to phonons at the stripe-ordering wavevectors at temperatures above the ordering temperature where stripes should be purely dynamic. This last question is especially important, because stripes in the cuprate superconductors are dynamic if they exist.

4.3.8. *Lessons Learned from Model Systems*

It follows from the investigations reviewed above as well as other similar studies that all types of structural and electronic instabilities have a similar impact on phonon dispersions as well as line-broadening. The following steps are essential to correctly identify the origin of anomalous phonon softening and broadening most of the time:

1. Measurements of phonon dispersions and linewidths.
2. Measurements of the temperature dependence of phonons at anomalous wavevectors.

3. DFT calculations of phonon dispersions for different values of the smearing of the Fermi surface.
4. ARPES measurements and/or DFT calculations of the electronic joint density of states in LDA or GGA approximation to determine the nesting properties of the Fermi surface.

*The phonon anomalies predicted by the DFT calculations that appear at the calculated and/or measured nesting wavevectors and are very sharp in **q**-space* result from the Fermi surface nesting. For example, the Fermi surface nesting-related anomalies in KCP, $ZrTe_3$ and Cr (near $\boldsymbol{q} = (0.95, 0, 0)$) cover about 5% of the Brillouin zone or less. In this case, the calculated anomalous phonon energies should be sensitive to the smearing of the Fermi surface.

*The anomalies predicted by DFT calculations that are broad in **q*** are not associated with nesting, but originate from structural instabilities or $\boldsymbol{q}$-dependence of electron–phonon coupling. For example, in La_2CuO_4, Cr (away from $\boldsymbol{q} = (0.95, 0, 0)$), and $NbSe_2$ the width is the anomalous phonon regions of about 15% of the Brillouin zone. If the calculated phonon dispersions are sensitive to the smearing of the Fermi surface, these structural instabilities originate from interactions with conduction electrons via an enhanced electron–phonon coupling matrix element. In this case soft phonon line-shapes can be deformed by the opening of a gap in the electronic spectra such as the superconducting gap. If the calculated phonon dispersions are not sensitive to the smearing of the Fermi surface (such as in Cr away from $\mathbf{q} = (0.95, 0, 0)$) or if there are no conduction electrons present (such as in La_2CuO_4), renormalization of the anomalous phonons by electrons near the Fermi surface can be ruled out. In this case, the relevant physics can be best described in terms of an anharmonic potential or coupling to electronic states far from the Fermi surface.

Phonon anomalies not explained by the DFT/LDA or DFT/GGA calculations originate from strong electronic correlations beyond the LDA/GGA (mean field) level such as the tendency to form stripes. The first step to identify such phonon anomalies is to look at features in the phonon spectra that are not predicted by the calculations and have the $\mathbf{q}$-dependence that may be associated with stripe formation. However, this prescription is insufficient to unambiguously identify signatures of dynamic stripes. More work on model systems where stripes are known to exist is necessary to make further progress. The study of Tranquada *et al.* [44] was the first step in this direction. Further investigations are under way [45].

4.4. Phonon Anomalies Possibly Related to Stripes in Doped La_2CuO_4

4.4.1. *Summary of Early Work*

Calculations performed soon after the discovery of high-temperature superconductors suggested that electron–phonon coupling is too weak to account for high-temperature superconductivity [46,47]. However, measurements performed in conjunction with shell model calculations showed that the bond-stretching branch softened strongly towards the zone boundary as doping increased from the insulating phase to the superconducting phase [48]. This behavior pointed towards strong electron–phonon coupling and a possible role of the zone boundary "half breathing" bond-stretching mode in the mechanism of high-temperature superconductivity [49]. It later became apparent that this trend continues into the over doped nonsuperconducting phase, which indicates that zone boundary softening is related to the increase of metallicity with doping rather than to the mechanism of superconductivity (Fig. 4.11) [50].

These experiments and related calculations are extensively covered in the previous review by Pintschovius [10].

A recent investigation of Park *et al.* [51] also revealed that the linewidth of the zone boundary bond-stretching phonon (around $\boldsymbol{q} = (0.5, 0, 0)$) broadens strongly when the doping, x, is reduced and then narrows abruptly at $x = 0$. They explained this effect in light of NQR/NMR results showing that the doped carrier concentration in the Cu–O planes is inhomogeneous, and since the zone boundary phonon frequency depends on doping, the experimental lineshape broadens. This sensitivity to doping increases towards small x, which results in the increased phonon broadening. The disappearance of the effect at $x = 0$, where there are no doped carriers, also naturally follows from this model.

Here, I will focus not on the zone boundary, but half-way to the zone boundary in the [100]-direction (along the Cu–O bond) where the most interesting physics has been observed. This work began with the INS experiments of McQueeney *et al.* [52] who reported anomalous lineshape and temperature dependence of the bond-stretching phonons in $La_{1.85}Sr_{0.15}CuO_4$ near $\boldsymbol{q} = (0.25, 0, 0)$ (Fig. 4.12) and interpreted these results in terms of line splitting due to unit cell doubling induced by stripe formation, which is similar to the branch splitting later reported for stripe-ordered $La_{1.69}Sr_{0.31}NiO_4$

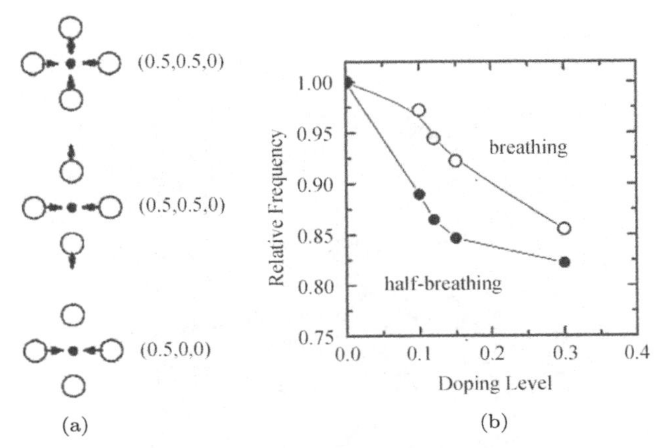

Fig. 4.11. (a) Displacement patterns of zone-boundary bond-stretching modes in cuprates. Top: longitudinal mode in the [110]-direction (breathing mode); middle: transverse mode in the [110]-direction (quadrupolar mode); bottom: longitudinal mode in the [100]-direction (half-breathing mode). Circles and full points represent oxygen atoms and copper atoms, respectively. Only the displacements in the Cu–O planes are shown. All other displacements are small for these modes. (b) Schematic of doping dependence of the breathing $\boldsymbol{q} = (0.5, 0.5, 0)$ and half-breathing $\boldsymbol{q} = (0.5, 0, 0)$ zone boundary mode frequencies. This behavior is probably not directly related to the mechanism of superconductivity, since the softening continues into the overdoped nonsuperconducting part of the phase diagram (from Ref. [50]).

by Tranquada *et al.* [44] (see above) (Fig. 4.10). It is important to emphasize here that long-range-ordered static stripes do not form in this compound, and the stripes must be dynamic if they exist.

This interpretation evolved considerably in recent years as a result of further measurements and calculations. Pintschovius and Braden repeated the experiment using different experimental conditions, which had a higher energy resolution due to their use of the Cu220 monochromator [53]. They also measured the interesting wavevector range between $\boldsymbol{q} = 0.1$ and 0.4 in the so-called focusing condition with the tilt of the resolution ellipsoid matching the phonon dispersion, which further improved the resolution compared to Ref. [52] (see Section 4.2 for a more detailed discussion). Pintschovius and Braden reported enhanced linewidth near the same wavevector (with the strongest broadening at $\boldsymbol{q} = (0.3, 0, 0)$), but did not see any splitting of the phonon line (Fig. 4.13).

The origin of these effects was not clear at the time, but phonon anomalies near half way to the zone boundary, where static charge stripes

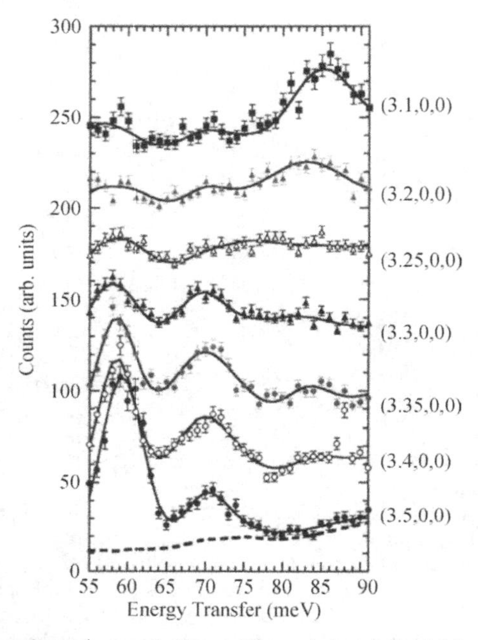

Fig. 4.12. Bond-bending (around 60 meV) and bond-stretching (above 65 meV) branches in optimally-doped LSCO [52].

appear in some cuprates, suggested a possible connection to incipient stripe formation.

Reznik *et al.* [54] investigated the same bond stretching branch in $La_{1.875}Ba_{0.125}CuO_4$ where static stripes appear at low temperatures. They performed the first set of measurements in the same scattering geometry as Pintschovius and Braden [53] (at wavevectors (5–4.5,0,0) or (5–4.5,0,1)) and found that the phonon dispersion could be described with two components (see Fig. 4.14): One had a "normal" dispersion following a cosine function (blue line), and the other softened and broadened abruptly at $\boldsymbol{q} = (0.25, 0, 0)$ (black line). The possible relationship between the phonon anomaly and stripe formation is further explored in Section 4.4.5.

Reznik *et al.* [54, 55] found that there was an overall hardening of the spectral weight on heating in both $La_{1.875}Ba_{0.125}CuO_4$ and $La_{1.85}Sr_{0.15}CuO_4$ at $\boldsymbol{q} = (0.25, 0, 0)$ (see Fig. 4.15). This indicates that the anomalous broadening does not originate from anharmonicity or structural inhomogeneity, since these have the opposite or no temperature dependence.

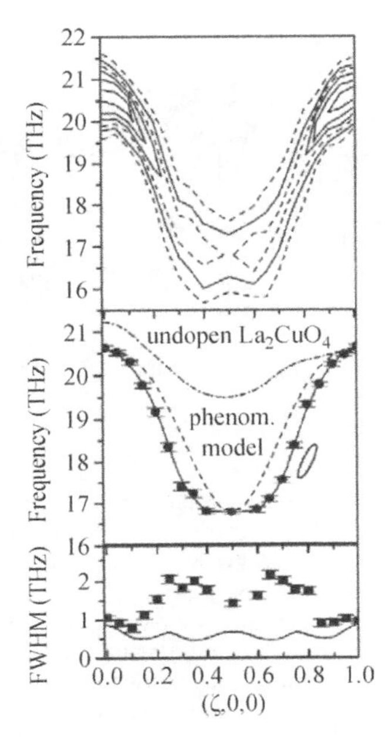

Fig. 4.13. Results of measurements on $La_{1.85}Sr_{0.15}CuO_4$ performed with a better energy resolution and similar in-plane wavevector resolution than in Fig. 4.12 (from Ref. [53]).

Pintschovius and Braden [53] observed no temperature dependence at $\boldsymbol{q} = (0.3, 0, 0)$, (it is equivalent to $\boldsymbol{q} = (0.7, 0, 0)$ if interlayer interactions are neglected) although the zone center phonon of the same branch softened on heating. This softening is due to increased anharmonicity and should affect the entire branch. The absence of softening at $\boldsymbol{q} = (0.3, 0, 0)$ that they report, indicates that there is a counterbalancing trend, which makes their results agree qualitatively with Refs. [54, 55]. I will explain the reason for the quantitative difference in the following section.

McQueeney *et al.* [52] reported the suppression of the anomalous behavior at room temperature. However, the room temperature data of Ref. [52] suffered from a much stronger background than the low temperature data and relatively large statistical error. Reference [55], which had a much better resolution and signal-to-background ratio, but was limited to only three wavevectors, also reported a suppression of the anomalous behavior at 330 K. In this regard the two studies are consistent, although Ref. [52] claims a much more radical change of the phonon dispersion than

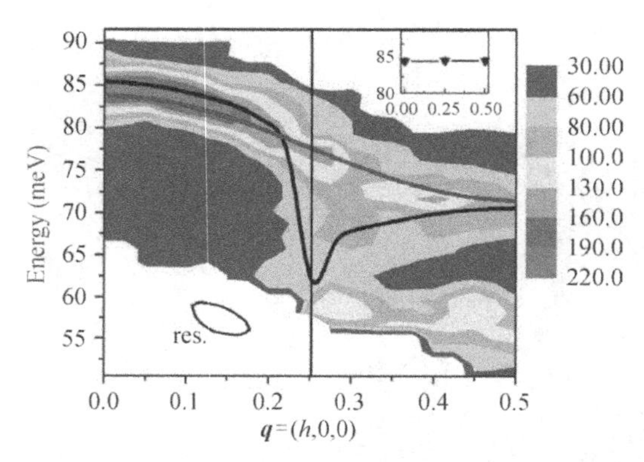

Fig. 4.14. Color-coded contour plot of the phonon spectra observed on $La_{1.875}Ba_{0.125}CuO_4$ at 10 K. The intensities above and below 60 meV are associated with plane-polarized Cu–O bond-stretching vibrations and bond-bending vibrations, respectively. Lines are dispersion curves based on two-peak fits to the data. The white area at the lower left corner of the diagram was not accessible in this experiment. The ellipse illustrates the instrumental resolution. The inset shows the dispersion in the [110]-direction. The vertical line represents the charge stripe ordering wavevector. Blue/black line represents the "normal"/"anomalous" component, respectively, in the original interpretation of the authors. Subsequent work showed that a substantial part of the intensity in the "normal" component is an artifact of finite wavevector resolution in the transverse-direction (out of the page) (from Ref. [54]). (For interpretation of the references to color in this figure legend, the reader is referred to the web version of this paper.)

reported in Ref. [55]. To resolve this disagreement it is necessary to perform measurements covering the entire BZ at 300 K with the experimental configuration of Ref. [55].

4.4.2. *Recent IXS Results*

Neutron scattering experiments have a relatively poor $\boldsymbol{Q}$ resolution. For the cuprates its full width half maximum (FWHM) is on the order of 15% of the in-plane Brillouin zone. The effects of finite $\boldsymbol{Q}$ resolution in the longitudinal direction have been carefully considered in early studies, but the finite resolution in the transverse direction has not. In this section, I will discuss recent IXS work and will show that some previous experiments need to be reinterpreted taking into account the finite transverse $\boldsymbol{Q}$.

More recent measurements using both INS [55] and IXS [56] yielded a somewhat surprising result that the anomalous softening/ broadening for $\boldsymbol{q} = (0.25, k, 0)$ occurs only very close to $k = 0$. For example in $La_{1.84}Nd_{0.04}Sr_{0.12}CuO_4$ the phonon anomaly significantly weakened at

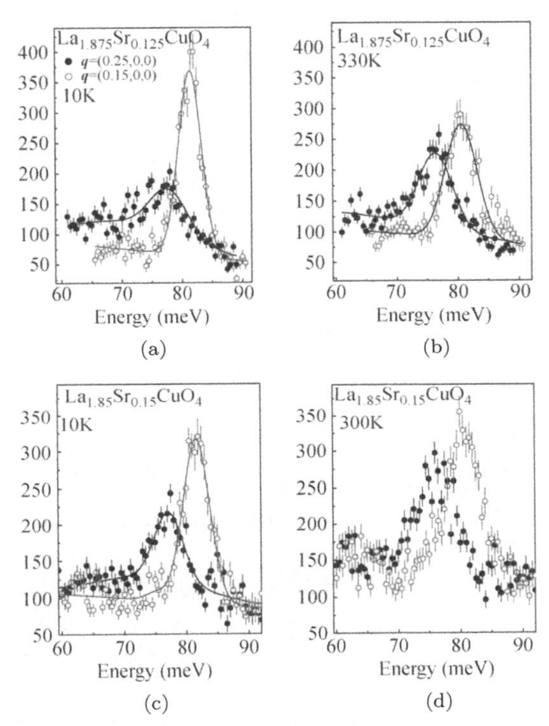

Fig. 4.15. Temperature dependence of the bond-stretching phonons at select wavevectors. Energy scans taken on $La_{1.875}Ba_{0.125}CuO_4$ (a) and (b) and on $La_{1.85}Sr_{0.15}CuO_4$ ((c)–(e)) at 10 K (a) and (c) and 330 K (b) and (d) (Ref. [55]). The phonon at $\boldsymbol{q} = (0.15, 0, 0)$ is "normal" in that it has a Gaussian lineshape on top of a linear background. This background results from multiphonon and incoherent scattering and has no strong dependence on $\boldsymbol{Q}$. The intensity reduction of this phonon in $La_{1.875}Ba_{0.125}CuO_4$ from 10 K (a) to 330 K (b) is consistent with the Debye–Waller factor. At $\boldsymbol{q} = (0.25, 0, 0)$, there is extra intensity on top of the background in the tail of the main peak. It originates from one-phonon scattering that extends to the lowest investigated energies, while the peak intensity is greatly suppressed as discussed in the text. The effect is reduced but does not disappear at 330 K. Note that 330 K in (b) is shown instead of 300 K in the same plot in Ref. [54] because of a typographical error in the latter. Integrated intensity of the phonon decreases from $\boldsymbol{q} = (0.15, 0, 0)$ to $\boldsymbol{q} = (0.25, 0, 0)$ due to the decrease of the structure factor (from Ref. [51]).

$|k| \approx 0.08$ compared to $k = 0$, and disappeared entirely at $|k| = 0.16$ (see Fig. 4.16) [56]. The "normal" component for $k \approx 0$ was significantly suppressed.

Neutron measurements have a much lower resolution in the k-direction, i.e., INS experiments nominally performed with $k = 0$ include a significant

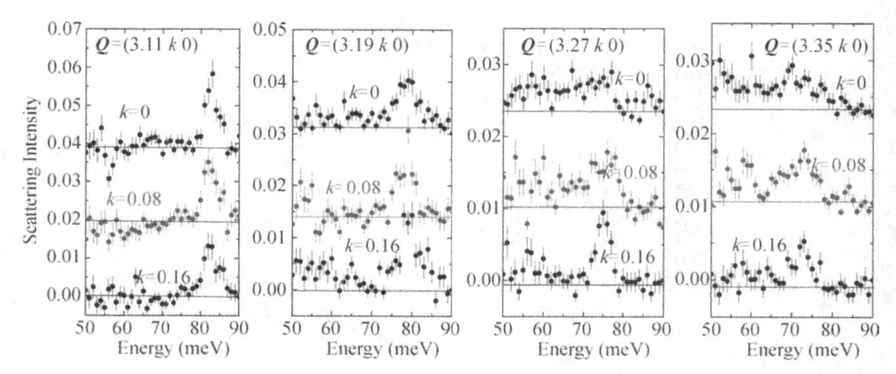

Fig. 4.16. IXS energy scans after subtraction of the elastic tail and a constant term corresponding to the stray radiation. The scans were taken with $\boldsymbol{q} = (h, k, 0)$ with $h = 0.11, 0.19, 0.27, 0.35$ (from the left to the right column) and k as indicated in the figure. The most interesting features are the suppression of the two-component behavior seen by INS at 77 meV near $\boldsymbol{Q} = (3.27, 0, 0)$ and the rapid narrowing and hardening of the phonon line from $k = 0$ to $k = 0.16$ for $\boldsymbol{Q} = (3.27, k, 0)$. From Ref. [56].

contribution from wavevectors with $|k| > 0.08$ even in the most optimal configuration ($a - b$ scattering plane). Thus most of the intensity in the"normal" component in the INS measurements probably comes from these phonons with $|k| > 0.08$.

With this information it now becomes possible to explain why the temperature effect in Ref. [53] was weaker than in Ref. [52]. The experiment of Pintschovius and Braden [53] was performed in the a-c scattering plane, which had poorer wavevector resolution in the k-direction, whereas the other study was performed in the a–b scattering plane, which had a better k-resolution.[a] Since the phonon anomaly is sharp in the k-direction, the anomalous behavior should be masked by the "normal" phonons with $|k| > 0$ in the a–c scattering plane more than in the a–b scattering plane. This "masking" would also reduce the phonon linewidth in Ref. [53] compared to the measurement in Ref. [57] performed with better k-resolution.

---

[a]In order to maximize flux neutron scattering instrument have a much larger beam divergence (factors 2–3) in the direction perpendicular to the scattering plane than in the scattering plane. Wavevector resolution scales with the beam divergence. The perpendicular direction for the a–c scattering plane is the b-direction, which corresponds to the k-direction in reciprocal space in my notation. In the a–b scattering plane has the poor resolution along l, and good resolution along h and k.

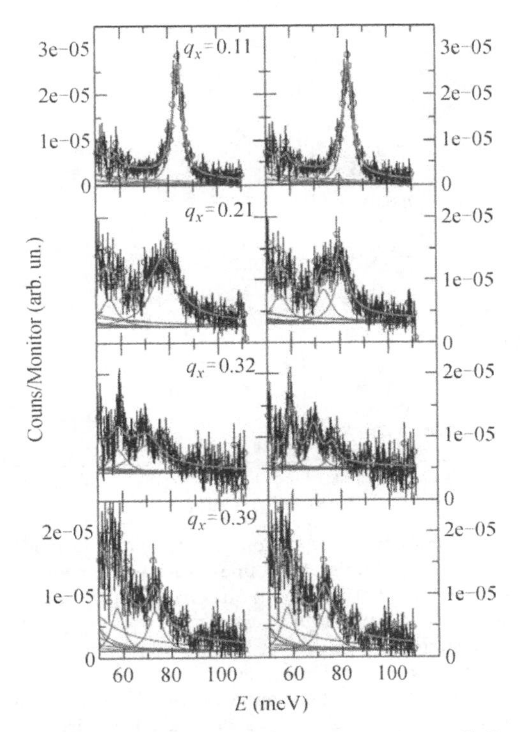

Fig. 4.17. Inelastic X-ray scattering spectra of $La_{1.86}Ba_{0.14}CuO_{4+\delta}$, at $\boldsymbol{Q} = (3 + q_x, q_y, 0)$ $(q_y < 0.04)$. In the left column a single peak is used to fit the data. In the right column the data are fitted with two Cu–O bond stretching modes (from Ref. [57]).

d'Astuto *et al.* [57] reported the two-branch behavior in the IXS spectra of $La_{1.86}Ba_{0.14}CuO_4$ with clearly resolved "normal" and "anomalous" components (Fig. 4.17). This result seems to be different from the $\boldsymbol{Q} = (3.27, 0, 0)$ data of Fig. 4.16 measured also by IXS with a much higher energy resolution, as well as with the results of Graf *et al.* [58] on $La_{1.92}Sr_{0.08}CuO_4$ and Sasagawa *et al.* [59], which are consistent with either a single broad peak or two strongly overlapping peaks. It is possible that since d'Astuto *et al.* measured a Ba-doped sample, and Reznik *et al.* investigated the Sr-doped systems, the difference may come from Ba vs. Sr doping. It is necessary to perform further experiments to clarify this potentially important issue as discussed by De Filippis *et al.* [60].

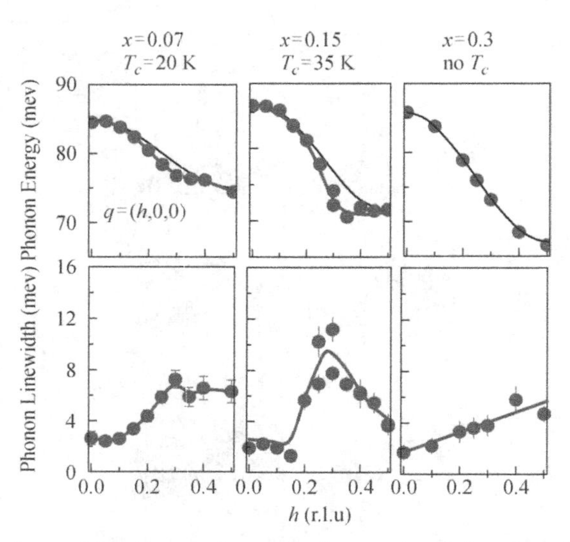

Fig. 4.18. Bond-stretching phonon dispersion (top row) and linewidth (bottom row) in $La_{2-x}Sr_xCuO_4$ at three doping levels. Black lines represent downward cosine dispersion. The overall increase of the bond-stretching mode linewidth towards the zone boundary appears to be doping-independent. Softening compared with the cosine dispersion as well as the linewidth enhancement half-way to the zone boundary do not shift with h between $x = 0.07$ and 0.15. Note that the enhancement of the linewidth at $x = 0.07$ near the zone boundary compared with the higher dopings is unrelated to the anomaly at $h = 0.3$ and can be explained by the inhomogeneous doping effect discussed in Section 4.4.1 and [51].

4.4.3. *Doping Dependence*

Figure 4.18 shows the bond-stretching phonon dispersion and line-width for $La_{2-x}Sr_xCuO_4$ with $x = 0.07$, 0.15, and 0.3. The dispersion is compared with the cosine function that typically comes out of DFT/LDA/GGA calculations (see for example Refs. [61–63]). Here, the data presented in Fig. 4.4 of Ref. [54] are combined with some unpublished results and refitted using the model that includes all phonon branches picked up by the spectrometer resolution as opposed to gaussian peaks [64]. Such an analysis provides more accurate values of intrinsic phonon linewidths. The strongest dip below the cosine function and the biggest peak of the linewidth is observed at optimal doping where the T_c is highest. These are smaller at $x = 0.07$ and disappear in the overdoped nonsuperconducting sample with $x = 0.3$. The position of the phonon anomaly does not change with doping.

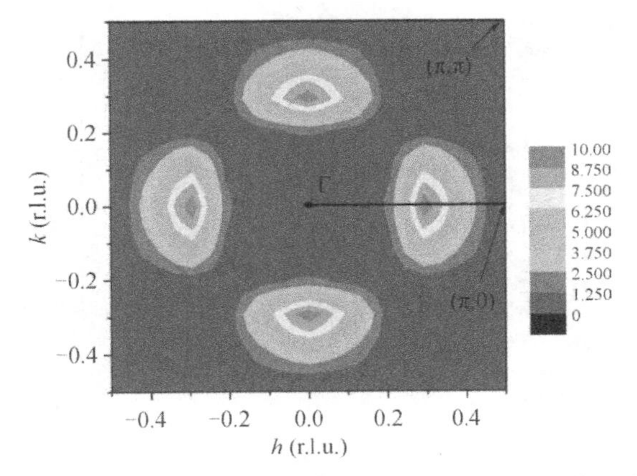

Fig. 4.19. Qualitative picture of the difference in the linewidths of the bond-stretching phonon in optimally-doped ($x = 0.15$) and overdoped ($x = 0.3$) $La_{2-x}Sr_xCuO_4$ as a function of wavevector in the *ab-basal* plane based on [50, 56]. The units of the color scheme are meV. The solid line indicates the [100] direction along which most of measurements were performed.

Comparison with the overdoped sample, where the physics are conventional, allows to identify the effects of electron–phonon coupling that are intrinsic to optimal doping. Figure 4.19 shows the schematic of the anomalous phonon broadening that appears on top of the broadening observed in the $x = 0.3$ sample. The anomalous broadening peaks at $\boldsymbol{q} = (0.3, 0, 0)$ and weakens rapidly in the longitudinal and transverse directions.

This effect is phenomenologically very similar to the renormalization of the acoustic phonons at specific wavevectors discussed in Sections 4.2 and 4.3. Next, I will show that profound differences exist between $La_{2-x}Sr_xCuO_4$ and conventional metals.

4.4.4. *Comparison with Density Functional Theory*

As discussed in Section 4.2, density functional theory gives a good description of phonon dispersions in metals where electron–electron interactions beyond the mean field level (as taken into account in LDA and GGA) can be neglected. Giustino *et al.* [62] performed such a calculation in the GGA for $La_{1.85}Sr_{0.15}CuO_4$, which approximately agreed with the early experimental data of Ref. [52]. In particular, they reproduced the overall downward dispersion of the longitudinal bond-stretching branch. However, the strong effect in the bond-stretching phonon half way to the zone boundary was

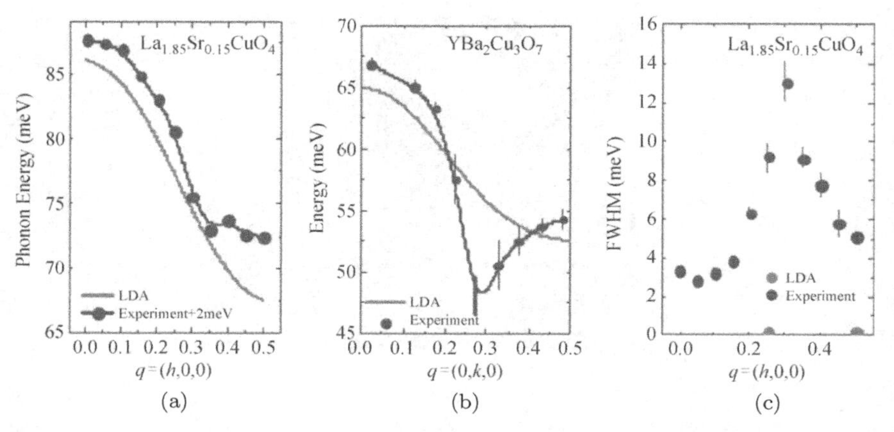

Fig. 4.20. Comparison of some LDA predictions with experimental results for $La_{1.85}Sr_{0.15}CuO_4$ [54] and $YBa_2Cu_3O_7$ [85] at 10 K. (a) and (b) Experimental bond-stretching phonon dispersions compared to LDA results. The data in (a) are shifted up by 2 meV. (c) Phonon linewidths in $La_{1.85}Sr_{0.15}CuO_4$ compared with LDA results on $YBa_2Cu_3O_7$. Reference [62] contains no linewidth results for $La_{1.85}Sr_{0.15}CuO_4$ but they should be similar.

not apparent in Fig. 4.1 of Ref. [62], because the 300 K results and 10 K results were plotted together. In the brief communication arising from the paper of Giustino *et al.*, Reznik *et al.* showed that DFT did not reproduce the phonon anomaly half way to the zone boundary that appears in Refs. [50, 52–58] as well as in later experiments discussed above in detail (Fig. 4.20 and [65]).

It is interesting that many-body calculations predict a substantial enhancement of the coupling to bond-stretching phonons compared to DFT (see, for example Refs. [49, 66]). t–J model-based calculations describe interesting doping dependence of the zone boundary phonons, suggesting that strong correlations might be relevant [67–69]. The idea that electronic correlations are responsible for the enhanced electron–phonon coupling is further reinforced on the qualitative level by theoretical investigations based on the Hubbard–Holstein model and similar models, which find an enhancement of phonon renormalization by electronic correlations not included in LDA [70].

Strong renormalization of the bond stretching phonons has been taken as evidence for a soft collective charge mode [71, 72] or an incipient instability [73] with respect to the formation of either polarons, biporarons [74–76], charge density wave order [77], phase separation [77–80], valence bond order [81], or other inhomogeneity [81]. These may or may not be related to the mechanism of stripe formation. A number of studies suggested that these instabilities may lead to superconductivity [73–75, 79].

Further evidence that the bond-stretching phonon anomaly results from electronic correlations comes from an excellent agreement between the GGA calculation and the experimental phonon dispersion in nonsuperconducting overdoped $La_{1.7}Sr_{0.3}CuO_4$ [50] and Fig. 4.18 in Ref. [11].

4.4.5. *Connection with Stripes and other Charge-inhomogeneous Models*

The bond-stretching phonon anomaly is strongest in $La_{1.875}Ba_{0.125}CuO_4$ and $La_{1.48}$ $Nd_{0.4}Sr_{0.12}CuO_4$, compounds that exhibit spatially modulated charge and magnetic order, often called stripe order. It appears when holes doped into copper–oxygen planes segregate into lines, which act as domain walls for an anti-ferromagnetically ordered background. Static long-range stripe order has been observed only in a few special compounds such as $La_{1.48}Nd_{0.4}Sr_{0.12}CuO_4$ and $La_{1.875}Ba_{0.125}CuO_4$ where anisotropy due to the transition to the low temperature tetragonal structure provides the pinning for the stripes while superconductivity is greatly suppressed [4]. In contrast, the more common low temperature orthorhombic (LTO) phase does not provide such a pinning and static stripes do not form. In the LTO phase, the stripes are assumed to be purely dynamic, which makes their detection extremely difficult [5]. Here, I discuss the possible relation between the phonon anomaly and dynamic stripes.

A detailed comparison between the bond-stretching phonon dispersion in stripe-ordered compounds and optimally-doped superconducting $La_{1.85}Sr_{0.15}CuO_4$ was performed by Reznik *et al.* [55]. They found the strongest phonon renormalization at $h = 0.25$ in the presence of static stripes and $h = 0.3$ at optimal doping (Fig. 4.21). It appears that static stripes pin the phonon anomaly at the stripe ordering wavevector.

Two mechanisms of the impact of dynamic stripes on phonons have been proposed: One is that the phonon eigenvector resonates with the charge component of the stripes; The other is that 1D nature of charge stripes makes them prone to a Kohn anomaly, which renormalizes the phonons. In the first scenario (2D picture) the propagation vector of the anomalous phonon must be parallel to the charge ordering wavevector,

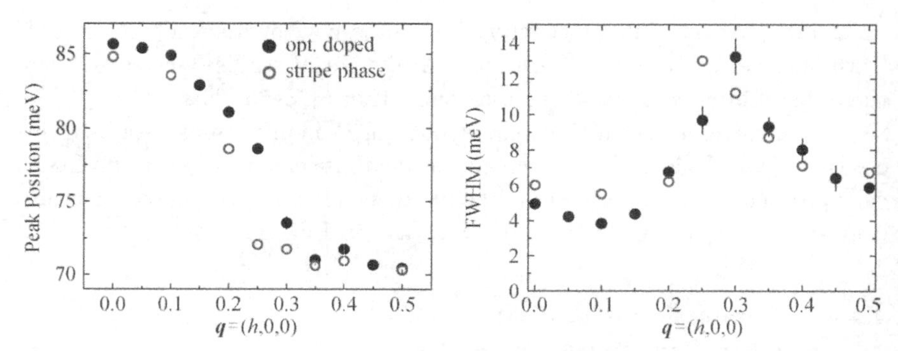

Fig. 4.21. Comparison of the phonon dispersions (a) and linewidth (b) of the bond-stretching branch in $La_{1.85}Sr_{0.15}CuO_4$ and $La_{1.48}Nd_{0.4}Sr_{0.12}CuO_4$ (from Ref. [55]).

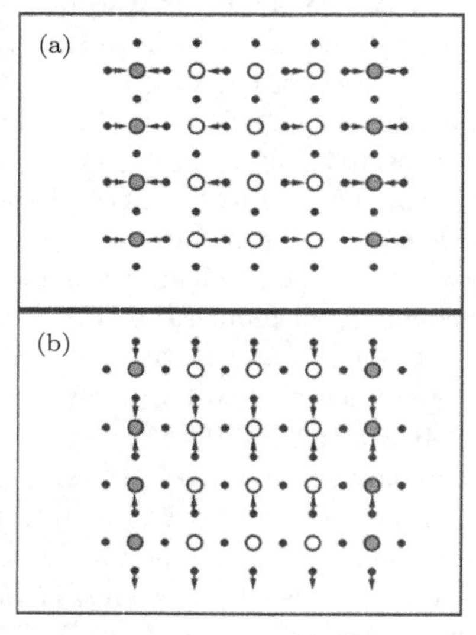

Fig. 4.22. Schematic of the eigenvectors for the phonons with $\boldsymbol{q} = (0.25, 0, 0)$ propagating perpendicular (a) and parallel (b) to the stripes. Open circles represent hole-poor anti-ferromagnetic regions, while the filled circles represent the hole-rich lines (from Ref. [56]).

whereas in the second scenario (1D picture) it must be perpendicular to the charge ordering wavevector (Fig. 4.22).

An important clue is that the phonon anomaly disappears quickly as one moves away from $k = 0$ along the line in reciprocal space: $\boldsymbol{q} = (0.25, k, 0)$ as shown in Refs. [51, 52] and discussed in Section 4.2.2. Such behavior is expected from the matching of the phonon wavevector and the stripe propagation vector. In contrast, a simple picture of a Kohn anomaly due to 1D physics inside the stripes predicts a phonon anomaly that only weakly depends on k. This observation favors the 2D picture, but an important caveat is that it may be possible to reconcile the 1D picture with experiment by including a decrease of the electron–phonon matrix element away from $k = 0$ [84].

Another way to distinguish between the two scenarios is to consider the doping dependence of the wavevector of the maximum phonon renormalization, $\boldsymbol{q}_{\text{max}}$. In the stripe picture, $\boldsymbol{q}_{\text{max}} = 2\boldsymbol{q}_{\text{in}}$ ($\boldsymbol{q}_{\text{in}}$ is the wavevector of incommensurability of low energy spin fluctuations) [5]. At doping levels of $x = 0.12$ and higher, $\boldsymbol{q}_{\text{in}} = 0.125$ which gives the charge ordering wavevector of 0.25. This value is indeed close to $\boldsymbol{q}_{\text{max}}$. At $x = 0.07, \boldsymbol{q}_{\text{in}} = 0.07$. This gives the charge stripe ordering wavevector of 0.14 whereas $\boldsymbol{q}_{\text{max}} = 0.3$. This discrepancy appears to contradict the 2D picture. But again there is a caveat: Anomalous phonons occur at a fairly high energy of about 75 meV, and a comparison to the dynamic stripe wavevector at low energies may not be appropriate.

Thus the question of which picture, 1D or 2D, agrees better with the data is not yet settled.

If the phonon renormalization is driven by static stripes, one may expect to see different behavior for phonons propagating parallel or perpendicular to the stripe propagation vector [43]. In this case, the phonon should split into two peaks. The dynamic stripes, according to Vojta *et al.* [84], may not break tetragonal symmetry, because fluctuations can occur in both directions simultaneously. Thus, a single-peak anomalous phonon lineshape is compatible with dynamic stripes.

4.5. Other Cuprates

4.5.1. *Bond-stretching Phonon Anomalies in $YBa_2Cu_3O_{6+x}$*

It is necessary to establish the universality of the phonon anomalies observed in the $La_{2-x}Sr_xCuO_4$ family. Until now much less work has been performed on other cuprates, because they are more difficult to measure either due to the higher background, no availability of large samples for INS, or low IXS scattering cross-section.

In the case of $YBa_2Cu_3O_{6+x}$ the orthorhombic structure combined with twinning complicates the interpretation of the results. Very little work has

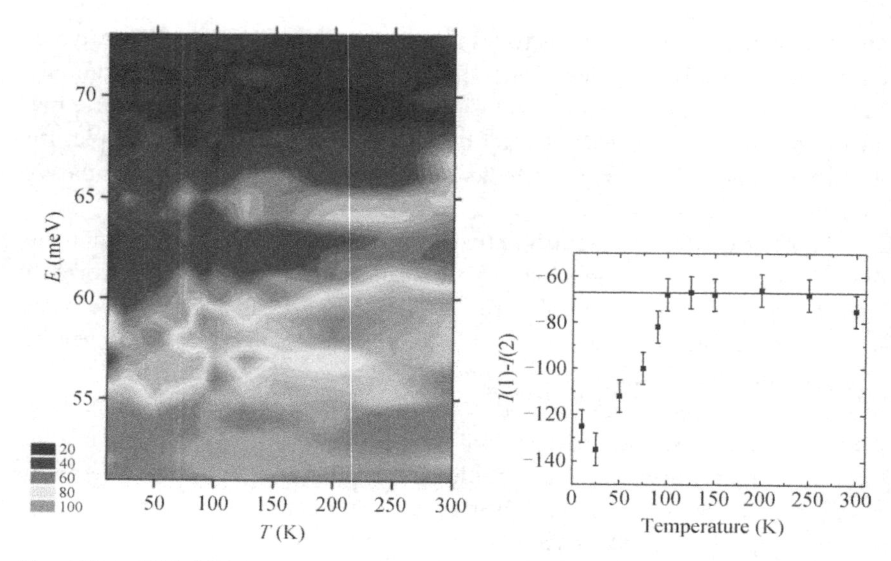

Fig. 4.23. (Right) Inelastic scattering intensity of $YBa_2Cu_3O_{6.95}$ at $\boldsymbol{Q} = (3.25, 0, 0)$ as a function of temperature, determined with the triple-axis spectrometer at the HFIR. Data were smoothed once to reduce noise. (Left) Temperature dependence of the intensity difference $I(1)$–$I(2)$, where $I(1)$ is the average intensity from 56 to 68 meV, $I(2)$ from 51 to 55 eV, at $\boldsymbol{Q} = (3.25, 0, 0)$. T_c of the sample was 93 K (from Ref. [87]).

been done so far on detwinned samples [85] because they are smaller than the twinned ones. Furthermore, two CuO_2 layers in the unit cell introduce two bond-stretching branches, of Δ_1 and Δ_4 symmetry.

At optimal doping, bond-stretching phonons propagating along the chain direction show an anomaly that is in many respects similar to the one in $La_{2-x}Sr_xCuO_4$ [86, 87]. It is absent at 300 K, and appears at low temperatures. Chung *et al.* [88] reported that the spectral weight of the bond-stretching phonons in the Δ_1 symmetry redistributes to lower energies below the superconducting transition temperature, $T_c = 93$ K (see Fig. 4.23).

Pintschovius *et al.* [86] and Reznik *et al.* [88] found that a similar transfer of spectral weight occurs for the Δ_4 phonons but starting close to 200 K, not at T_c (Fig. 4.24(a)). They interpreted this transfer of spectral weight as arising from softening of the bond-stretching phonon polarized along b^*, which transfers its eigenvector to the branches that are lower in energy. This interpretation could explain the observed behavior with some important caveats, but more work is necessary to better understand this

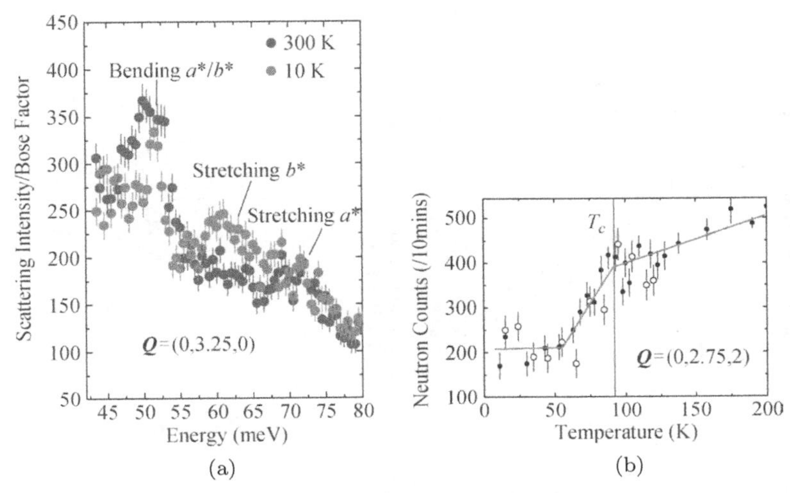

Fig. 4.24. Data of Ref. [88] for the Δ_4 symmetry in $YBa_2Cu_3O_{6.95}$. (a) Comparison of the 300 K and 10 K spectra. (b) Background-subtracted intensity at $\boldsymbol{Q} = (0, 2.75, -2)$ and $E = 60$ meV (see text). Open and solid circles represent different datasets.

effect. Figure 4.24(b) shows that this transfer of spectral weight accelerates below T_c saturating near 50 K. While clearly related to the onset of superconductivity, this effect is not understood.

Reznik *et al.* also showed that the transfer of spectral weight in the Δ_1 symmetry also begins well above T_c with the most pronounced change below T_c (Fig. 4.25). This result seems to contradict the observation of Chung *et al.* [88] who reported that the phonon intensity shift in Δ_1 symmetry occurs only below T_c. According to Reznik *et al.* the effect would also appear only below T_c if they excluded the intensity below 50 meV from their analysis [86] as was done in Ref. [87]. So in this respect the two studies are consistent.

The phonon anomaly in $YBa_2Cu_3O_{6.95}$ seems to extend far in the transverse direction (Fig. 4.26), i.e., it may be consistent with the 1D picture [86] (also see Section 4.4.5). However, twinning of the sample made the data difficult to interpret. Otherwise, the phonon anomaly in optimally-doped $YBa_2Cu_3O_{6.95}$ is similar to the effect in $La_{2-x}Sr_xCuO_4$.

Much less is known about $YBa_2Cu_3O_{6+x}$ at lower doping levels. Stercel *et al.* [89] reported splitting of the bond-stretching branch arguing in favor of charge inhomogeneity, whereas Pintschovius *et al.* [90] explained similar results in terms of the difference between the dispersion of the stretching

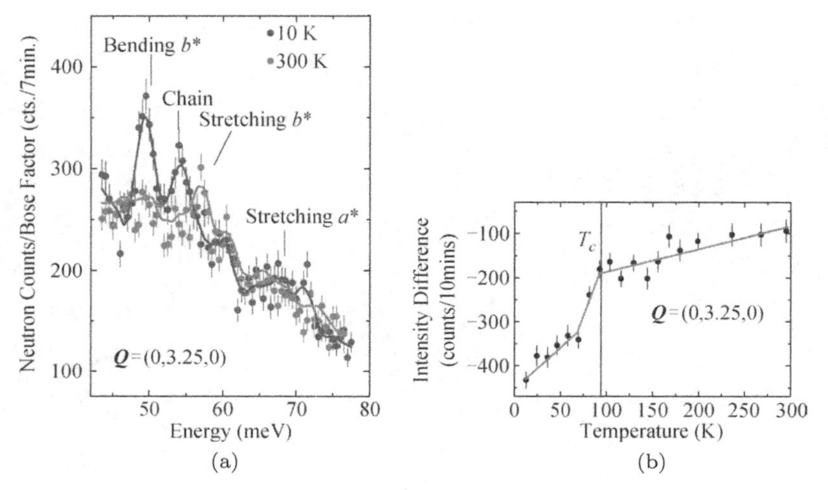

Fig. 4.25. Data of Ref. [88] for the Δ_1 symmetry. (a) Phonon spectra at 300 K and 10 K. The main difference with Ref. [87] and Fig. 4.23 is a bigger energy range here: 42–75 meV in Ref. [67] vs. 51–72 meV in [66]. (b) The difference between the intensity at 57 meV and the average of intensities at 53 meV and 49 meV at $\boldsymbol{Q} = (0, 3.25, 0)$. Temperature dependence above T_c not seen in Fig. 4.23 comes from including the 49 meV phonon, which falls outside the energy range investigated in Ref. [66].

phonons propagating parallel and perpendicular to Cu–O chains. This disagreement needs to be settled by measurements on detwinned samples.

4.5.2. *Bond-buckling Phonon Anomalies in* $YBa_2Cu_3O_{6+x}$

Early Raman scattering experiments on optimally doped cup-rates have shown that the bond-buckling mode exhibits a superconductivity-induced softening of ~1.5% at the wavevector $\boldsymbol{q} = 0$ [91]. This work was followed by neutron scattering measurements on a large twinned single crystal, which showed that the softening persists at nozero $\boldsymbol{q}$ in the directions along the Cu–O bonds (100 and 010-directions also called a^* and b^*, respectively), but not in the 110 direction [92]. This result was interpreted in terms of the interaction of the phonon branch with the d-wave superconducting gap. Recently, Raichle *et al.* [93] investigated a detwinned sample and found that the softening of the phonon decreased away from the zone center along a^*, but showed a pronounced enhancement along b^* around $k = 0.3$ (Fig. 4.27).

Raichle *et al.* concluded that the anomalous phonon with such a large a–b anisotropy cannot mediate d-wave pairing, but may contribute to

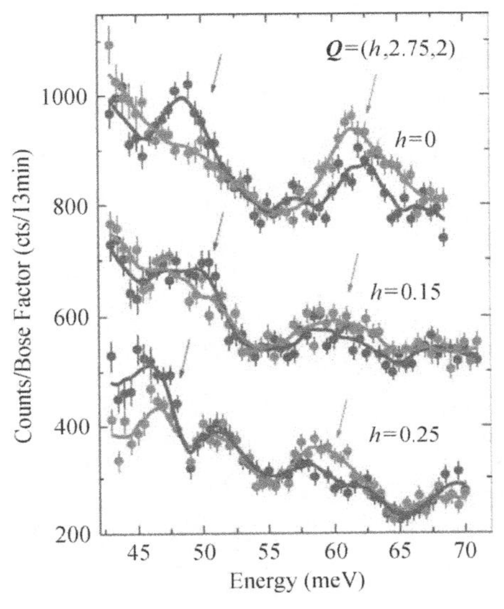

Fig. 4.26. Energy (E) scans taken at 200 K (red) and 10 K (blue). Data were taken with the final energy $E_F = 13.4$ meV for $E > 46$ meV and with $E_F = 12.5$ meV for $43 < E < 48$ meV. The 12.5-meV data were corrected for the different resolution volume by multiplying by $(13.4/12.5)^2$. The resulting intensities were averaged in the overlapping energy range (46.5–48 meV). The 200 K data were divided by the Bose factor and 23 counts were subtracted to correct for the temperature dependence of the background. Blue/red arrows indicate intensity gain/loss (from Ref. [88]). (For interpretation of the references to color in this figure legend, the reader is referred to the web version of this paper.)

the gap anisotropy and the hotly-debated kink in the electronic dispersion observed by angle-resolved photoemission. More importantly, both the bond-stretching and bond-buckling modes indicate that there is a charge fluctuation in $YBa_2Cu_3O_{6+x}$ with the wavevector close to $\boldsymbol{q} = (0.3, 0, 0)$, which indicates nematic behavior. It is possibly associated with stripe formation with stripes running along the a-axis (perpendicular to the chains). An alternative explanation also proposed in Ref. [93] is that it could be related to CDW-type instability in the copper oxygen chains.

4.5.3. $HgBa_2CuO_{4+x}$

Bond-stretching phonons in $HgBa_2CuO_{4+x}$ have been measured by Uchiyama *et al.* [94]. These measurements showed that the bond-stretching phonons soften similarly to $La_{2-x}Sr_xCuO_4$ and $YBa_2Cu_3O_{6+x}$ (Fig. 4.28).

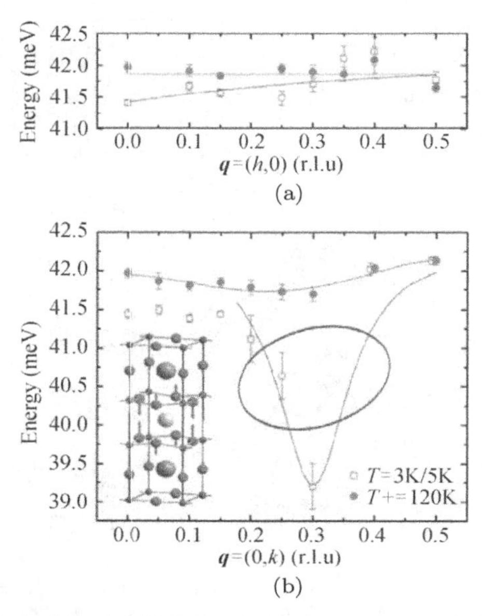

Fig. 4.27. Dispersion of the buckling mode in optimally-doped $YBa_2Cu_3O_{6+x}$ along the a^* and b^*. The black line is the dispersion used for the resolution calculation. A projection of the 4D resolution ellipsoid is shown for comparison. The data point at in-plane $\boldsymbol{q}_{\text{in}} = (0, 0.3)$ is the result of the resolution convolution; the remaining points were determined by fits to standard Voigt functions. The inset shows the eigenvector of the buckling mode at $\boldsymbol{q}_{\text{in}} = (0, 0.3)$. The elongations of the apical oxygen atoms and of the in-plane oxygen atoms along b were enlarged by a factor of 4 for clarity [93].

It is important to extend this study to different dopings, temperatures and nonzero transverse wavevectors.

4.5.4. $Bi_2Sr_{1.6}La_{0.4}Cu_2O_{6+x}$

Graf *et al.* [95] measured phonon dispersions by IXS and electronic dispersions by ARPES in a single-layer Bi-based cuprate, $Bi_2Sr_{1.6}La_{0.4}Cu_2O_{6+x}$. They reported a similar phonon anomaly as in other cuprates (Fig. 4.29) and argued in favor of a correlation between this phonon anomaly, the kink observed in photoemission and the Fermi arc that characterizes the pseudogap phase. They related the sudden onset of phonon broadening near $\boldsymbol{q} = (0.2, 0, 0)$ to coupling of the phonon to the Fermi arc region of the Fermi surface, but not to the pseudogap region. The Fermi arc region is not nested, so exceptionally large electron–phonon coupling for the stretching branch is necessary for this interpretation to be valid (as in Cr as described

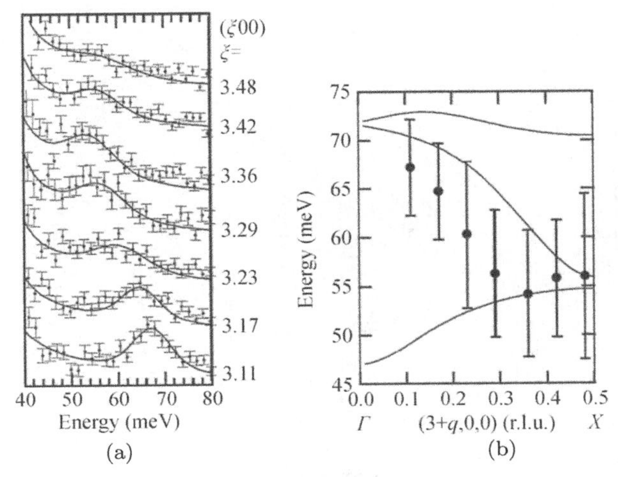

Fig. 4.28. Bond stretching phonons in $HgBa_2CuO_{4+x}$. (a) Enlarged spectra taken close to the bond stretching mode plotted on a linear scale. (b) Data points represent frequencies of the bond-stretching phonons. The lines show the shell model calculation in which the interaction between the next-nearest neighbor oxygens in the CuO_2 plane is added. The lines indicate (top to bottom) the c-polarized apical oxygen mode, the a-polarized Cu–O bond stretching mode, and the a-polarized in-plane Cu–O bending mode, respectively. The vertical bars indicate the FWHM of the peaks determined in fitting data shown in (a) (from Ref. [94]).

in Section 4.3.6). In addition one needs to consider that in other families of cuprates, where the doping dependence has been investigated, the wavevector of the onset of the phonon effect is doping independent, whereas the length of the Fermi arc strongly depends on doping. More detailed studies of this compound, especially as a function of doping, are necessary to clarify these issues.

4.5.5. *Electron-doped Cuprates*

Bond-stretching phonons have been investigated in electron-doped cuprates only in $Nd_{2-x}Ce_xCuO_4$. Phonon density of states measurements on powder samples showed that electron doping softens the highest energy oxygen phonons as occurs in the case of hole-doping [96]. The first single crystal experiment has been performed by d'Astuto *et al.* by IXS [97] who found that the bond-stretching phonon branch dispersed steeply downwards beyond $h = 0.15$. This work was, in fact, the first IXS experiment on the high T_c cuprates. These measurements, however were complicated by the

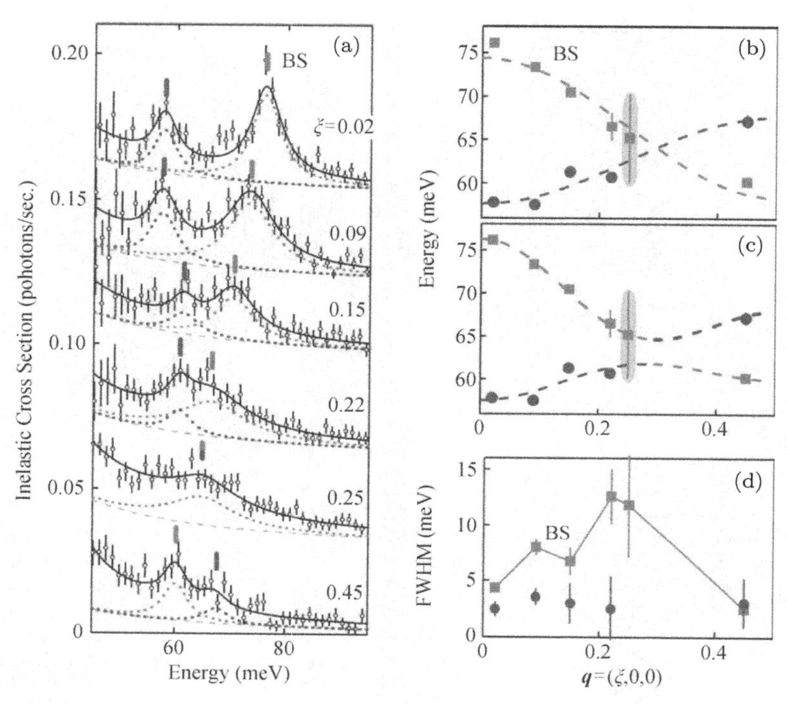

Fig. 4.29. LO phonon dispersions in $Bi_2Sr_{1.6}La_{0.4}Cu_2O_{6+x}$ [95]. (a) IXS spectra for $\boldsymbol{Q} = (3+\xi, 0, 0)$ with ξ from the BZ center (top spectrum, $\xi = 0.02$) to the BZ boundary (bottom spectrum, $\xi = 0.45$). The spectra are vertically shifted. The solid lines show the harmonic oscillator fit, the dashed lines show the elastic tail and the dotted lines show the two modes used in the fit. (b) and (c) Phonon dispersions and linewidths. The cosine dashed lines are guides for the eyes illustrating the crossing (b) and anti-crossing (c) scenarios. (d) Full width at half maximum. The error bars are an estimate of the standard deviation of the fit coefficients.

anti-crossing of the bond-stretching branch with another branch due to Nd–O vibrations that dispersed sharply upwards. The anti-crossing occurs near $h = 0.2$ complicating the interpretation of the data near these wavevectors. Another difficulty came from low IXS scattering cross sections for the oxygen vibrations.

A neutron scattering investigation has been performed by Braden *et al.* [98] once large single crystals became available. Oxygen phonons have a higher scattering cross section in the INS than in the IXS experiments, allowing a more accurate determination of the phonon dispersions.

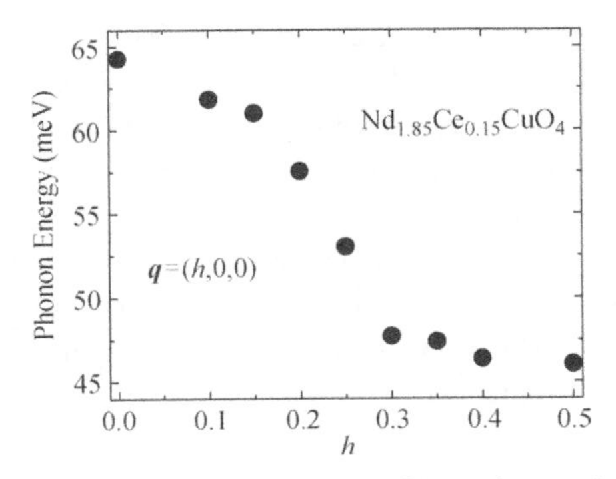

Fig. 4.30. Dispersion of the Cu–O bond-stretching phonon in $Nd_{1.85}Ce_{0.15}CuO_4$ adapted from Ref. [98].

The two studies showed that the bond-stretching phonon dispersion in $Nd_{1.85}Ce_{0.15}CuO_4$ was similar to that in the hole-doped compounds (Fig. 4.30). This similarity points at a commonality between the tendencies to charge inhomogeneity between the hole-doped and electron-doped compounds as discussed in Ref. [98].

4.6. Conclusions

A lot of progress has been made in recent years in understanding of phonon anomalies in a variety of compounds including the ones with static or dynamic stripes. It was possible to distinguish between different mechanisms behind the phonon softening and broadening: bond-length mismatch, $\boldsymbol{q}$-dependence of electronphonon coupling to conduction electrons, Fermi surface nesting, and electronic correlations.

In copper oxide superconductors, where the dynamic stripes are suspected, the bond-stretching phonons around $\boldsymbol{q}$ =(0.3,0,0) are softer and broader than expected from conventional theory. This effect may be related to incipient instability with respect to the formation of dynamic stripes or another charge-ordered or inhomogeneous state.

Much more experimental and theoretical work is necessary to understand the interplay between stripes and phonons. Up to now very little experimental work has been done on compounds where static stripes are clearly present, such as the nickelates and the cobaltates. Understanding the phenomenology of phonon anomalies in these compounds is essential to be able to clearly distinguish effects of stripes from other causes of phonon anomalies discussed in Sections 4.2 and 4.3. In the cuprates, new evidence emerged that makes it unlikely that anomalous phonons, that may be interacting with stripes, directly mediate superconductivity (Section 4.5.2) [93]. However, phonon renormalization in optimally-doped YBCO accelerates at or near the superconducting T_c [86–88, 93]. Furthermore, doping dependence of phonon anomalies in LSCO suggests that it is indirectly associated with the mechanism of superconductivity. Thus the relationship between phonon anomalies and the mechanism of high-temperature superconductivity needs to be explored further.

Acknowledgments

Work on this article was supported by the DOE, Office of Basic Energy Sciences under Contract No. DE-SC0006939. I greatly benefited from interactions with many people over the years without which this work would not have been possible. In particular, I would like to acknowledge discussions with L. Pintschovius, R. Heid, K. -P. Bohnen, W. Reichardt, H. von Löhneysen, F. Weber, D. Lamago, A. Hamann, J. M. Tranquada, S. A. Kivelson, T. Egami, Y. En-doh, M. Arai, K. Yamada, P. B. Allen, I. I. Mazin, J. Zaanen, D. J. Singh, G. Khaliullin, B. Keimer, D.A. Neumann, J. W. Lynn, A. Mischenko, N. Nagaosa, F. Onufrieva, P. Pfeuty, P. Bourges, Y. Sidis, O. Gunnarson, S. I. Mukhin, P. Horsch, T. P. Devereaux, Z.-X. Shen, A. Q. R. Baron, D. S. Dessau, P. Böni, M. d'Astuto, A. Lanzara, M. Greven, and M. Hoesch.

References

[1] J. Zaanen, O. Gunnarsson, *Phys. Rev. B* **40** (1989), 7391.
[2] K. Machida, *Physica C* **158** (1989), 192.
[3] J. M. Tranquada, D. J. Buttrey, V. Sachan, and J. E. Lorenzo, *Phys. Rev. Lett.* **73** (1994) 1003.
[4] J. M. Tranquada, B. J. Sternlieb, J. D. Axe, Y. Nakamura, and S. Uchida, *Nature* **375** (1995) 561.
[5] S. A. Kivelson, I. P. Bindloss, E. Fradkin, V. Oganesyan, J. M. Tranquada, A. Kapitulnik, and C. Howald, *Rev. Mod. Phys.* **75** (2003) 1201.
[6] C. Castellani, C. Di Castro, and M. Grilli, *Phys. Rev. Lett.* **75** (1995) 4650.
[7] V. J. Emery and S. A. Kivelson, *Physica C* **189** (1994), 235–240.
[8] S. A. Kivelson and V. J. Emery. *Strongly Correlated Electronic Materials* in: The Los Alamos Symposium 1993, vol. 619. (Adison-Wesley, Reading, MA, 1994).

[9] V. I. Yukalov and E. P. Yukalova, *Phys. Rev. B* **70** (2004), 224516.
[10] L. Pintschovius, *Phys. Status Solidi B* **242** (2005), 30.
[11] D. Reznik, *Adv. Condens. Matter Phys.* **2010** (2010), 24, http://dx.doi.org/10.1155/2010/523549 (Article ID 523549).
[12] M. Braden, W. Schnelle, W. Schwarz, N. Pyka, G. Heger, Z. Fisk, K. Gamayunov, I. Tanaka, and H. Kojima, *Z. Phys. B* **94** (1994) 29.
[13] W. Kohn, *Phys. Rev. Lett.* **2** (1959), 393.
[14] B. Renker, L. Pintschovius, W. Gläser, H. Rietschel, R. Comès, L. Liebert, and W. Drexel, *Phys. Rev. Lett.* **32** (1974), 836.
[15] H. G. Smith and W. Gläser, *Phys. Rev. Lett.* **25** (1970), 1161.
[16] W. Weber, *Phys. Rev. B* **8** (1973), 5082.
[17] S. N. Sinha and B. N. Harmon, *Phys. Rev. Lett.* **35** (1975), 1515.
[18] S. Baroni, S. de Gironcoli, A. Dal Corso, and P. Giannozzi, *Rev. Mod. Phys.* **73** (2001), 515.
[19] R. Heid, L. Pintschovius, W. Reichardt, and K.-P. Bohnen, *Phys. Rev. B* **61** (2000), 12059.
[20] C. M. Varma and A. L. Simons, *Phys. Rev. Lett.* **51** (1983), 138.
[21] M. D. Johannes, I. I. Mazin, and C. A. Howells, *Phys. Rev. B* **73** (2006), 205102.
[22] M. Calandra, I. I. Mazin, and F. Mauri, *Phys. Rev. B* **80** (2009) 241108.
[23] F. Weber, S. Rosenkranz, J.-P. Castellan, R. Osborn, R. Hott, R. Heid, K.-P. Bohnen, T. Egami, A. H. Said, and D. Reznik, *Phys. Rev. Lett.* **107** (2011), 107403.
[24] M. Hoesch, A. Bosak, D. Chernyshov, H. Berger, and M. Krisch, *Phys. Rev. Lett.* **102** (2009) 086402.
[25] D. S. Inosov, V. B. Zabolotnyy, D. Evtushinsky, A. Kordyuk, B. Buechner, R. Follath, H. Berger, and S. Borisenko, *New J. Phys.* **10** (2008), 125027.
[26] A. Shukla, M. Calandra, M. d'Astuto, M. Lazzeri, F. Mauri, C. Bellin, M. Krisch, J. Karpinski, S. M. Kazakov, J. Jun, D. Daghero, and K. Parlinski, *Phys. Rev. Lett.* **90** (2003), 095506.
[27] A. Q. R. Baron, H. Uchiyama, Y. Tanaka, S. Tsutsui, D. Ishikawa, S. Lee, R. Heid, K.-P. Bohnen, S. Tajima, and T. Ishikawa, *Phys. Rev. Lett.* **92** (2004), 197004.
[28] P. B. Allen, M. L. Cohen, *Phys. Rev. Lett.* **29** (1972), 1593.
[29] B. Klein, D. A. Papaconstantopoulos, *Phys. Rev. Lett.* **32** (1974), 1193.
[30] E. I. Isaev, R. Ahuja, S. I. Simak, A. I. Lichtenstein, Yu. Kh. Vekilov, B. Johansson, and I. A. Abrikosov, *Phys. Rev. B* **72** (2005), 064515.
[31] J. Noffsinger, F. Giustino, S. G. Louie, and M. L. Cohen, *Phys. Rev. B* **77** (2008), 180507.
[32] S. Y. Savrasov, *Phys. Rev. B* **54** (1996), 16470.
[33] H. Kawano, H. Yoshizawa, H. Takeya, and K. Kadowaki, *Phys. Rev. Lett.* **77** (1996), 4628.
[34] W. Reichardt, R. Heid, and K.-P. Bohnen, *J. Supercond.* **18** (2005), 759.
[35] L. Pintschovius, F. Weber, W. Reichardt, A. Kreyssig, R. Heid, D. Reznik, O. Stockert, and K. Hradil, *Pramana J. Phys.* **71** (2008), 687.

[36] P. B. Allen, V. N. Kostur, N. Takesue, and G. Shirane, *Phys. Rev. B* **56** (1997), 5552.
[37] F. Weber, A. Kreyssig, L. Pintschovius, R. Heid, W. Reichardt, D. Reznik, O. Stockert, and K. Hradil, *Phys. Rev. Lett.* **101** (2008), 237002.
[38] F. Weber and L. Pintschovius, *Phys. Rev. B* **82** (2010) 024509.
[39] P. Aynajian, T. Keller, L. Boeri, S. M. Shapiro, K. Habicht, and B. Keimer, *Science* **319** (2008), 1509.
[40] E. Fawcett, *Rev. Mod. Phys.* **60** (1988), 209.
[41] W. M. Shaw and L. D. Muhlestein, *Phys. Rev. B* **4** (1971), 969.
[42] D. Lamago, M. Hoesch, M. Krisch, R. Heid, P. Böni, and D. Reznik, *Phys. Rev. B* **82** (2010), 195121.
[43] Eiji Kaneshita, Masanori Ichioka, and Kazushige Machida, *Phys. Rev. Lett.* **88** (2002), 115501.
[44] J. M. Tranquada, K. Nakajima, M. Braden, L. Pintschovius, and L. Pintschovius, *Phys. Rev. Lett.* **88** (2002), 075505.
[45] D. Reznik *et al.*, unpublished.
[46] W. Weber and L. F. Mattheiss, *Phys. Rev. B* **37** (1988), 599.
[47] P. B. Allen, W. E. Pickett, and H. Krakauer, *Phys. Rev. B* **37** (1988), 7482.
[48] L. Pintschovius and W. Reichardt, *Neutron scattering in layered copper-oxide superconductors*, in: Physics and Chemistry of Materials with Low-Dimensional Structures, vol. 20 (Kluwer Academic, Dordrecht, 1998).
[49] S. Ishihara, T. Egami, and M. Tachiki, *Phys. Rev. B* **55** (1997), 3163.
[50] L. Pintschovius, D. Reznik, and K. Yamada, *Phys. Rev. B* **74** (2006), 174514.
[51] S. R. Park, A. Hamann, L. Pintschovius, D. Lamago, G. Khaliullin, M. Fujita, K. Yamada, G. D. Gu, J. M. Tranquada, and D. Reznik, *Phys. Rev. B* **84** (2011), 214516.
[52] R. J. McQueeney, Y. Petrov, T. Egami, M. Yethiraj, G. Shirane, and Y. Endoh, *Phys. Rev. Lett.* **82** (1999), 628.
[53] L. Pintschovius and M. Braden, *Phys. Rev. B* **60** (1999), 15039.
[54] D. Reznik, L. Pintschovius, M. Ito, S. Iikubo, M. Sato, H. Goka, M. Fujita, K. Yamada, G. D. Gu, and J. M. Tranquada, *Nature* **440** (2006), 1170.
[55] D. Reznik, L. Pintschovius, M. Fujita, K. Yamada, G. D. Gu, and J. M. Tranquada, *J. Low Temp. Phys.* **147** (2007), 353.
[56] D. Reznik, T. Fukuda, A. Baron, M. Fujita, and K. Yamada, *J. Phys. Chem.* Solids **69** (2008), 3103.
[57] M. d'Astuto, G. Dhalenne, J. Graf, M. Hoesch, P. Giura, M. Krisch, P. Berthet, A. Lanzara, and A. Shukla, *Phys. Rev. B* **78** (2008), 140511.
[58] J. Graf, M. d'Astuto, P. Giura, A. Shukla, N. L. Saini, A. Bossak, M. Krisch, S. W. Cheong, T. Sasagawa, and A. Lanzara, *Phys. Rev. B* **76** (2007), 172507.
[59] T. Sasagawa, H. Yuia, S. Pyonb, H. Takagib, and A. Q. R. Baron, *Physica C* **470** (2010), S51.
[60] G. De Filippis, V. Cataudella, R. Citro, C. A. Perroni, A. S. Mishchenko, and N. Nagaosa, *EPL* **91** (2010), 47007.
[61] C. Falter, T. Bauer, and F. Schnetgöke, *Phys. Rev. B* **73** (2006), 224502.

[62] F. Giustino, M. L. Cohen, and S. G. Louie, *Nature* **452** (2008), 975–978.
[63] K.-P. Bohnen, R. Heid, and M. Krauss, *Europhys. Lett.* **64** (2003), 104.
[64] A. Hamann *et al.*, unpublished results.
[65] D. Reznik, G. Sangiovanni, O. Gunnarsson, and T. P. Devereaux, *Nature* **452** (2008), E6.
[66] P. Piekarz and T. Egami, *Phys. Rev. B* **72** (2005), 054530.
[67] O. Rösch and O. Gunnarsson, *Phys. Rev. B* **70** (2004), 224518.
[68] P. Horsch and G. Khaliullin, *Physica B* **359–361** (2005), 620–622.
[69] S. Ishihara and N. Nagaosa, *Phys. Rev. B* **67** (2004), 144520.
[70] E. von Oelsen, A. Di Ciolo, J. Lorenzana, G. Seibold, and M. Grilli, *Phys. Rev. B* **81** (2010), 155116.
[71] S. Cojocaru, R. Citro, and M. Marinaro, *Phys. Rev. B* **75** (2007), 014516.
[72] S. Cojocaru, R. Citro, and M. Marinaro, arXiv: 0905 (2009).
[73] C. J. Zhang and H. Oyanagi, *Phys. Rev. B* **79** (2009), 064521.
[74] T. Mertelj, V. V. Kabanov, and D. Mihailovic, *Phys. Rev. Lett.* **94** (2005) 147003; J. Miranda, T. Mertelj, V. V. Kabanov, D. Mihailovic, *J. Supercond. Nov. Magn.* **22** (2009), 281.
[75] J. Ranninger and A. Romano, *Phys. Rev. B* **78** (2008), 054527.
[76] H. Keller, A. Bussmann-Holder, and K. A. Muller, *Mater. Today* **11** (2008), 38.
[77] S. Sykora, A. Hübsch, and K. W. Becker, *Europhys. Lett.* **85** (2009), 57003.
[78] A. Di Ciolo, J. Lorenzana, M. Grilli, and G. Siebold, *Phys. Rev. B* **79** (2009), 085101.
[79] A. S. Alexandrov, *J. Supercond. Nov. Magn.* **22** (2009), 95.
[80] B. V. Fine and T. Egami, *Phys. Rev. B* **77** (2008), 014519.
[81] M. Vojta and O. Rösch, *Phys. Rev. B* **77** (2008), 094504.
[82] J. Röhler, *Physica C* **374** (2007), 460–462.
[83] S. I. Mukhin, A. Mesaros, J. Zaanen, and F. V. Kusmartsev, *Phys. Rev. B* **76** (2007), 174521.
[84] M. Vojta, T. Vojta, and R. K. Kaul, *Phys. Rev. Lett.* **97** (2006), 097001.
[85] D. Reznik, L. Pintschovius, W. Reichardt, Y. Endoh, H. Hiraka, J. M. Tranquada, S. Tajima, H. Uchiyama, and T. Masui, *J. Low Temp. Phys.* **131** (2003), 417.
[86] L. Pintschovius, D. Reznik, W. Reichardt, Y. Endoh, H. Hiraka, J. M. Tranquada, H. Uchiyama, T. Masui, and S. Tajima, *Phys. Rev. B* **69** (2004) 214506.
[87] J.-H. Chung, Th. Proffen, S. Shamoto, A. M. Ghorayeb, L. Croguennec, W. Tian, B. C. Sales, R. Jin, D. Mandrus, and T. Egami, *Phys. Rev. B* **67** (2003), 014517.
[88] D. Reznik, L. Pintschovius, J. M. Tranquada, M. Arai, T. Masui, and S. Tajima, *Phys. Rev. B* **78** (2008), 094507.
[89] F. Stercel, T. Egami, H. A. Mook, M. Yethiraj, J.-H. Chung, M. Arai, C. Frost, and F. Dogan, *Phys. Rev. B* **77** (2008), 014502.
[90] L. Pintschovius, W. Reichardt, M. Kläser, T. Wolf, and H. V. Löhneysen, *Phys. Rev. Lett.* **89** (2002), 037001.

[91] B. Friedl, C. Thomsen, and M. Cardona, *Phys. Rev. Lett.* **65** (1990), 915.
[92] D. Reznik, B. Keimer, F. Dogan, and I. A. Aksay, *Phys. Rev. Lett.* **75** (1995), 2396.
[93] M. Raichle, D. Reznik, D. Lamago, R. Heid, Y. Li, M. Bakr, C. Ulrich, V. Hinkov, K. Hradil, C. T. Lin, and B. Keimer, *Phys. Rev. Lett.* **107** (2011), 177004.
[94] H. Uchiyama, A. Q. R. Baron, S. Tsutsui, Y. Tanaka, W.-Z. Hu, A. Yamamoto, S. Tajima, and Y. Endoh, *Phys. Rev. Lett.* **92** (2004), 197005.
[95] J. Graf, M. d'Astuto, C. Jozwiak, D. R. Garcia, N. L. Saini, M. Krisch, K. Ikeuchi, A. Q. R. Baron, H. Eisaki, and A. Lanzara, *Phys. Rev. Lett.* **100** (2008), 227002.
[96] H. J. Kang, S. Pengcheng Dai, D. Mandrus, R. Jin, H. A. Mook, D. T. Adroja, S. M. Bennington, S. H. Lee, and J. W. Lynn, *Phys. Rev.* **66** (2002), 064506.
[97] M. d'Astuto, P. K. Mang, P. Giura, A. Shukla, P. Ghigna, A. Mirone, M. Braden, M. Greven, M. Krisch, and F. Sette, *Phys. Rev. Lett.* **88** (2002), 167002.
[98] M. Braden, L. Pintschovius, T. Uefuji, and K. Yamada, *Phys. Rev. B* **72** (2005), 184517.

Part II

Isotopic Effect

5
Oxygen Isotope Effect in Cuprates Results from Polaron-Induced Superconductivity*

S. Weyeneth and K. A. Müller

Physik-Institut der Universität Zürich, Winterthurerstrasse 190, 8057 Zürich, Switzerland

The planar oxygen isotope effect coefficient measured as a function of hole doping in the Pr- and La-doped $YBa_2Cu_3O_7$ (YBCO) and the Ni-doped $La_{1.85}Sr_{0.15}CuO_4$ (LSCO) superconductors quantitatively and qualitatively follows the form originally proposed by Kresin and Wolf [*Phys. Rev. B* **49**, (1994) 3652], which was derived for polarons perpendicular to the superconducting planes. Interestingly, the inverse oxygen isotope effect coefficient at the pseudogap temperature also obeys the same formula. These findings allow the conclusion that the superconductivity in YBCO and LSCO results from polarons or rather bipolarons in the CuO_2 plane. The original formula, proposed for the perpendicular direction only, is obviously more generally valid and accounts for the superconductivity in the CuO_2 planes.

Copper oxides are the only compounds that show superconductivity at T_c above the boiling point of nitrogen. When not doped, these materials are anti-ferromagnetic insulators due to the strong correlation splitting of the $d_{x^2-y^2}$-type band into an upper empty and a lower occupied Mott–Hubbard band. When the latter is doped and holes are present, very high T_c's are found. A two-decade-long controversy resulted from the question as

*Reprinted with the permission from S. Weyeneth, K. A. Müller. Orignial published in *J. Supercond. Nov. Magn.* **24** (2011), 1235–1239.

to whether the superconducting phenomenon still results from the electronic correlations present or whether lattice dynamics play a role. In favor of the latter, Kresin and Wolf wrote a colloquium [1] in which they review couplings from an extended lattice to phonons, plasmons, exitons and polarons, both theoretically and experimentally. For the case of polarons, i.e., a local vibronic lattice deformation, they quote a formula for the oxygen isotope coefficient $\alpha(n)$, originally derived by them [1, 2] in 1994 for lattice distortions along the c-direction:

$$\alpha(n, T_c) = \nu(n)\frac{n}{T_c(n)}\frac{\partial T_c(n)}{\partial n}. \tag{5.1}$$

Here, n denotes the superconducting carrier density. The parameter $\nu(n)$ entering (5.1) depends only weakly on n.

Equation (5.1) resembles the formula proposed earlier by Schneider and Keller, derived by analysing the universal proporties of cuprates [3, 4]. Already there, an important ingredient for mapping the experimentally obtained $\alpha(n)$ was the detailed $T_c(n)$ dependence. In the original report by Kresin and Wolf [2], the applicability of the model is discussed on the basis of the early data for $\alpha(n)$. The result of their fitting of (5.1) to the data yielded $\nu = 0.13$. However, it is not really discussed how and in which manner the gradient of $T_c(n)$ is incorporated into (5.1) during fitting.

Experimentally, it was observed by Zech *et al.* [5] that site-selective oxygen substitution $^{16}O \rightarrow ^{18}O$ in optimally doped $YBa_2Cu_3O_7$ (YBCO) results in more than 80% of the oxygen isotope effects being due to the planar oxygen. This is because the hole density present was shown to reside mainly in the planar CuO_2 [6]. Accordingly, the carrier density is mainly planar $n_{\|}$. Note that for the perpendicular direction $n_{\perp} = 0$ and therefore also $\alpha_{\perp} = 0$. Various groups [7–10] carefully studied $\alpha_{\|}(n)$ for YBCO and doped $La_{1.85}Sr_{0.15}CuO_4$ (LSCO) as shown in Fig. 5.1, taken from Ref. [11]. In it, $\alpha_{\|}(n)$ is compared with the results of a vibronic theory. The latter comprises two electronic bands present in the cuprates, a lower t–J-type d-band and a nearby, higher-lying s-band coupled by a linear vibronic term [6, 11, 12]. Within this model, the oxygen isotope effects on T_c result from considering in-plane bands with first t_1, second t_2, third neighbor t_3, and interplanar hopping integrals t_4 [6, 11]. The polaronic renormalizations of the second-nearest-neighbor hopping integral t_2 and the interplanar hopping term t_4 yield the correct trend for $\alpha_{\|}(n)$ close to optimal doping. From Fig. 5.1 it can be seen that the agreement is rather good at and

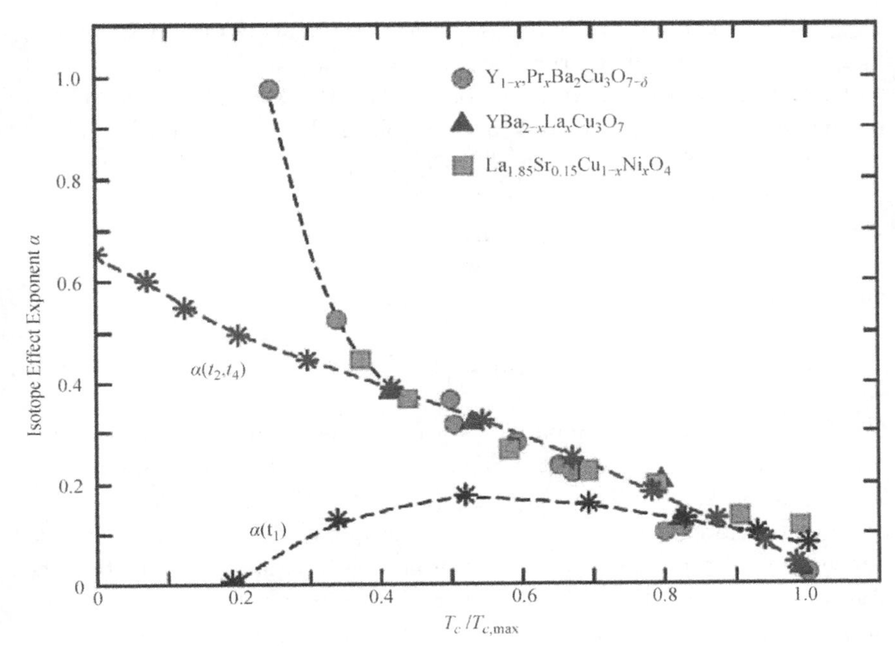

Fig. 5.1. The oxygen isotope effect exponent α as a function of $T_c/T_{c,\max}$. The *black stars* refer to the calculated α when only the nearest-neighbor hopping integral (t_1) is renormalized. The *purple stars* are theoretically derived, with both the second-nearest-neighbor and the interplanar hopping integrals (t_2, t_4) renormalized. *Red, blue* and *green* data are experimental values for α for $Y_{1-x}Pr_xBa_2Cu_3O_{7-\delta}$ [7, 8], $YBa_{2-x}La_xCu_3O_{7-\delta}$ [9], and $La_{1.85}Sr_{0.15}Cu_{1-x}Ni_xO_4$ [10], respectively. Taken from Ref. [11].

below optimal doping, but clearly deviates for very low doping. Being based on a mean-field theory, this may be expected because at very low doping individual polarons exist. A result of that theory is that the local coupling mode is of Jahn–Teller-type t_2/t_4 motion. From this we may infer that for a polaronic scenario the formal $\alpha(n)$ [see (5.1)] may also be valid for the planar polarons. Thus we identify in (5.1) $n = n_{\|}$ and $\alpha(n) = \alpha_{\|}(n)$, as described in the next paragraph.

The dependence of $T_c(n)$ can be parameterized according to early work [13–15] as

$$\frac{T_c(n)}{T_{c,\max}} = 1 - A\left(\frac{n}{n_{\max}} - 1\right)^2 . \tag{5.2}$$

$T_{c,\max}(n)$ denotes the maximum transition temperature of a given family, $n_{\max}$ the corresponding carrier density, and A an empirical constant found to be $\approx$ 2–3 [13–15]. In the inset of Fig. 5.2 the dependence in (5.1) is presented. Combining (5.1) and (5.2), we obtain for the generic formula of $\alpha_{\|}(n)$ in terms of $z = n/n_{\max}$ and $t = T_c/T_{c,\max}$

$$\alpha_{\|}(z) = 2\nu(n)\frac{z(z-1)}{(z-1)^2 - A^{-1}} \tag{5.3}$$

and within the underdoped region:

$$z = 1 - \sqrt{\frac{1-t}{A}}. \tag{5.4}$$

This result can be compared with experimental data for $\alpha_{\|}(n)$ from the literature. In Fig. 5.2, we depict data for $\alpha_{\|}(n)$ as a function of $T_c/T_{c,\max}$ for $Y_{1-x}Pr_xBa_2\ Cu_3O_{7-\delta}$ [7, 8], for $YBa_{2-x}La_xCu_3O_{7-\delta}$ [9], and for $La_{1.85}Sr_{0.15}Cu_{1-x}\ Ni_xO_4$ [10]. Clearly, (5.3) describes the universal dependence of the measured $\alpha_{\|}(n)$ rather well, assuming $\nu(n)$ to be constant. By fitting (5.3) and (5.4) to the data, we derive a value of $A = 2.5(3)$, in good agreement with literature values [13–15], and $\nu = 0.146(8)$, slightly larger than the value of $\nu = 0.13$ reported in Ref. [2] based on early results for $\alpha(n)$.[a] Globally, this calculation is similar to those previously discussed in Refs. [2] and [3].[b]

The agreement between the measured oxygen isotope effect $\alpha_{\|}(n)$ as a function of doping n and the curve calculated with (5.3) and (5.4), but with a constant parameter ν, is really remarkable. From optimal doping $n_{\max}$ down to near-vanishing superconductivity, where the measured $\alpha_{\|}(n)$ exhibits a characteristic upturn, the data follow the curve obtained with (5.1) very well. This allows the conclusion that this simple expression is more generally valid than only for polarons axed along the crystallographic c-direction, for which it was originally obtained [1, 2]. It quantitatively yields the correct behavior *for polarons sited in the* CuO_2 *plane.* Now the vibronic theory of Bussmann-Holder and Keller [6, 11] reproduces $\alpha_{\|}(n)$ quite well near optimum doping, see Fig. 5.1. In their work, the

[a]Note that for this analysis newer experimental data were used, obtained from a larger range of doping, in combination with the early data analyzed in Ref. [2].

[b]Note that an obvious difference in the calculation presented in Refs. [2, 3] and the work presented here is the functional form of $T_c(n)$ used for calculation. The generality of this form, as presented in (5.1) by the freedom of parameter A, allows an unambiguous calculation of the isotope coefficient α.

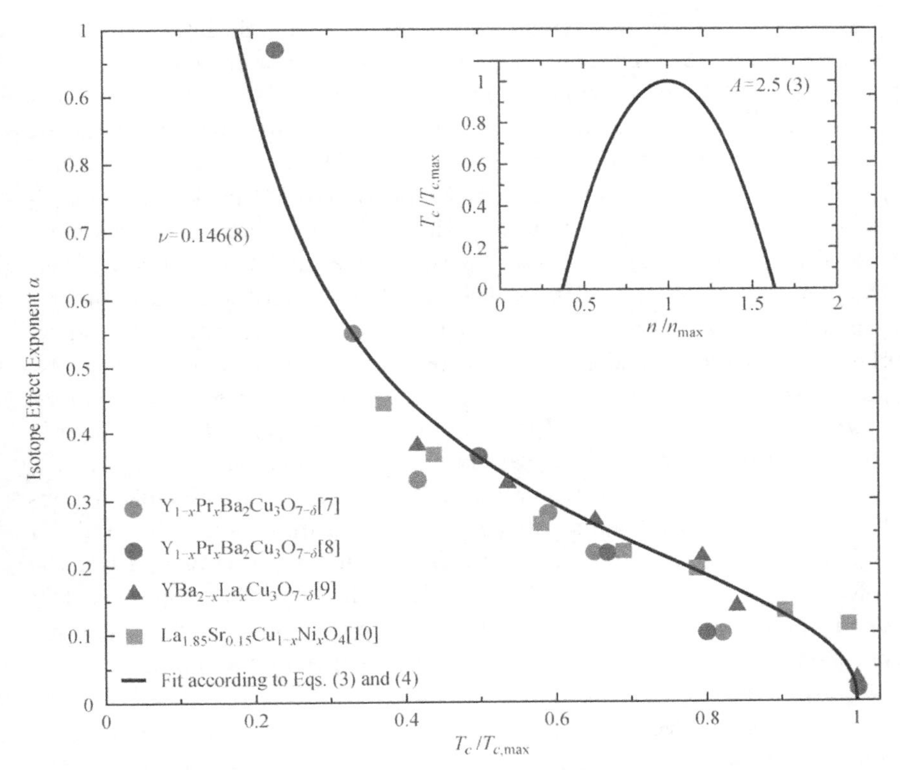

Fig. 5.2. The same data for α as in Fig. 5.1 as a function of $T_c/T_{c,\max}$ for $Y_{1-x}Pr_xBa_2Cu_3O_{7-\delta}$ [7, 8], $Y_{1-x}Pr_xBa_2Cu_3O_{7-\delta}$ [9], and $La_{1.85}Sr_{0.15}Cu_{1-x}Ni_x$ O_4 [10]. Clearly, (5.3) describes the universal dependence of α rather well. We can derive a value of $\nu = 0.146(8)$ from the fit. The *inset* shows the doping-dependent T_c according to (5.3) and (5.4) using the fitted $A = 2.5\,(3)$.

authors deduced that the *local* polaronic lattice deformation is of t_2/t_4, i. e., Jahn–Teller type. From this fact, we may assume that this is also the case in the entire range shown in Fig. 5.2.[c] In other words, the effects of doping-dependent oxygen isotope and the theories shown in Figs. 5.1 and 5.2 are evidence of the planar polaronic origin of the high-temperature superconductivity.

The above conclusion agrees with the ones reached in the viewpoint published by the second author [16] four years ago. In this viewpoint, the

[c]This behavior is similar to those in critical phenomena, where the local conformations near structural phase transitions that occur in the mean field regime subsist symmetry-wise in the critical regime.

possibility for superconductivity to be due to magnetic interactions was also addressed. At that time, no electronic theories were known to yield finite $T_c's$. Furthermore, experiments were discussed which indicated that magnetic interactions could not be responsible for the high-temperature superconductivity. However, the group of Keimer at the MPI in Stuttgart and others investigated the anti-ferromagnetic (AFM) peaks near the M-point of the Brillouin zone with inelastic neutron scattering [17] for a number of compounds. These peaks are very weak, and the group was able to enhance the signal by first exciting electrons from the $2p$ level of oxygen with X-rays. The signal obtained in this way quantitatively follows the temperature-dependent superconducting gap. From this, a magnetic mechanism was deduced by *assuming an intrinsic homogeneous state of the material. However, this is not the case*! This has been pointed out recently [16] and also earlier [18]. For example, X-ray [19] and early Tm NMR [20] investigations yielded near-insulating AFM regions and metallic-like regions forming clusters. Close to a superconducting cluster or stripe, weak AFM signals exist at the border, owing to the continuous electronic part of the wave function [21]. Indeed, weak spin ordering has been detected by nuclear quadrupole resonance [22]. It is thus possible that this weak AFM signal is coherent by interaction with the coherent superconducting part, and is responsible for the AFM signal.

The largest oxygen isotope effect at T_c shown in Fig. 5.1, occurring at low doping, is about α =1.0, i.e., much lower than those observed at the pseudogap temperature T^* reported for LSCO [23] and $HoBa_2Cu_4O_8$ [24]. For the latter, $\alpha^* = -2.2$. It is sign-inverted, as expected from (1), which also applies to the pseudogap isotope effect [1]. With the nearly linear $T^*(n)$, we arrive at $\nu^* = 1.7$ from these data. The existence of an even greater α^* for copper in YBCO in terms of local lattice conformations [24] has been commented on in Refs. [6, 16]. In the latter paper, the occurrence of T^* has been assigned to the formation of inter-site Jahn–Teller bipolarons [16]. That these giant isotope effects on T^*, which were observed and confirmed a decade ago [23, 24], were ignored by the proponents of pure electronic mechanisms for the occurrence of superconductivity in the cuprates is not acceptable from a scientific point of view.

The isotope effect with $\alpha = 0.5$ in the classical superconductors was substantial for the acceptance of the BCS theory. The carrier-dependent $\alpha_{\|}(n)$ at $T_c(n)$ quantitatively follows the Kresin–Wolf formula [see (5.1)] derived for polarons along the c-direction over the entire doping range (see Fig. 5.2). We regard this as a strong clue that, for the materials shown, superconductivity is induced by polarons, or rather bipolarons, with spin $S = 0$, but that they occur *in* the CuO_2 plane [16]. The latter follows from the agreement of the calculated planar isotope effects at $T_c(n)$ with the vibronic theory near optimum doping, see Fig. 1. Comparing the observed isotope effects with the vibronic theory indicates that the *local polaronic conformations* are of the t_2/t_4 Jahn–Teller type. This theory also yields the large inverted oxygen isotope effect of $\alpha^* = -2.2$ observed at the pseudo-gap temperature T^* and other behaviors [11]. Furthermore, already at very low doping, intersite Jahn–Teller bipolarons are formed, which cluster in probably ramified, conducting entities, such as stripes [25].[d]

Acknowledgments

The authors would like to thank H. Keller for various helpful discussions and the hospitality at the Physics Institute of the University of Zurich, and one of us (K. A. M.) thanks A. Bishop for correspondence and M. Mali and B. Keimer for discussions. This work was partly supported by the Swiss National Science Foundation.

References

[1] V. Z. Kresin and S. A. Wolf, *Rev. Mod. Phys.* **81** (2009), 481.
[2] V. Z. Kresin and S. A. Wolf, *Phys. Rev. B* **49** (1994), 3652(R).
[3] T. Schneider and H. Keller, *Phys. Rev. Lett.* **69** (1992), 3374.
[4] T. Schneider and H. Keller, *Phys. Rev. Lett.* **86** (2001), 4899.
[5] D. Zech, H. Keller, K. Conder, E. Kaldis, E. Liarokapis, N. Poulakis and K. A. Müller, *Nature (London)* **371** (1994), 681.
[6] A. Bussmann-Holder and H. Keller, *Eur. Phys. J. B* **44** (2005), 487.
[7] J. P. Franck, J. Jung, M. A. -K. Mohamed, S. Gygax and G. I. Sproule, *Phys. Rev. B* **44** (1991), 5318.
[8] R. Khasanov, A. Shengelaya, E. Morenzoni, K. Conder, I. M. Savic and H. Keller, *J. Phys. Condens. Matter* **16** (2004), S4439.

[d]Note that inelastic neutron scattering and photoemission spectroscopy *integrate over the volume of the sample used.* For the former, this has already been alluded to in the text. For the latter, this means that because of the *intrinsic* heterogeneity at a particular instant, emissions *from AFM near insulating parts of the sample*, close to the M-point of the Brillouin zone, as well as *from other conducting Fermi-like ones*, near X-point directions, are detected, i.e., *the observed excitations result from different parts of the sample.* Assuming homogeneity leads to questionable conclusions as in Ref. [26].

[9] H. J. Bornemann and D. E. Morris, *Phys. Rev. B* **44** (1991), 5322.
[10] N. Babushkina, A. Inyushkin, V. Ozhogin, A. Taldenkov, I. Kobrin, T. Vorobeve, L. Molchanova, L. Damyanets, T. Uvarova and A. Kuzakov, *Physica C* **185** (1991), 901.
[11] H. Keller, A. Bussmann-Holder and K. A. Müller, *Mater. Today* **11** (2008), 38.
[12] A. Bussmann-Holder, H. Keller, A. R. Bishop, A. Simon, R. Micnas and K. A. Müller, *Europhys. Lett.* **72** (2005), 423.
[13] M. R. Presland, J. L. Tallon, R. G. Buckley, R. S. Liu and N. E. Flower, *Physical C* **176** (1991), 95.
[14] J. L. Tallon, C. Bernhard, H. Shaked, R. L. Hitlerman and J. D. Jorgenen, *Phys. Rev. B* **51** (1995), 12911(R).
[15] T. Schneider, *The Physics of Superconductors* (Springer, Berlin, 2004).
[16] K. A. Müller, *J. Phys. Condens. Matter* **19** (2007), 251002.
[17] H. He, P. Bourges, Y. Sidis, C. Ulrich, I. P. Regnault, S. Paihes, N. S. Berzigiarova, N. N. Kolesnikov and B. Keimer, *Science* **295** (2002), 1045.
[18] D. Mihailovic and K. A. Müller, *High-T_c Superconductivity 1996: Ten Years After the Discovery*, NATO ASI Ser. E. vol. 343. (Kluwer, Dordrecht, 1997).
[19] A. Bianconi, N. L. Saini, A. Lanzara, M. Missori, T. Rosetti, H. Oyanagi, H. Yamaguchi, K. Oka and T. Ito, *Phys. Rev. Lett.* **76** (1996), 3412.
[20] M. A. Teplov *et al.*, *High-T_c Superconductivity 1996: Ten Years After the Discovery*, NATO ASI. Ser. E, vol. 343 (Kluwer, Dordrecht, 1997).
[21] I. Martin, E. Kaneshita, A. R. Bishop, R. J. McQueeney and Z. G. Yu, *Phys. Rev. B* **70** (2004), 224514.
[22] F. Raffa, T. Ohno, M. Mali, J. Roos, D. Brinkmann, K. conder and M. Eremin, *Phys. Rev. Lett.* **81** (1998), 5912.
[23] A. Lanzara, G. M. Zhao, N. L. Saini, A. Bianconi, K. Conder, H. Keller and K. A. Müller, *J. Phys. Condens. Matter* **11** (1999), L541.
[24] A. Furrer, *Superconductivity in Comples Systems, Structure and Bonding Series*, vol. 114, p.171 (Springer, Berlin, 2005).
[25] A. Shengelaya, M. Bruun, B. J. Kochelaev, A. Safina, K. Conder and K. A. Müller, *Phys. Rev. Lett.* **93** (2004), 017001.
[26] I. M. Vishik, W. S. Lee, R.-H. He, M. Hashimoto, Z. Hussain, T. P. Devereaux, and Z. X. Shen, *New J. Phys.* **12** (2010), 105008.

6

Oxygen Isotope Effect on the Effective Mass of Carriers from Magnetic Measurements on $La_{2-x}Sr_xCuo_4$*

Guo-Meng Zhao, K. K. Singh, A. P. B. Sinha and D. E. Morris

Morris Research, Inc., 1918 University Avenue, Berkeley, California 94704, USA

Oxygen isotope effects on T_c and the Meissner fraction f have been investigated in fine-grained, decoupled $La_{2-x}Sr_xCuO_4$ with $x = 0.105, 0.110$, and 0.115. We find that these oxygen isotope effects are related to each other: $d\ln T_c/d\ln M \approx d\ln f/d\ln M$ (where M is the mass of oxygen). We also show that the trapped flux due to intergrain weak-link and intragrain flux pinning is negligible in our samples, and that the Meissner fractions for the samples with the same ^{16}O isotope are nearly equal The observed large oxygen isotope effects on T_c and the Meissner fraction can be explained as due to the oxygen-mass dependence of m^* (effective mass of carriers). The results may suggest that the conducting carriers in the cuprate superconductors are of polaronic type.

The discovery and development of the high-T_c cuprate superconductors have prompted a burst of experimental and theoretical investigations on these systems. However, the microscopic pairing mechanism for high-T_c superconductivity remains controversial. Opinions on the role of

*Reprinted with the permission from Guo-meng Zhao, K. K. Singh, A. P. B. Sinha, D. E. Morris. Original published in *Phys. Rev. B* **52** (1995), 6840–6844.

electron–phonon interaction vary widely [1–3]. One way to assess the importance of phonons in the pairing is the isotope effect on T_c. Very small oxygen isotope effect on T_c was observed in the optimally doped cuprates [4–7] while large oxygen isotope effect was found in the underdoped cuprates [8–11]. The large isotope effect may point to the phonon-mediated pairing mechanism for high-T_c superconductivity. However, since T_c is sensitive to the hole concentration in the underdoped range, it has been suggested [12] that the observed large oxygen isotope effect could be due to a slight difference in the mobile-hole concentrations of the ^{16}O and ^{18}O samples. Therefore it is essential to address this issue before one can draw any conclusion about pairing mechanism from these isotope effect experiments.

It is important to note that the ^{16}O-containing samples of $La_{2-x}Sr_xCuO_4$ have a local minimum in T_c at $x \sim 0.110$, as shown by Takagi *et al.* [13] and Xiong *et al.* [14] One possible explanation to the local minimum in T_c is that the mobile-hole concentration has a local minimum at this composition. However, several substitution experiments [15, 16] indicate that the mobile-hole concentration n is nearly equal to x when $x < 0.15$. The Hall effect measurements [15] further show that the Hall coefficient is a monotonic function of x. These results consistently suggest that the local minimum in T_c for the ^{16}O-containing samples occurs at the mobile-hole concentration $n \approx x \approx 0.110$. Since the T_c values of the ^{16}O-containing samples have a minimum at $n \approx x \approx 0.110$, any isotopeinduced increase or decrease in the hole concentration at this composition will always raise T_c. Therefore, the isotope-induced decrease in T_c must be caused by variables other than the change in the hole concentration.

The muon-spin-relaxation experiments [17] have shown a universal linear relationship: $T_c \propto 1/\lambda_{ab}(0)^2 \propto n/m^*$ for underdoped p-type cuprates, where $\lambda_{ab}(0)$ is the in-plane penetration depth at zero temperature and m^* is the effective mass of carriers along a–b plane. This result implies that the observation of oxygen isotope effect on T_c must be accompanied by the observation of oxygen isotope effect on the penetration depth $\lambda_{ab}(0)$. Since the Meissner fraction $f(0)$ for a fine-grained sample is a function of grain size and penetration depth, the observation of oxygen isotope effect on $\lambda_{ab}(0)$ will be equivalent to the observation of oxygen isotope effect on $f(0)$.

Here, we report the detailed results of oxygen isotope effects on T_c and Meissner fraction f for the fine-grained and decoupled $La_{2-x}Sr_xCuO_4$

with $x = 0.105, 0.110$, and 0.115. We find that the oxygen isotope effects on T_c and Meissner fraction f are related to each other: $d\ln T_c/d\ln M \approx d\ln f/d\ln M$ (where M is the mass of oxygen). The results can be explained as due to the oxygen-mass dependence of m^*.

Samples of $La_{2-x}Sr_xCuO_4$ with $x = 0.100, 0.105, 0.110, 0.115$, and 0.120 were prepared by conventional solid-state reaction using La_2O_3 (99.99%), $SrCO_3$ (99.999%), and CuO (99.999%). The La_2O_3 was dried for 6 h at 900°C prior to weighing. The powders were mixed and ground thoroughly under isopropyl alcohol, and then were pressed into pellets of 6 mm diam with ~2 ton/cm^2 pressure. These pellets were then fired in air at 1000°C for ~70 h with three intermediate grindings. X-raydiffraction (XRD) results indicate that the samples are very good single phase without any observable trace of the Sr-rich phase $La_{1.67}Sr_{0.33}Cu_2O_5$ To obtain samples with small grains and enough porosity, we reground the samples thoroughly, pressed them into pellets, and annealed them in air at 900°C for 12 h. The cooling time to room temperature was 4 h.

Each pellet was broken in half, and the halves were then subjected to ^{16}O and ^{18}O isotope diffusion, which was conducted in two parallel quartz tubes separated by about 2 cm (see Ref. [7]). The purities of the ^{16}O and ^{18}O gases used for isotope exchange are as follows: $^{16}O_2$ gas contains 99.993% ^{16}O, 0.005% ^{17}O, 0.002% ^{18}O, CO < 14 ppm, $CO_2 < 7$ppm, $CH_4 < 5$ ppm, H < 39 ppm, He < 13 ppm, and N < 18 ppm; $^{18}O_2$ gas contains 96.9% ^{18}O, 2.6% ^{16}O, 0.5% ^{17}O, CO < 18 ppm, $CO_2 < 18$ ppm, $CH_4 < 7$ ppm, N < 21 ppm, H < 18 ppm, and He < 15 ppm. The diffusion was carried out for 40 h at 900°C and oxygen pressure of about 0.7 bar. The cooling time to room temperature was 4 h. The oxygen isotope enrichment was determined from the weight changes of the ^{16}O and ^{18}O samples. The ^{18}O samples had ~85–90% ^{18}O and ~15–10% ^{16}O. Back exchange was carried out at 900°C for 40 h in flowing $^{16}O_2$ on the samples with $x = 0.115$.

The susceptibility was measured with a Quantum Design superconducting quantum interference device magnetometer. The field-cooled, measured-on-warming (FCW) susceptibility was measured in a nominal field of 5 Oe. All samples were carefully aligned in the same direction (along the axis parallel to the field) during each measurement. The samples were held at 40 K in a field of ~5 Oe for ~10 min, and then cooled directly to 5 K. The cooling rates for all measurements were similar. The data were collected upon warming. The magnetic field was kept unchanged throughout each series of measurements. The before-diffusion samples had been measured at an earlier time in the nominal field of 5 Oe while the back-exchanged samples were measured at a later time in the same nominal field. The zero-field-cooled (ZFC) susceptibility was measured in a field of 4.8 Oe after the sample was zero-field (~0.1 Oe) cooled from normal state to 5 K, and the field-cooled (FC) susceptibility was measured in the same field as that for ZFC measurement.

In Figs. 6.1(a)–6.1(c), we show the FCW susceptibility for the ^{16}O and ^{18}O samples with $x = 0.105, 0.110$, and 0.115. The nominal magnetic field for the measurement is 5 Oe. From Fig. 6.1, we find that the Meissner fractions for the ^{18}O samples are ~5–8% smaller than for the ^{16}O samples at low temperatures, in agreement with results reported by Franck *et al.* [18]. The detailed isotope-effect results are summarized in Table 6.1 and Fig. 6.2. From Table 6.1, one can see that the oxygen isotope effect on T_c ($\alpha_0 = -d\ln T_c/d\ln M$) is correlated with the oxygen isotope effect on the Meissner fraction f as $d\ln T_c/d\ln M \sim d\ln f/d\ln M$ (where M is the mass of oxygen).

Since the T_c values of the ^{16}O-containing samples have a local minimum at $n \approx x \approx 0.110$ (see Fig. 6.2), any increase or decrease in the hole concentration upon isotope substitution will raise the T_c of the sample at this composition. Therefore, a lowering in T_c of the ^{18}O sample at this composition is not possibly due to a lower or higher hole concentration in the ^{18}O sample. From Fig. 6.2, one can also see that $dT_c/dn < 0$ at $n = 0.105$, and $dT_c/dn > 0$ at $n = 0.115$. So there would be a positive oxygen isotope effect on T_c at $n \approx x \approx 0.115$, and a negative oxygen isotope effect on T_c at $n \approx x \approx 0.105$, if there were only a positive oxygen isotope effect on the hole concentration. The observed large positive oxygen isotope effect on T_c at all the three compositions implies that the oxygen isotope effect on the hole concentration, if it exists, will be small.

The observed oxygen isotope effect on the Meissner fraction is possibly caused by the differences in grain-size, remanent magnetization (trapped fiux) and demagnetization factor of the ^{16}O and ^{18}O samples. To rule out this possibility, we show, in Figs. 6.3(a) and 6.3(b), the results of FCW susceptibility (a) for $x = 0.110$ before isotope exchange; (b) for $x = 0.115$ after back exchange. The nominal magnetic field for the measurements is 5 Oe. It is clear that the Meissner fractions of the ^{16}O and ^{18}O samples are the same within ±1.5% before isotope exchange (see Fig. 6.3(a)). After back exchange, the T_c of the $x = 0.115$ sample which previously contained ^{18}O is ~0.2 K lower than the original ^{16}O sample as shown in Fig. 6.3(b),

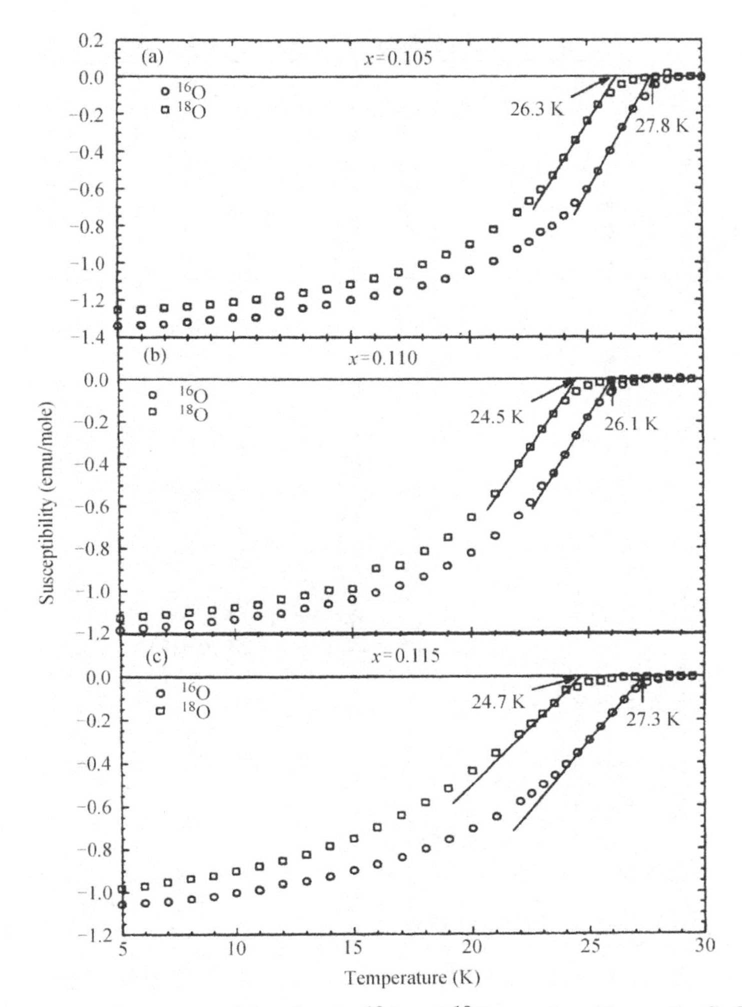

Fig. 6.1. The FCW susceptibility for the ^{16}O and ^{18}O samples of $La_{2-x}Sr_xCuO_4$ with (a) $x = 0.105$; (b) $x = 0.110$; (c) $x = 0.115$. The data shown were divided by a nominal field of 5 Oe and not corrected by demagnetization factors. The real magnetic field for measurements is 5.9 ± 0.2 Oe.

possibly due to incomplete substitution of ^{16}O for ^{18}O ($\sim$6% ^{18}O remains). The reversibility of the Meissner fraction upon isotope exchange suggests that the observed isotope effect is not due to the differences in grain size, demagnetization factor, and remanent magnetization of the ^{16}O and ^{18}O samples.

Table 6.1. Oxygen isotope effects in $La_{2-x}Sr_xCuO_4$.

x	^{18}O (%)	T_c (K)	α_O[a]	$\Delta T_c/T_c$ (%)	f (%)[b]	$\Delta f/f$ (%)
0.105	0	27.8	0.50	−5.4	16.6	−6.8
0.105	86	26.3			15.5	
0.110	0	26.1	0.54	−6.1	14.7	−5.0
0.110	90	24.5			14.0	
0.115	0	27.3	0.90	−9.5	13.1	−7.8
0.115	85	24.7			12.1	

Notes:[a]$\alpha_O = -d\ln T_c/d\ln M$.
[b]f is the Meissner fraction at 5 K. The data shown were divided by a real field of 5.9 Oe and corrected by a demagnetization factor of 1/3.

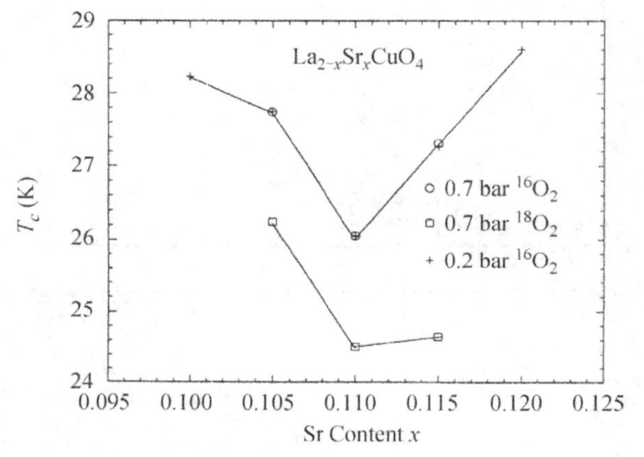

Fig. 6.2. The superconducting transition temperatures T_c vs. x for the ^{16}O and ^{18}O samples of $La_{2-x}Sr_xCuO_4$. Since the T_c values of the ^{16}O-containing samples have a local minimum at $n \approx x \approx 0.110$, any increase or decrease in the hole concentration upon isotope substitution will raise the T_c of the sample at this composition. Therefore, a lowering in T_c of the ^{18}O sample at this composition is not due to a lower or higher hole concentration in the ^{18}O sample.

Note that the apparent differences in the absolute susceptibility of the ^{16}O samples as shown in Figs. 6.1(b) and 6.3(a), or in Figs. 6.1(c) and 6.3(b) are due to resetting the magnetic field in the different series of measurements which were conducted in different periods. We usually set a magnetic field in no overshoot mode after the field is reduced from − or +5.0 T to zero in oscillate mode [which leads to a remanent field of + or $-(0.9 \pm 0.2)$ Oe]. Then the real magnitude of the field will be 4.1 ± 0.2 or 5.9 ± 0.2 Oe if one sets a nominal field of 5 Oe.

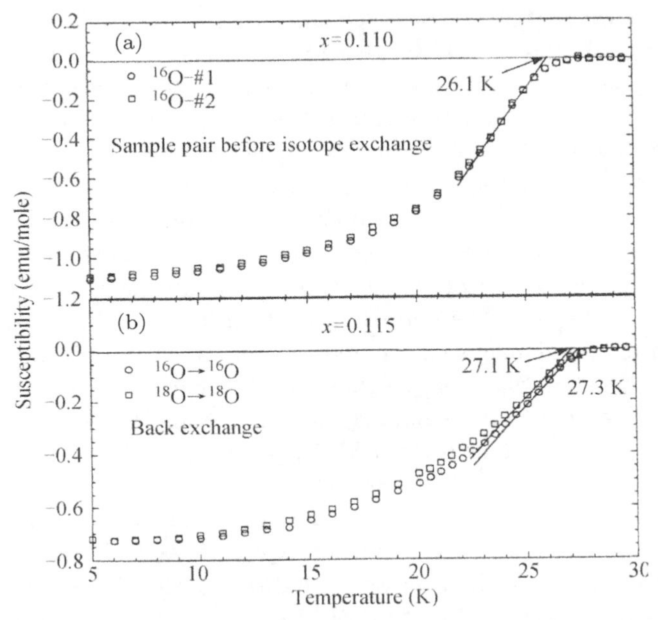

Fig. 6.3. The FCW susceptibility for the samples of $La_{2-x}Sr_xCuO_4$ with (a) $x = 0.110$, before isotope exchange; (b) $x = 0.115$, after back exchange. The data shown were divided by a nominal field of 5 Oe and not corrected by demagnetization factors. The real magnetic field for measurements is 5.9 ± 0.2 Oe in (a), and 4.1 ± 0.2 Oe in (b).

It was also suggested that the observed oxygen isotope effects on T_c and f could be explained by a picture in which the ^{18}O samples contain more nonsuperconducting phase ($La_{1.67}Sr_{0.33}Cu_2O_5$) than the ^{16}O samples [19]. However, this is not consistent with our XRD results, which do not show any evidence of a second phase (e.g., $La_{1.67}Sr_{0.33}Cu_2O_5$) in either the ^{16}O or the ^{18}O samples. In our XRD patterns (the wavelength of X ray is 1.54060 Å), the weak peak at $2\theta = 36.277°$ for $La_{2-x}Sr_xCuO_4$ (3% of the most intense peak) can be clearly identified, but the most intense peak for $La_{1.67}Sr_{0.33}Cu_2O_5$ is not visible. This indicates that the Sr-rich phase is below 5% in our samples, and cannot account for the 5%–8% difference in the Meissner fractions of the ^{16}O and ^{18}O samples. In addition, the isotope-exchange temperature is 900°C, so any impurity phase, once formed, would not be expected to convert back to the superconducting phase. The reversible change of the Meissner fraction upon oxygen isotope substitution further rules out this possibility.

It is known that the field-cooled susceptibility (or Meissner effect) for fine-grained samples is reduced due to the penetration-depth, intergrain weak links, and intragrain flux pinning. If the intergrain link is not well established, the flux cannot be trapped in the intergrain boundary. If the samples have negligible defects, and the magnetic field is low, the flux will not be trapped inside the grain. The magnitude of the trapped flux can be obtained by the ZFC and FC measurements. The difference of ZFC and FC magnetization is the remanent magnetization due to the trapped flux [20]. In Fig. 6.4, we show the ZFC and FC susceptibility for $x = 0.115$ sample (after back exchange). The real magnetic field of "zero field" was checked by measuring the diamagnetic signal of Pb, and was determined to be $\sim$0.1 Oe. From Fig. 6.4, one can see that the difference of the ZFC and FC signals is very small, suggesting negligible flux trapping. Further evidence for negligible flux trapping is the field independence of the FC susceptibility below 7 Oe.

Since there is negligible flux trapping in our finegrained and decoupled samples, the Meissner fraction can be expressed as $f(0) = f(\{r\}, \lambda(0))$, where $\{r\}$ is the radius distribution of grains and $\lambda(0)$ is the effective penetration depth at low temperatures. For $\Delta\lambda(0)/\lambda(0) \ll 1$, we can make a Taylor expansion for the Meissner fraction $f(0)$, leading to

$$-\Delta f(0)/f(0) = B[\Delta\lambda(0)/\lambda(0)], \tag{6.1}$$

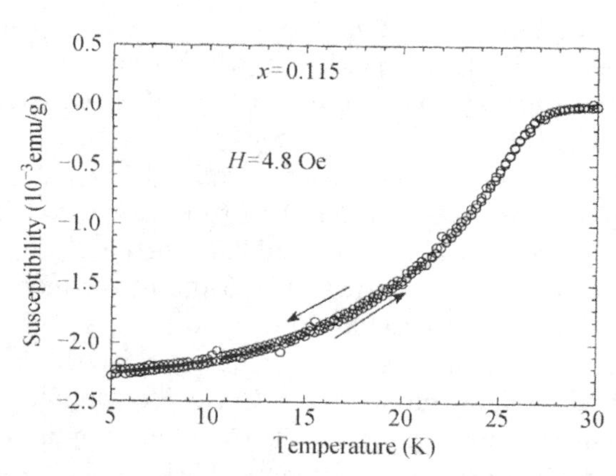

Fig. 6.4. The temperature dependence of ZFC and FC susceptibility for the $x = 0.115$ sample. The difference of the ZFC and FC signals is small, suggesting negligible flux trapping.

where $\Delta\lambda(0)$ and $\Delta f(0)$ mean isotope-induced changes of $\lambda(0)$ and $f(0)$, and $B = -[\lambda(0)/f(0)][\partial f(\{r\}, \lambda(0))/\partial\lambda(0)]$. If $\lambda(0)/r << 1, f(0) = [1 - \lambda(0)/R]^3$ (where R is the average radius of grains), then $B \approx 3[1 - f(0)^{1/3}]/f(0)^{1/3}$. The above relation can give a rough estimation for the coefficient B [e.g., $B \sim 2$ for $f(0) \sim 20\%$].

Equation (6.1) indicates that the observation of oxygen isotope effect on the Meissner fraction is equivalent to the observation of oxygen isotope effect on the penetration depth. Since $\lambda \propto (m^*/n)^{1/2}$, Eq. (6.1) implies that the observed oxygen-mass dependence of $f(0)$ originates from the oxygen-mass dependence of n and/or m^*. As discussed above, the results shown in Fig. 6.2 imply that the oxygen isotope effect on the hole concentration, if it exists, will be small. Then the observed large oxygen isotope effect on the Meissner fraction is mainly caused by the oxygen-mass dependence of m^* (effective mass of carriers).

Since the hole concentration n is closely related to a ratio $g = c/a$ (where a and c are the lattice constants along ab plane and c-axis direction, respectively), any isotopeinduced change in the hole concentration (e.g., a possible difference in the oxygen contents of the ^{16}O and ^{18}O samples) will lead to a change of g. Crawford *et al.* [9] have precisely determined the lattice constants for the ^{16}O and ^{18}O samples of $La_{2-x}Sr_xCuO_4$ from synchrotron X-ray diffraction measurements, and found that the values of g for the ^{16}O and ^{18}O samples are the same within $\pm 10^{-6}$ [for orthorhombic symmetry, g is defined as $g = c/(ab/2)^{1/2}$]. Using a relation $\Delta n \sim 10\Delta g/g$ (Refs. [21–24]), one can estimate that the hole concentrations of the ^{16}O and ^{18}O samples are the same within $\pm 10^{-5}$ Therefore, the observed large oxygen isotope effect on the penetration depth (or Meissner fraction) is due to the oxygen-mass dependence of m^* or the density of states $N(E)$.

There are several theoretical explanations to the oxygen-mass dependence of the $N(E)$. Pickett *et al.* [25] have shown that the strong anharmonicity of phonon modes will make $N(E)$ depend on the ion mass due to structure fluctuation. Alternatively, the ionmass dependence of $N(E)$ could be due to the breakdown of Migdal adiabatic approximation. Engelesberge and Schrieffer [26] have shown that the Migdal approximation does not hold if there is a reasonably strong interaction between electrons and long wavelength optical phonons. The breakdown of the Migdal approximation will lead to the formation of polarons [3]. The interaction of the electrons with the long wavelength optical phonons (or the local phonons as in the

Holstein model) narrows the electronic bandwidth and enhances the density of states by a factor $A(\omega)\exp(B/\omega)$ which is ion-mass dependent, where $A(\omega)$ has a weak dependence on ω and B is a constant [3].

In conclusion, we have observed oxygen isotope effects on T_c and the Meissner fraction f in the fine-grained, decoupled $La_{2-x}Sr_xCuO_4$ with $x = 0.105, 0.110$, and 0.115. The observed large oxygen isotope effects on T_c and the Meissner fraction can be explained as due to the oxygen-mass dependence of m^* (effective mass of carriers). The present results may suggest that the conducting carriers in the cuprate superconductors are of polaronic type.

Acknowledgments

We would like to thank M. L. Cohen and V. H. Crespi for stimulating discussion.

References

[1] V. J. Emery, *Phys. Rev. Lett.* **58** (1987), 2794.
[2] P. W. Anderson, *Science* **235** (1987), 1196.
[3] A. S. Alexandrov and N. F. Mott, *Int. J. Mod. Phys. B* **8** (1994), 2075.
[4] B. Batlogg *et al.*, *Phys. Rev. Lett.* **58** (1987), 2333.
[5] L. C. Bourne *et al.*, *Phys. Rev. Lett.* **58** (1987), 2337.
[6] D. E. Morris *et al.*, *Phys. Rev. B* **37** (1988), 5936.
[7] J. H. Nickel, D. E. Morris, and J. W. Ager III, *Phys. Rev. Lett.* **70** (1993), 81.
[8] M. K. Crawford *et al.*, *Phys. Rev. B* **41** (1990), 282.
[9] M. K. Crawford *et al.*, *Science* **250** (1990), 1390.
[10] H. J. Bornemann and D. E. Morris, *Phys. Rev. B* **44** (1991), 5322.
[11] J. P. Franck *et al.*, *Phys. Rev. B* **44** (1991), 5318.
[12] T. Schneider and H. Keller, *Phys. Rev. Lett.* **69** (1992), 3374.
[13] H. Takagi *et al.*, *Phys. Rev. B* **40** (1989), 2254.
[14] Q. Xiong *et al.*, *Phys. Rev. B* **46** (1992), 581.
[15] T. Nishikawa, H. Harashina, and M. Sato, *Physica C* **211** (1993), 127.
[16] Y. Maeno *et al.*, *Physica C* **185–189** (1991), 909
[17] Y. J. Uemura *et al.*, *Phys. Rev. Lett.* **62** (1989), 2317.
[18] J. P. Franck *et al.*, *Phys. Rev. Lett.* **71** (1993), 283.
[19] Maria Ronay *et al.*, *Phys. Rev. B* **45** (1992), 355.
[20] J. R. Clem and Z. D. Hao, *Phys. Rev. B* **48** (1993), 13774.
[21] G. M. Zhao *et al.*, *Phys. Rev. B* **50** (1994), 4112.
[22] P. Ganguly *et al.*, *Phys. Rev. B* **47** (1993), 991.

[23] K. Sreedhar and P. Ganguly, *Phys. Rev. B* **41** (1990), 371.
[24] J. C. Grenier *et al.*, *Physica C* **202** (1992), 209.
[25] W. E. Pickett, R. E. Cohen, and H. Krakauer, *Phys. Rev. Lett.* **67** (1991), 228.
[26] S. Engelesberge and J. R. Schrieffer, *Phys. Rev.* **131** (1963), 993.

7

Isotope Effects and Possible Pairing Mechanism in Optimally Doped Cuprate Superconductors*

Guo-Meng Zhao[†,‡], Vidula Kirtikar[‡] and Donald E. Morris[‡]

[†] *Physik-Institut der Universität Zürich, CH-8057 Zürich, Switzerland*
[‡] *Morris Research, Inc., 44 Marguerito Road, Berkeley, California 96707, USA*

We have studied the oxygen-isotope effects on T_c and in-plane penetration depth $\lambda_{ab}(0)$ in an optimally doped three-layer cuprate $Bi_{1.6}Pb_{0.4}Sr_2Ca_2Cu_3O_{10+y}$ ($T_c \sim 107$ K). We find a small oxygen-isotope effect on T_c ($\alpha_O = 0.019$), and a substantial effect on $\lambda_{ab}(0)[\Delta\lambda_{ab}(0)/\lambda_{ab}(0) = 2.5 \pm 0.5\%]$. The present results along with the previously observed isotope effects in single-layer and double-layer cuprates indicate that the isotope exponent α_O in optimally doped cuprates is small while the isotope effect on the in-plane effective supercarrier mass is substantial and nearly independent of the number of the CuO_2 layers. A plausible pairing mechanism is proposed to explain the isotope effects, high-T_c superconductivity, and tunneling spectra in a consistent way.

The pairing mechanism responsible for high-T_c superconductivity is still controversial. In conventional superconductors, a strong effect of changing ion mass M on the transition temperature T_c implies that lattice vibrations (phonons) play an important role in the microscopic mechanism of superconductivity. An isotope exponent $\alpha(= -d\ln T_c/d\ln M)$ of about 0.5

*Reprinted with the permission from Guo-meng Zhao, Vidula Kirtikar, Donald E. Morris. Original published in *Phys. Rev. B* **63**, (2001), 220506(R).

is consistent with the phonon-mediated BCS theory. A nearly zero oxygen-isotope effect ($\alpha_O \simeq 0.03$) was earlier observed in a double-layer cuprate superconductor $YBa_2Cu_3O_{7-y}$ which is optimally doped ($T_c \sim 92$ K) [1, 2]. Such a small isotope effect might suggest that phonons should not be important to the pairing mechanism. On the other hand, large oxygen-isotope shifts were later observed in several underdoped cuprate superconductors [3–9]. Further, three indirect experiments have consistently demonstrated that the difference in the hole densities of the ^{16}O and ^{18}O samples is smaller than 0.0002 per Cu site [7–9]. Moreover, a quantitative data analysis on the isotope-exchanged $YBa_2Cu_3O_{6.94}$ (Ref. [10]) suggested that there is a negligible oxygen-isotope effect on the supercarrier density n_s.

Since muon-spin rotation experiments [11] showed that T_c is approximately proportional to n_s/m^{**}_{ab} in deeply underdoped cuprates (where m^{**}_{ab} is the in-plane effective supercarrier mass), a large oxygen-isotope shift of T_c observed in this doping regime should arise from a large oxygen-isotope effect on m^{**}_{ab}. Indeed, several independent experiments [7–9, 12] have consistently demonstrated that both the average super-carrier mass m^{**} and the in-plane supercarrier mass m^{**}_{ab} strongly depend on the oxygen isotope mass in underdoped cuprates. Such an unconventional isotope effect suggests that there exist polaronic charge carriers, which are condensed into supercarriers in the superconducting state. This appears to give support to a theory of (bi)polaronic superconductivity [13]. On the other hand, within this theory, it is difficult to explain a small isotope shift of T_c and a large reduced energy gap (i.e., $2\Delta(0)/k_BT_c > 6$) observed in optimally doped cuprates where the single-particle excitation gap vanishes above T_c [14]. Therefore, an alternative theoretical approach is required to explain superconductivity in optimally doped and overdoped cuprates. A possibly correct pairing mechanism should be able to consistently explain a small isotope shift of T_c, a substantial isotope effect on the supercarrier mass [8, 10], and a large reduced energy gap.

Here, we report the observation of the oxygen-isotope effects on T_c and in-plane penetration depth $\lambda_{ab}(0)$ in a three-layer cuprate $Bi_{1.6}Pb_{0.4}Sr_2Ca_2Cu_3O_{10+y}$ ($T_c \sim 107$ K). We find a small oxygen-isotope effect on T_c and a substantial effect on $\lambda_{ab}(0)$. We propose a possible theoretical model which is able to consistently explain these isotope effects and the large reduced energy gap.

Samples of $Bi_{1.6}Pb_{0.4}Sr_2Ca_2Cu_3O_{10+y}$ were prepared from high purity Bi_2O_3, PbO, $SrCO_3$, $CaCO_3$, and CuO. The samples were ground, pelletized and fired at 865/855/845°C for 37/40/44 h in air with two intermediate grindings. Two samples were prepared under nearly the same heat treatment. To ensure that the samples have small grain size and enough porosity, they were reground thoroughly, pelletized, and annealed in flowing oxygen at 600°C for 10 h.

Two pelletized samples were broken into halves (producing two sample pairs), and the halves were subject to the ^{16}O and ^{18}O isotope diffusion, which was conducted in two parallel quartz tubes separated by about 2 cm [5, 7, 10]. The diffusion was carried out for 68 h at 600°C and oxygen pressure of 0.8 bar for sample pair I, and for 40 h at 650°C and oxygen pressure of 1 bar for sample pair II. The cooling rate was 30°C/h. The oxygen isotope enrichments were determined from the weight changes of both ^{16}O and ^{18}O samples. The ^{18}O samples had $85 \pm 5\%$ ^{18}O and $15 \pm 5\%$ ^{16}O.

The susceptibility was measured with a quantum design superconducting quantum interference device (SQUID) magnetometer. The field-cooled, measured-on-warming susceptibility was measured in a field of 15 Oe for sample pair I, and 10 Oe for sample pair II. The temperature measurements were performed with a platinum resistance thermometer (Lakeshore PT-111) placed in direct contact with the sample and driven by a microprocessor controlled ac bridge in the SQUID magnetometer. The resolution is 2.5 mK and reproducibility is 10 mK at 77 K after cycling to room temperature [5, 10].

In Fig. 7.1, we show the susceptibility near T_c for the ^{16}O and ^{18}O samples of $Bi_{1.6}Pb_{0.4}Sr_2Ca_2Cu_3O_{10+y}$: (a) pair I; (b) pair II. In all the cases, the T_c for the ^{18}O samples is 0.22 ± 0.01 K lower than for the ^{16}O samples. Extrapolating to 100% exchange, we calculate the isotope exponent $\alpha_O = -d\ln T_c/d\ln M_O = 0.019 \pm 001$. We also note that there is a well-defined linear portion on the transition curve ~1 K below the diamagnetic onset temperature. It is evident that ^{18}O the slope of the linear portion (denoted by P_1) for the ^{18}O samples is $5.0 \pm 1.0\%$ smaller than for the ^{16}O samples, that is, $\Delta P_1/P_1 = -5.0 \pm 1.0\%$.

For comparison, the results for single-layer $La_{1.85}Sr_{0.15}CuO_4$ (LSCO) and double-layer $YBa_2Cu_3O_{6.94}$ (YBCO) compounds are reproduced in Fig. 7.2. It is clear that $\Delta P_1/P_1 = -5.5 \pm 1.0\%$ for LSCO and $-6.8 \pm 1.0\%$ for YBCO. Comparing Fig. 7.2 with Fig. 7.1, one can see that the isotope effect on P_1 is nearly the same for all three compounds within the experimental uncertainty. It is also interesting to note that the isotope effect on T_c decreases monotonically with increasing T_c.

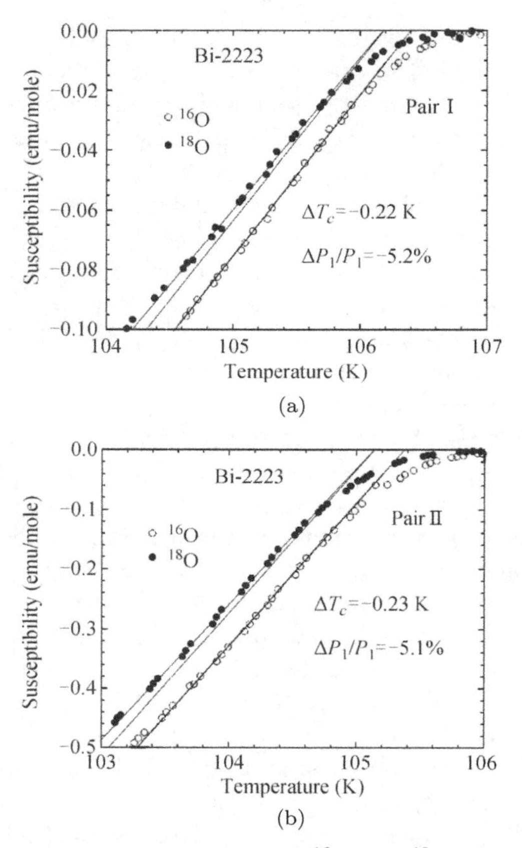

Fig. 7.1. The susceptibility near T_c for the ^{16}O and ^{18}O samples of $Bi_{1.6}Pb_{0.4}Sr_2Ca_2Cu_3O_{10+y}$ (Bi-2223): (a) Pair I; (b) Pair II.

The observed oxygen-isotope effect on the slope P_1 is caused by the dependence of the penetration depth on the oxygen mass. For nearly isotropic materials with $\lambda_{ab}(0) \sim \lambda_c(0) \sim \lambda(0)$, it was shown that: [10] $\Delta P_1/P_1 = -\Delta T_c/T_c - 2\Delta\lambda(0)/\lambda(0)$. For highly anisotropic materials such as cuprates, a relation $\lambda_c(T) \gg \lambda_{ab}(T) > R$ (where R is the maximum particle size) holds near T_c so that the diamagnetic signal is proportional to $1/\lambda_{ab}^2(T)$ [15]. Then one readily finds that

$$\Delta P_1/P_1 \simeq -\Delta T_c/T_c - 2\Delta\lambda_{ab}(0)/\lambda_{ab}(0). \tag{7.1}$$

From Figs. 7.1 and 7.2, we obtain $\Delta\lambda_{ab}(0)/\lambda_{ab}(0) = 3.2 \pm 0.7\%$ using Eq. (7.1). This is in remarkably good agreement with the recent muon

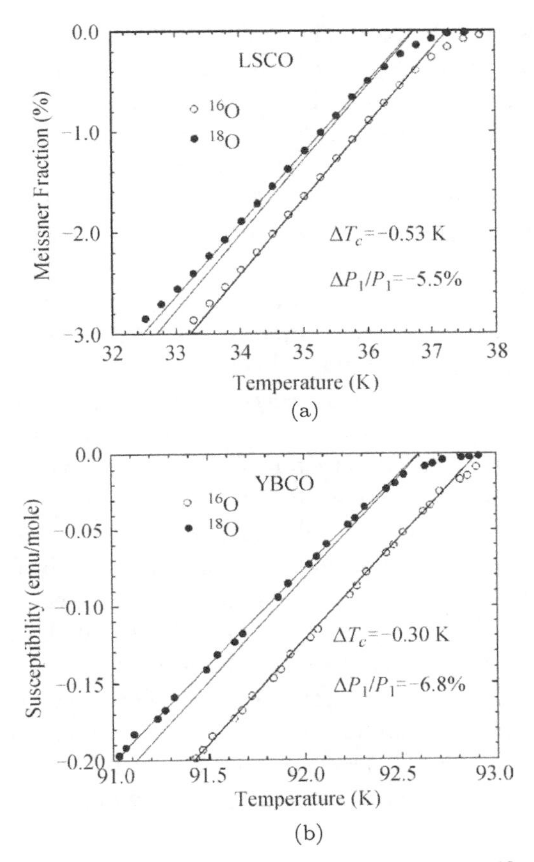

Fig. 7.2. The susceptibility data near T_c for the ^{16}O and ^{18}O samples of (a) $La_{1.85}Sr_{0.15}CuO_4$ (LSCO) and (b) $YBa_2Cu_3O_{6.94}$ (YBCO) (after Refs. [8, 10]).

spin rotation experiments on the oxygen-isotope exchanged $YBa_2Cu_3O_{6.96}$, which show that $\Delta\lambda_{ab}(0)/\lambda_{ab}(0) = 2.5 \pm 0.5\%$ [16].

Since both n and n_s are independent of the isotope mass as discussed above, the observed oxygen-isotope effect on the in-plane penetration depth is caused by the isotope dependence of m^{**}_{ab}. The substantial isotope effect on m^{**}_{ab} may suggest that charge carriers in the optimally doped cuprates remain polaronic nature, and that those polaronic carriers are condensed into supercarriers in the superconducting state.

Now a question arises: why is the isotope effect on m^{**}_{ab} substantial while the isotope shift of T_c is very small in these optimally doped cuprates?

In order to answer this question, we should first find out which phonon modes are strongly coupled to doped holes in these materials. Inelastic neutron scattering experiments [17, 18] show that the Cu–O bond stretching modes in the CuO_2 planes are strongly coupled to the doped holes. The average frequency of this mode is about 75 meV in $La_{1.85}Sr_{0.15}CuO_4$ [17], and about 60 meV in $YBa_2Cu_3O_{6.92}$ [18]. Such a strong electron–phonon coupling should be also manifested in tunneling spectra, as is the case in conventional superconductors. Although it is difficult to obtain reliable tunneling spectra for cuprates due to a short coherent length, there are two high-quality tunneling spectra for slightly overdoped YBCO (Ref. [19]) and $Bi_2Sr_2CaCu_2O_{8+y}$ (BSCCO) [20].

In Fig. 7.3, we show normalized conductance data for scanning tunneling microscopy (STM) on a slightly overdoped YBCO crystal with a Pt–Ir tip at 4.2 K [19]. The crystal has $T_c \simeq 90$ K with ~ 1 K transition width [19]. Since the presence of oxygen vacancies in the CuO chains can lead to residual density of states and to zero-bias conductance in the super-conducting state, a negligible zero-bias conductance in the spectrum suggests that the spectrum represents the intrinsic density of states contributed only from the CuO_2 planes. It is striking that the strong coupling features similar to that in conventional superconductors can be clearly seen in the spectrum (as indicated by the arrows). The strong-coupling features correspond to a strong coupling between charge carriers and the phonon mode with $\omega_{ph} = 19$ meV. The phonon density of states in an optimally doped YBCO also reveals a large peak at about 20 meV [21].

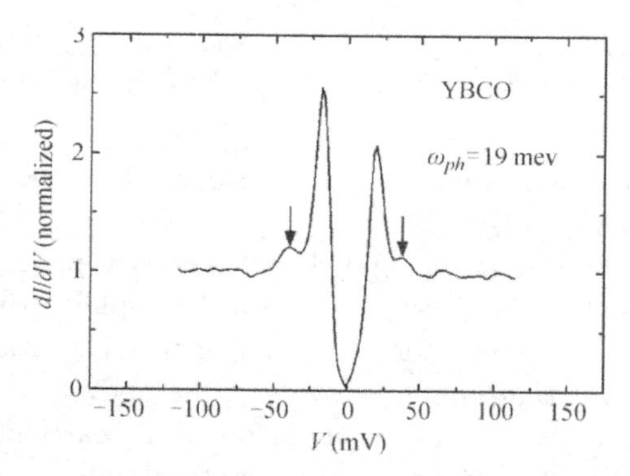

Fig. 7.3. The normalized conductance data for a scanning tunneling microscopy (STM) tunnel junction on a slightly overdoped $YBa_2Cu_3O_{7-y}$ (YBCO) crystal with a Pt–Ir tip at 4.2 K (after Ref. [19]).

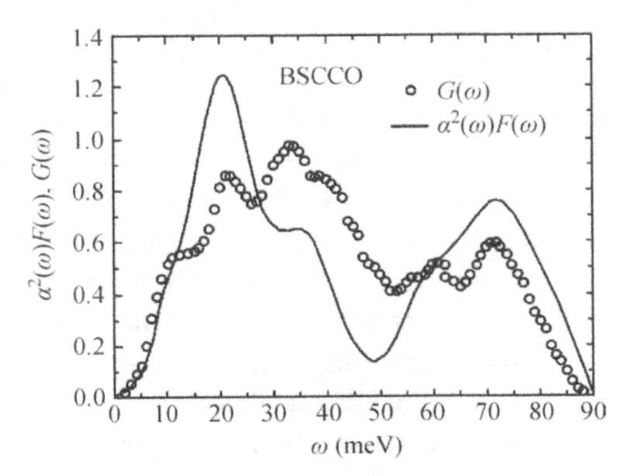

Fig. 7.4. The electron–phonon spectral density $\alpha^2F(\omega)$ for an optimally doped $Bi_2Sr_2CaCu_2O_{8+y}$ (BSCCO) crystal, which was deduced from an SIS break-junction spectrum (Ref. [20]).

In Fig. 7.4, we plot the electron–phonon spectral density $\alpha^2F(\omega)$ for an optimally doped BSCCO crystal, which was extracted from a superconductor-insulator-superconductor (SIS) break-junction spectrum [20]. A strong coupling feature at an energy of about 20 meV is clearly seen. This feature also corresponds to the large peak in the phonon density of states at about 20 meV (see open circles). In addition to a strong coupling feature at 20 meV, there is another strong coupling feature at about 73 meV, which corresponds to the phonon energy of the Cu-O bond stretching mode discussed above. Therefore, these tunneling spectra (Figs. 7.3 and 7.4) consistently suggest that both the low-energy phonon mode at about 20 meV and the high-energy phonon mode at about 73 meV are strongly coupled to conduction electrons in these double-layer compounds.

A theoretical approach to a strong electron–phonon coupling system depends not only on the adiabatic ratio ω/E_F but also on the coupling strengths with different phonon modes. When the Fermi energy E_F is smaller than these strong-coupling phonon energies, the phonon-induced effective interaction between carriers is nonretarded so that the real-space pairing (e.g., intersite bipolaron formation) becomes possible [13, 22]. This should be the case for the doping level $x \leq 0.10$ in $La_{1-x}Sr_xCuO_4$ [9]. In contrast, E_F in doped oxygen-hole bands may lie in between 20 and 73 meV in the optimally doped and overdoped regimes, so that the pairing interaction becomes retarded for the low-energy phonons, and remains nonretarded for the high-energy phonons. The retarded electron–phonon coupling for the low-energy phonons could be treated within the Migdal approximation, while the nonretarded electron–phonon coupling for the high-energy phonons should be modeled separately within the polaron theory. This theoretical approach has been successfully applied to fullerenes [23]. The strong coupling between doped holes and the high-energy phonons leads to a polaronic mass enhancement and to an attractive nonretarded potential between doped holes. Effectively, the polaronic holes could then form k-space Cooper pairs by interacting with the low-energy phonons. The problem could thus be solved within Eliashberg equations with an effective electron–phonon spectral density for the low-energy phonons and a negative Coulomb pseudopotential produced by the high-energy phonons and other high-energy bosonic excitations of purely electronic origin (e.g., spin fluctuations, excitons, and plasmons). Within this simplified approach, the effective electron-phonon coupling constant λ_{ep} for the low-energy phonons is enhanced by a factor of $f_p = \exp(g^2)$. Here $g^2 = A/\omega H$, A is a constant, and ω_H is the frequency of the high-energy phonon mode. The effective Coulomb pseudopotential μ^* is negative and also proportional to f_p.

For slightly overdoped BSCCO, g^2 can be evaluated from the midinfrared optical conductivity which exhibits a maximum at $E_m \simeq 0.12$ eV [26]. With $E_m = 0.12$ eV, $\hbar\omega_H = 75$ meV, we find $g^2 = E_m/(2\hbar\omega_H) = 0.8$, leading to $f_p = 2.2$. From the spectral density shown in Fig. 7.4, we can extract the effective electron–phonon coupling constant λ_{ep} for the low-energy phonon mode, that is, $\lambda_{ep} \simeq 2.6$. If there were no polaronic mass enhancement due to the high-energy phonons, the coupling constant contributed from the low-energy phonons would be about 1.2. With $\mu^* \simeq 0.1$ and $\lambda_{ep} = 1.2$, we calculate $T_c = 18$ K according to a T_c formula

$$k_B T_c = 0.25\hbar\sqrt{\langle\omega^2\rangle}[\exp(2/\lambda_{eff}) - 1]^{-1/2}, \tag{7.2}$$

where

$$\lambda_{eff} = (\lambda_{ep} - \mu^*)/[1 + 2\mu^* + \lambda_{ep}\mu^* t(\lambda_{ep})]. \tag{7.3}$$

The function $t(\lambda_{ep})$ is plotted in Fig. 7.2 of Ref. [27]. In the present case, $\hbar\sqrt{\langle\omega^2\rangle}$ is contributed only from the low-energy phonons and equal to 20 meV. Therefore, without the high-energy phonons, T_c would not be higher than 20 K. The high-energy phonons not only enhance λ_{ep} by a factor of 2.2, but also reduce μ^* substantially [23]. It has recently been shown that μ^* in cuprates becomes negative (i.e., $\mu^* \simeq -0.05$) due to the presence of low-energy electronic collective modes (acoustic plasmons) in layered conductors [24]. Since the high-energy phonon mode reduces μ^* further, it is likely that the value of μ^* should be in the range of -0.1 to -0.2. If we take $\mu^* = -0.15$, $\lambda_{ep} = 2.6$ (see above), we can get $T_c = 95$ K. This leads to $k_B T_c/(\hbar\sqrt{\langle\omega^2\rangle}) = 0.41$, and $2\Delta(0)/k_B T_c \simeq 7$ according to the known relation between $k_B T_c/(\hbar\sqrt{\langle\omega^2\rangle})$ and $2\Delta(0)/k_B T_c$ [25]. The calculated reduced energy gap is in good agreement with experiment. We would like to mention that the present calculation is valid for an isotropic s-wave gap, which should be the case for the polaronic oxygen-hole bands near $(0, \pm\pi)$ and $(\pm\pi, 0)$ regions. In addition to the polaronic oxygen holes, there are electronlike free carriers which could condense into super-carriers through interband scattering. The pairing symmetry of the two-carrier system may be an extended s-wave with eight line nodes [28].

Now we can calculate the total isotope exponent α using Eqs. (7.2) and (7.3), and the relations $\lambda_{ep} \propto f_p, \mu^* \propto f_p, t(\lambda_{ep}) \simeq 1.8/\lambda_{ep}$. The calculated total isotope exponent is $\alpha \simeq 0$. The nearly zero isotope exponent is due to the fact that the isotope dependencies of λ_{ep} and μ^* (arising from the polaronic effect) cancel out the isotope effect on the prefactor of Eq. (7.2). Moreover, the formation of polaronic Cooper pairs naturally explains the sizable isotope effect on the supercarrier mass.

In summary, we report the observation of the oxygen-isotope effects in the three-layer cuprate $Bi_{1.6}Pb_{0.4}Sr_2Ca_2Cu_3O_{10+y}$. The present results along with the previously observed isotope effects in single-layer and double-layer cuprate superconductors indicate that the isotope effect on T_c in optimally doped cuprates is small while the isotope effect on the effective supercarrier mass is substantial. These isotope effects and tunneling spectra observed in optimally doped cuprates can be consistently explained within a scenario where polaronic oxygen holes are bound into Cooper pairs.

References

[1] B. Batlogg *et al.*, *Phys. Rev. Lett.* **58** (1987), 2333.
[2] D. E. Morris, R. M. Kuroda, A. G. Markelz, J. H. Nickel, and J. Y. T. Wei, *Phys. Rev. B* **37** (1988), 5936.
[3] M. K. Crawford, M. N. Kunchur, W. E. Farneth, E. M. McCarron III, and S. J. Poon, *Phys. Rev. B* **41** (1990), 282.
[4] M. K. Crawford, W. E. Farneth, E. M. McCarron III, R. L. Harlow, and A. H. Moudden, *Science* **250** (1990), 1390.

[5] H. J. Bornemann, D. E. Morris, H. B. Liu, and P. K. Narwankar, *Physica C* **191** (1992), 211.
[6] J. P. Franck, S. Harker, and J. H. Brewer, *Phys. Rev. Lett.* **71** (1993), 283.
[7] G. M. Zhao, K. K. Singh, A. P. B. Sinha, and D. E. Morris, *Phys. Rev. B* **52** (1995), 6840.
[8] G. M. Zhao, M. B. Hunt, H. Keller, and K. A. Müller, *Nature (London)* **385** (1997), 236.
[9] G. M. Zhao, K. Conder, H. Keller, and K. A. Müller, *J. Phys.: Condens. Matter* **10** (1998), 9055.
[10] G. M. Zhao, and D. E. Morris, *Phys. Rev. B* **51** (1995), R16 487.
[11] Y. J. Uemura *et al.*, *Phys. Rev. Lett.* **62** (1989), 2317.
[12] J. Hofer, K. Conder, T. Sasagawa, G. M. Zhao, M. Willemin, H. Keller, and K. Kishio, *Phys. Rev. Lett.* **84** (2000), 4192.
[13] A. S. Alexandrov, and N. F. Mott, *Polarons and Bipolarons* (World Scientific, Singapore, 1995).
[14] N. Miyakawa, P. Guptasarma, J. F. Zasadzinski, D. G. Hinks, and K. E. Gray, *Phys. Rev. Lett.* **80** (1998), 157.
[15] A. Buzdin, A. Neminsky, P. Nikolaev, and C. Baraduc, *Physica C* **227** (1994), 365.
[16] R. Khasanov *et al.*, unpublished.
[17] R. J. McQueeney, Y. Petrov, T. Egami, M. Yethiraj, G. Shirane, and Y. Endoh, *Phys. Rev. Lett.* **82** (1999), 628.
[18] Y. Petrov, T. Egami, R. J. McQueeney, M. Yethiraj, H. A. Mook, and F. Dogan, cond-mat/0003414, unpublished.
[19] J. Y. T. Wei, N.-C. Yeh, D. F. Garrigus, and M. Strasik, *Phys. Rev. Lett.* **81** (1998), 2542.
[20] R. S. Gonnelli, G. A. Ummarino, and V. A. Stepanov, *Physica C* **275** (1997), 162.
[21] B. Renker, F. Gompf, E. Gering, D. Ewert, H. Rietschel, and A. Dianoux, *Z. Phys. B: Condens. Matter* **73** (1988), 309.
[22] J. Bonca, T. Katrasnik, and S. A. Trugman, *Phys. Rev. Lett.* **84** (2000), 3153.
[23] A. S. Alexandrov, and V. V. Kabanov, *Phys. Rev. B* **54** (1996), 3655.
[24] A. Bill, H. Morawitz, and V. Z. Kresin, *J. Supercond.* **13** (2000), 907.
[25] J. P. Carbotte, *Rev. Mod. Phys.* **62** (1990), 1027.
[26] M. A. Quijada, D. B. Tanner, R. J. Kelley, M. Onellion, H. Berger, and G. Margaritondo, *Phys. Rev. B* **60** (1999), 14917.
[27] V. Z. Kresin, *Phys. Lett. A* **122** (1987), 434.
[28] G. M. Zhao, *Phys. Rev. B* (2001) to be published.

8
Isotopic Fingerprint of Electron–Phonon Coupling in High-T_c Cuprates*

H. Iwasawa[†,‡], J. F. Douglas[§], K. Sato[‡,¶], T. Masui[||], Y. Yoshida[‡], Z. Sun[§], H. Eisaki[‡], H. Bando[‡], A. Ino[**], M. Arita[††], K. Shimada[††], H. Namatame[††], M. Taniguchi[**,††], S. Tajima[||], S. Uchida[‡‡], T. Saitoh[†], D. S. Dessau[§] and Y. Aiura[‡,††]

[†]*Department of Applied Physics, Tokyo University of Science, Shinjuku-ku, Tokyo 162-8601, Japan*
[‡]*National Institute of Advanced Industrial Science and Technology, Tsukuba, Ibaraki 305-8568, Japan*
[§]*Department of Physics, University of Colorado, Boulder, Colorado 80309-0390, USA*
[¶]*Faculty of Science, Ibaraki University, Mito, Ibaraki 310-8512, Japan*
[||]*Department of Physics, Osaka University, 1-1 Machikaneyama, Toyonaka, Osaka 560-0043, Japan*
[**]*Graduate School of Science, Hiroshima University, Higashi-Hiroshima 739-8526, Japan*
[††]*Hiroshima Synchrotron Radiation Center, Hiroshima University, Higashi-Hiroshima 739-8526, Japan*
[‡‡]*Department of Physics, University of Tokyo, Tokyo 113-8656, Japan*

*Reprinted with the permission from H. Iwasawa, J. F. Douglas, K. Sato, T. Masui, Y. Yoshida, Z. Sun, H. Eisaki, H. Bando, A. Ino, M. Arita, K. Shimada, H. Namatame, M. Taniguchi, S. Tajima, S. Uchida, T. Saitoh, D. S. Dessau, and Y. Aiura. Original published in *Phys. Rev. Lett.* **101** (2008), 157005.

Angle-resolvedphotoemission spectroscopy with low-energy tunable photons along the nodal direction of oxygen isotope substituted $Bi_2Sr_2CaCu_2O_{8+\delta}$ reveals a distinct oxygen isotope shift near the electronboson coupling "kink" in the electronic dispersion. The magnitude (afew meV) and direction of the kink shift are as expected due to the measured isotopic shift of phonon frequency, and are also in agreement with theoretical expectations. This demonstrates the participation of the phonons as dominant players, as well as pinpointing the most relevant of the phonon branches.

The effects of electron–boson interactions on electronic self-energies show up as sudden changes in the electron dispersion, or "kinks," via angle-resolved photoemission spectroscopy (ARPES) [1, 2]. While the kinks are now indisputable, their origin as arising from electronic coupling to phonons [3–5], magnetic excitations [6–9], or both [10], remains unclear. The best way to identify the origin of an interaction is to slightly modify the interaction in a controlled manner. For phonons, this can be done via an isotopic exchange, which varies particle masses and hence the vibrational energies. Here, we substitute the oxygen isotopes ($^{16}O \rightarrow {}^{18}O$) leading to a softening of phonon energies of a few meV [$= \Omega(1 - \sqrt{16/18})$] where Ω are the bare phonon frequencies of 40–70 meV [11, 12]. Recent STM experiments showed an isotopic shift of a second derivative feature of 3.7 meV [13], though whether the observed shift is evidence for electron-phonon coupling has been disputed by a number of groups in terms of the inelastic tunneling barrier [14–16]. Even earlier than that, Gweon *et al.* used ARPES to study isotope effects on the electronic structure [17]. They found that the primary effect was at high energy and was unusually strong (up to 30 meV), though these results were not reproduced by more modern experiments [18, 19]. For the effect on the kink scale they presented a roughly 5–10 meV isotopic softening but about 2–3 times larger than one of observed in STM. Therefore, a direct and clear "fingerprint" of strong electron–phonon interactions in the high-T_c superconductors has been missing up to now.

The momentum selectivity of ARPES brings an additional tool to bear on the study of the isotope effect which is not available from tunneling. This information is very helpful in separating out the energy and momenta of the bosonic modes which couple most strongly to the electronic degrees of freedom. Here we focus on the "nodal" spectra, i.e., those along the $(0,0)-(\pi,\pi)$ direction in the Brillouin zone where both the superconducting gap and pseudogap are minimal or zero [1, 2]. This is the portion of the Brillouin zone where the kinks or self-energy effects have been most strongly studied both experimentally [3, 20, 21] and theoretically [22, 23]. As described very recently by Giustino *et al.* [24] and Heid *et al.* [25], there is still great controversy about the nature of the coupling in this direction; these two theoretical arguments imply that the phonon coupling could only be a minor player in the overall electron-boson coupling of these states because the calculations indicate a significantly weaker kink than what is found in experiment.

Analysis and interpretation of the nodal data are also more straightforward than the data throughout the Brillouin zone, as the energies of any bosonic modes in the spectra are expected to be shifted by the energies of the gaps. Assuming any isotope shifts go roughly as $[\Omega(1-\sqrt{16/18})]$, these are at most a few meV, requiring an unusually high-precision ARPES experiment. To verify such a tiny effect in the most convincing way, we utilized bulk-sensitive low photon energy ARPES (LEARPES), which has greatly improved spectral resolution compared to "conventional" ARPES [26]. In this paper, we report an isotopic fingerprint of electron–phonon coupling as a clear isotope shift of the kink energy taking advantages of LEARPES. Present results provide straightforwardly that the origin of the nodal kink is due to the electron–phonon interactions.

High-quality optimally doped $Bi_2Sr_2CaCu_2O_{8+\delta}$ (Bi2212) single crystals were prepared by the traveling solvent floating-zone technique [27]. Oxygen isotope substitution was performed by annealing procedures (R-Dec Co. Ltd., ASF-11T),[a] yielding a slight decrease of T_c from 92.1 to 91.1 K. A softening of the oxygen vibration modes and the high oxygen isotope substitution rate (more than 80%) was confirmed by Raman spectroscopy.[b] Further, LEARPES gives us unparalleled precision in determining the Fermi surface areas directly from the Fermi surface maps, and we have confirmed that these are the optimal and the same for the two isotopes to the level below 2% (a doping level uncertainty $\Delta x = 0.003$).

The present data were collected at BL-9A of Hiroshima Synchrotron Radiation Center using a Scienta R4000 electron analyzer. We used our newly developed high-precision 6-axis sample manipulator to remove the extrinsic effect due to a sample misalignment. The clean and fiat surface

[a]We performed annealing under the same thermal conditions in either $^{16}O_2$ or $^{18}O_2$ gas: 700°C for 168 h at 1.0 bar, followed by 700°C for 168 h at 0.2 bar.
[b]The isotopic shifts were observed in the B_{1g} mode in the CuO_2 layer, and also in oxygen vibrational modes in the BiO layer and the SrO layer. These shifts give the estimation of isotope exchange rates.

of the samples was obtained by cleaving *in situ* in ultrahigh vacuum better than 4×10^{-11} Torr below 10 K. The total instrumental energy and angular (momentum) resolutions were better than 5 meV and 0.4° (0.005 Å^{-1}) at 7.0 eV photons, respectively.

First, we optimized the excitation energy to avoid the complications in extracting accurate band dispersions due to the very small but finite (0.01 Å^{-1}) bilayer splitting, which is now known to exist even along the nodal line (Figs. 8.1(a) and 8.1(b)) [28]. By tuning the photon energy to 7.0 eV, we can see the complete isolated and individual anti-bonding band dispersion (Figs. 8.1(c) and 8.1(d)), giving the extremely fine quasi-particle dispersion with the momentum full width (Δk) below 0.005 Å^{-1} (black (blue) line in Fig. 8.1(e)). This is in sharp contrast to the cases of the dispersion with bilayer splitting (Fig. 8.1(b)), and also to the conventional

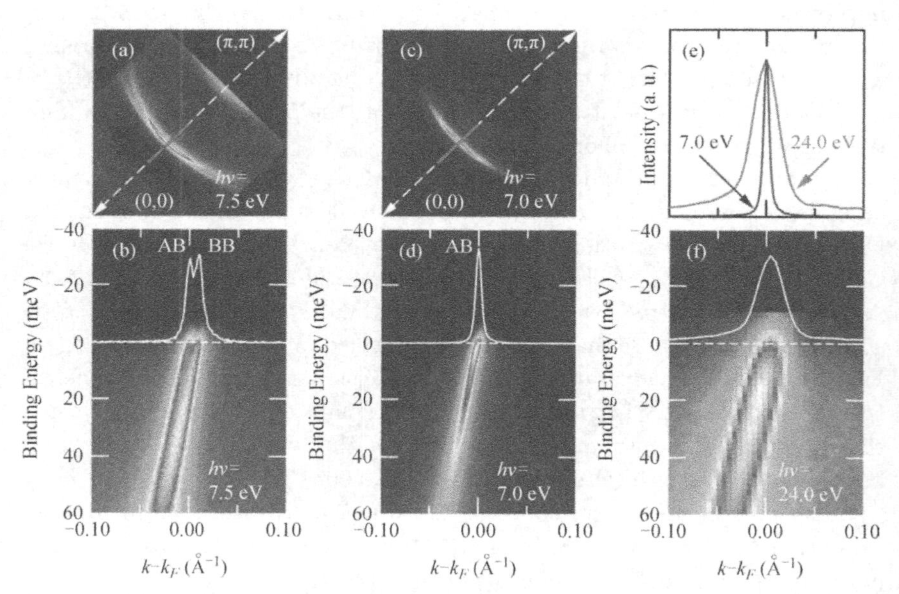

Fig. 8.1. (a) and (c) Fermi surface maps taken with 7.5 and 7.0 eV photons, respectively. (b), (d) and (f) Energy–momentum distribution maps taken with 7.5, 7.0, and 24.0 eV along the nodal direction, indicated by the gray (red) line in (a) and (c). MDCs at the Fermi level (E_F) are also shown by the white line, where the BB and AB denote bonding band and anti-bonding band, respectively. The Fermi momentum is set to the momentum of the AB crossing E_F. (e) Comparison of MDCs at E_F between 7.0 eV (blue line) and 24.0 eV (red line). Note that ARPES data using a conventional photon-energy of 24.0 eV were taken at BL5-4 of the Stanford Synchrotron Radiation Laboratory.

ARPES spectra broadened by a limit of resolutions as well as the unresolved bilayer splitting (Fig. 8.1(f)). These advantages of LEARPES, including the band selectivity, are mandatory to verify the subtle change in the ARPES spectra with isotope substitution.

Figure 8.2(a) compares the ^{16}O (blue lines) and ^{18}O (red lines) energy-momentum dispersions derivedfrom the momentum distribution curves

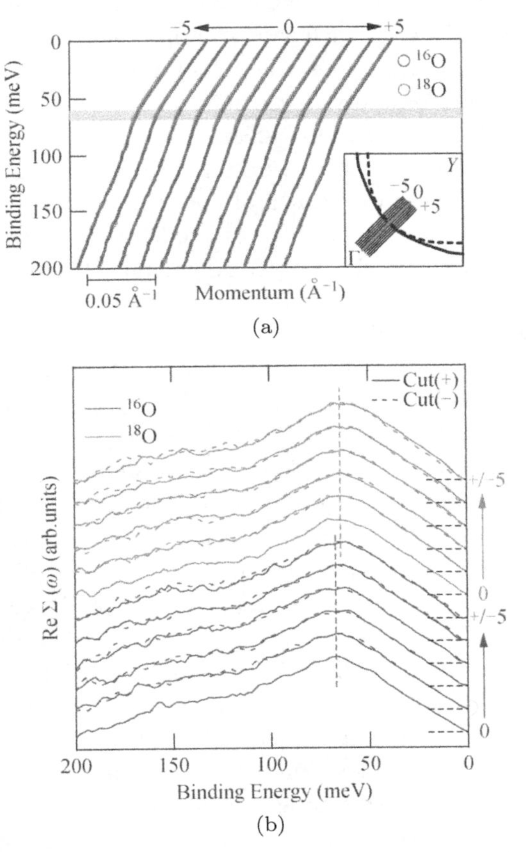

Fig. 8.2. (a) Energy–momentum dispersions near the nodal region both for ^{16}O (blue lines) and ^{18}O (red lines) from optimally doped Bi2212. Those measured cuts are labeled −5 to +5, displayed in the inset. Gray shaded area roughly indicates the kink in all the dispersions. (b) Real parts of the selfenergy both for ^{16}O [blue (lower) lines] and ^{18}O (red (upper) lines), showing an isotope shift of ~70 meV peak. Blue dashed lines and red dashed lines indicates the kink energy, averaged from cut −5 to cut +5, for ^{16}O and ^{18}O, respectively. Positive and negative cuts are represented by solid lines and dashed lines, respectively.

(MDCs) near the nodal region taken with 7.0 eV photons. We found a very small isotope effect around the kink in the dispersions, independent of the cut position. We emphasize here that each of the five positive dispersions is exactly identical with each of the five negative dispersions for both ^{16}O and ^{18}O (as seen in Fig. 8.2(b)), which rules out the possibility of an extrinsic effect due to a sample misalignment. It should be noted that the present results are consistent with our previous ARPES [18, 19], though we cannot see a convincing evidence of such a tinyisotope effect in there. This should be attributed to the fact that the experimental resolutions and/or accuracy were still not enough to observe a few meV order effect.

To visualize the subtle isotope effect more clearly, we deduced the real part of the self-energy $\mathrm{Re}\Sigma(\omega)$. We extracted $\mathrm{Re}\Sigma(\omega)$ by subtracting a bare band dispersion from the experimental one using three different forms of bare band in this analysis: linear (presented in Fig. 8.2(b)) [29], second order polynomial [30], and third order polynomial. It is worth noting that the present result is robust in regard to this choice. As seen in Fig. 8.2(b), clearly observed is a few meV isotopic shift of the ~70 meV peak of $\mathrm{Re}\Sigma(\omega)$. Here we estimated the kink energy as the energy giving the peak maximum of $\mathrm{Re}\Sigma(\omega)$, which was obtained by fitting the top portion of $\mathrm{Re}\Sigma(\omega)$ with a Gaussian. To quantify the energy scale of the kink as well as the isotopic kink shift, we have studied multiple samples systematically.

Figure 8.3(a) shows the real part of the self-energy $\mathrm{Re}\Sigma(\omega)$ both for ^{16}O (blue lines) and ^{18}O (red lines) multiple samples. From these $\mathrm{Re}\Sigma(\omega)$, we obtained kink energy plotted as a function of five different samples each for both ^{16}O and ^{18}O in Fig. 8.3(c). Error bars for each sample were determined from a statistical analysis, and are smaller than the overall (sample-to-sample) spread which is more dominated by systematics.We found a clear isotopic softening of the kink energy from about 69.0 meV to about 65.6 meV, or a softening of 3.4 ± 05 meV.[c] Additionally, we used a completely independent analysis method using the widths of the ARPES peaks. This analysis has the advantage of not having any assumptions about a bare band, such that the isotope effect should appear more straightforward. Thus, we see a 3.2 ± 0.6 meV shift in the imaginary part of the self-energy $\mathrm{Im}\Sigma(\omega)$ (Figs. 8.3(b) and 8.3(d)). By studying ten samples as

[c]The isotope shift was found to be insensitive to the choice of the bare velocity; the isotope shift shown here of 3.4 ± 0.5 meV utilized a linear bare dispersion, while a 3.3 ± 0.5 meV shift was found in the case of a second order polynomial.

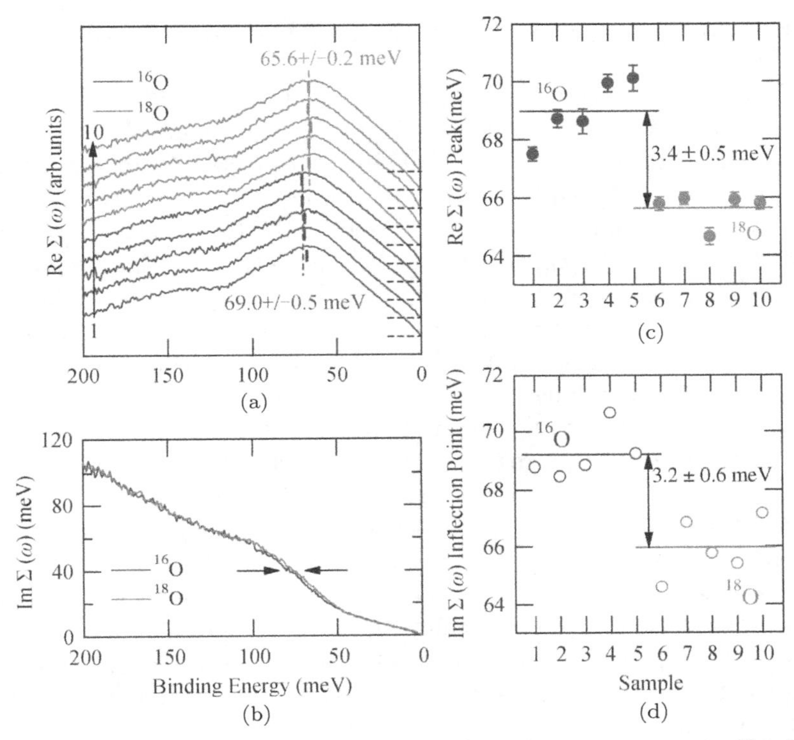

Fig. 8.3. (a) Real part of the self-energy Re$\Sigma(\omega)$ from five samples both for ^{16}O (blue lines) and ^{18}O (red lines) along the nodal direction indicated by the gray (red) line in Fig. 8.1(c). All Re$\Sigma(\omega)$ are deduced by subtracting a bare band dispersion from the experimental one, where ω is the energy relative to the Fermi energy, and normalized by the peak maximum, and are also offset for clarity. (b) Imaginary part of the self-energy Im $\Sigma(\omega)$ determined from MDC full widths. An impurity scattering term at $\omega = 0$ is subtracted as an energy independent constant background. (c), (d) Obtained kink energy as a function of sample numbers both for ^{16}O (blue line) and ^{18}O (red line) from Re$\Sigma(\omega)$ and Im $\Sigma(\omega)$, respectively.

well as by using multiple independent analysis methods, we compensated for possible systematic errors which might come into play when trying to determine energies to such a great precision. Further, we have also found a similar 3 meV isotope shift in the real and imaginary parts of the self-energy derived from the energy distribution curves. Therefore, we can state with confidence that the ~70 meV feature in the nodal electron self-energy is due to the coupling of the electrons with phonons. That this is the dominant feature in the electron self-energy, as is seen from both the real

(Fig. 8.3(a)) and imaginary (Fig. 8.3(b)) parts of the spectrum, is clear and significant.

Then, which phonons are responsible for this coupling? Neutron scattering experiments [31] as well as first-principles phonon calculations [32] indicate a few phonon modes that are likely to be most relevant for the coupling: the in-plane "half-breathing" phonon mode ($\Omega \sim 70$ meV) and the "buckling or stretching" modes ($\Omega \sim 36$ meV). The apical oxygen stretching mode ($\Omega \sim 50$ meV) could also be considered, though calculations indicate that the number of allowed final states for these phonons is negligible [24]. By coupling the nodal electrons with momentum, $\boldsymbol{k}$, to other parts of the Fermi surface ($\boldsymbol{k}'$), the electron self-energy can in principle pick up the energy of the superconducting gap $\Delta(\boldsymbol{k}')$ weighted over the Brillouin zone (and hence gap size) by the electron–phonon matrix elements $|g(\boldsymbol{k};\boldsymbol{k}')|$, though there is still theoretical disagreement about how to do this gap referencing in a *d*-wave superconductor [33]. On the one hand, it is suggested that the kink energy appears at the mode energy plus the maximum gap energy, or $\Omega + \Delta_{\text{max}}$ [34]. This scenario would indicate that the nodal kink is caused by the 36 meV buckling phonon. Alternatively, it is suggested that the kink represents simply the mode energy, indicating that the half-breathing phonon is dominant and couples electrons strongly along the node [24, 25, 35]. It is here that the present isotope effect is uniquely capable to distinguish the two.

For the buckling mode with $\Omega \sim 36$ meV, the isotope shift would be expected to be 2.1 meV, more than two error bars outside of the results published here. On the other hand, the breathing mode with $\Omega \sim 69$ meV would have an isotope shift of 3.9 meV, within the error bar of the isotopic shift seen here. This serves as a strong indication that the breathing mode is responsible for the nodal kink in ARPES data, consistent with expectations from recent theory [24]. However, the moderately strong coupling that we observe (a coupling parameter $\lambda \sim 0.6$[d]) is significantly stronger than that calculated by theory [24, 25], indicating that these calculations are missing an important ingredient to the electron–phonon coupling in the cuprates. Nominally we expect such enhancements in the electronphonon coupling to originate in the strong electron correlation effects [36] — a problem which

[d]The ratio of the "bare" dispersion in the absence of the kink to the renormalized dispersion due to the kink is $v_{\text{bare}}/v_F = 1 + \lambda \sim 1.6$.

is interesting in its own right, but which also may have particular relevance to the mechanism of superconductivity in the cuprates.

In summary, we reported the isotopic fingerprint of the nodal kink probed by high-precision LEARPES. The present isotope shift of the kink energy provides the first convincing and direct evidence that the electron–phonon interactions are responsible for the origin of the nodal kink.

Acknowledgments

This work was supported by KAKENHI (19340105), DOE Grant No. DE-FG02-03ER46066, and Grant-in-Aid for COE research (No. 13CE2002) of MEXT Japan. The synchrotron radiation experiments have been done under the approval of HSRC (Proposal No. 06-A-15). H. I. acknowledges financial support from JSPS.

References

[1] A. Damascelli, Z. Hussain, and Z.-X. Shen, *Rev. Mod. Phys.* **75** (2003), 473.
[2] J. C. Campuzano, M. R. Norman, and M. Randeria, *Physics of Superconductors* (Springer, Berlin, 2004).
[3] A. Lanzara *et al.*, *Nature (London)* **412** (2001), 510.
[4] Z.-X. Shen *et al.*, *Philos. Mag. B* **82** (2002), 1349.
[5] T. Cuk *et al.*, *Phys. Rev. Lett.* **93** (2004), 117003.
[6] D. J. Scalapino, *Science* **284** (1999), 1282.
[7] J. P. Carbotte, E. Schachinger, and D. N. Basov, *Nature (London)* **401** (1999), 354.
[8] H. He *et al.*, *Science* **295** (2002), 1045.
[9] K. Terashima *et al.*, *Nature Phys.* **2** (2006), 27.
[10] A. D. Gromko *et al.*, *Phys. Rev. B* **68** (2003), 174520.
[11] D. Reznik *et al.*, *Phys. Rev. Lett.* **75** (1995), 2396.
[12] R. J. McQueeney *et al.*, *Phys. Rev. Lett.* **87** (2001), 077001.
[13] Jinho Lee *et al.*, *Nature (London)* **442** (2006), 546.
[14] S. Pilgram, T. M. Rice, and M. Sigrist, *Phys. Rev. Lett.* **97** (2006), 117003.
[15] D. J. Scalapino, *Nature Phys.* **2** (2006), 593.
[16] J. Hwang, T. Timusk, and J. P. Carbotte, *Nature (London)* **446** (2007), E4.
[17] G.-H. Gweon *et al.*, *Nature (London)* **430** (2004), 187.
[18] J. F. Douglas *et al.*, *Nature (London)* **446** (2007), E5.
[19] H. Iwasawa *et al.*, *Physica (Amsterdam)* **463–465C** (2007), 52.
[20] T. Valla *et al.*, *Science* **285** (1999), 2110.
[21] X. J. Zhou *et al.*, *Nature (London)* **423** (2003), 398.
[22] P. Zhang, S. G. Louie, and M. L. Cohen, *Phys. Rev. Lett.* **98** (2007), 067005.
[23] M. L. Kulić and O. V. Dolgov, *Phys. Rev. B* **76** (2007), 132511.

[24] F. Giustino, M. L. Cohen, and S. G. Louie, *Nature (London)* **452** (2008), 975.
[25] R. Heid *et al.*, *Phys. Rev. Lett.* **100** (2008), 137001.
[26] J. D. Koralek *et al.*, *Phys. Rev. Lett.* **96** (2006), 017005.
[27] H. Eisaki *et al.*, *Phys. Rev. B* **69** (2004), 064512.
[28] T. Yamasaki *et al.*, *Phys. Rev. B* **75** (2007), 140513(R).
[29] P. D. Johnson *et al.*, *Phys. Rev. Lett.* **87** (2001), 177007.
[30] W. Meevasana *et al.*, *Phys. Rev. Lett.* **96** (2006), 157003.
[31] L. Pintschovius, D. Reznik, and K. Yamada, *Phys. Rev. B* **74** (2006), 174514.
[32] K.-P. Bohnen, R. Heid, and M. Krauss, *Europhys. Lett.* **64** (2003), 104.
[33] A. Balatsky, private communication.
[34] W. S. Lee *et al.*, *Phys. Rev. B* **77** (2008), 140504(R).
[35] T. P. Devereaux *et al.*, *Phys. Rev. Lett.* **93** (2004), 117004.
[36] S. Koikegami and Y. Aiura, *Phys. Rev. B* **77** (2008), 184519.

9
Kink Structure in the Electronic Dispersion of High-T_c Superconductors from the Electron–Phonon Interaction*

Shigeru Koikegami[†,‡] and Yoshihiro Aiura[‡]

[†]*Second Lab, LLC, 10-7-204 Inarimae, Tsukuba 305-0061, Japan*

[‡]*Nanoelectronics Research Institute, AIST Tsukuba Central 2, Tsukuba 305-8568, Japan*

We investigate the electronic dispersion of high-T_c superconductor on the basis of the 2D three-band Hubbard model with the electron–phonon interaction (EPI) together with the strong electron–electron interaction (EEI). In our model, it is shown across the hole-doped region of high-T_c superconductor that the EPI makes a dispersion kink, observed along the nodal direction, and that the small isotope effect appears on the electronic dispersion.

9.1. Introduction

For the past two decades, extensive studies of high-T_c cuprates have spotlighted many curious phenomena. It has been argued that most phenomena are attributable to the strong correlations among electrons, which play significant roles in these materials. However, since the discovery of sudden changes in the electron dispersion or "kinks" shown by the angle-resolved

*Reprinted with the permission from Shigeru Koikegami, and Yoshihiro Aiura. Original published in *Phys. Rev. B* **77** (2008), 184519.

photoemission spectroscopy (ARPES) [1, 2], effects of electron–boson interactions on electronic self-energies have been recognized. While the kinks are now indisputable in cuprates [3], their origin as arising from electronic coupling to phonons [4] or magnetic excitations [5–8] remains unclear.

Recent scanning tunneling microscope study showed that the statistical distribution of energy of bosonic modes (Ω) has meaningful difference between ^{16}O-and ^{18}O-materials. Thus, it should be hard to exclude the possibility that electron–phonon interaction (EPI) significantly affects the electronic states in cuprates.

In this study, we investigate the analysis upon the EPI together with the electron–electron interaction (EEI) on the basis of the 2D three-band Hubbard–Holstein (HH) model. With the use of our three-band HH model, we can reproduce the situation of the real high-T_c materials, in which EPI mainly works on p electrons at O sites.

9.2. Formulation

Our model Hamiltonian H is composed of d electrons at each Cu site, p electrons at O site, and lattice vibrations of O atoms. We consider only the on-site Coulomb repulsion U between d electrons at each Cu site as our EEI. Let us define that μ and N_0 represent the chemical potential and the number of all electrons, respectively. Then, $H - \mu N_0$ is divided into the non-interacting part H_0, the electron–electron interacting part $H_{\text{el-el}}$, the phonon part H_{ph}, and the electron–phonon interacting part $H_{\text{el-ph}}$ as

$$H - \mu N_0 = H_0 + H_{\text{el-el}} + H_{\text{ph}} + H_{\text{el-ph}},$$
$$N_0 = \sum_{\mathbf{k}\sigma}(d^\dagger_{\mathbf{k}\sigma} d_{\mathbf{k}\sigma} + p^{x\dagger}_{\mathbf{k}\sigma} p^x_{\mathbf{k}\sigma} + p^{y\dagger}_{\mathbf{k}\sigma} p^y_{k\sigma}). \tag{9.2.1}$$

Here, $d_{\mathbf{k}\sigma}(d^\dagger_{\mathbf{k}\sigma})$ and $p^{x(y)}_{\mathbf{k}\sigma}(p^{x(y)\dagger}_{\mathbf{k}\sigma})$ are the annihilation (creation) operator for d and $p^{x(y)}$ electrons of momentum $\mathbf{k}$ and spin σ, respectively. The noninteracting part H_0 is represented by

$$H_0 = \sum_{\mathbf{k}\sigma}(d^\dagger_{\mathbf{k}\sigma} p^{x\dagger}_{\mathbf{k}\sigma} p^{y\dagger}_{\mathbf{k}\sigma}) \begin{pmatrix} \Delta_{dp} & \zeta^x_{b_\mathbf{k}} & \zeta^y_\mathbf{k} \\ -\zeta^x_\mathbf{k} & 0 & \zeta^p_\mathbf{k} \\ -\zeta^y_\mathbf{k} & \zeta^p_\mathbf{k} & 0 \end{pmatrix} \begin{pmatrix} d_{\mathbf{k}\sigma} \\ p^x_{\mathbf{k}\sigma} \\ p^y_{\mathbf{k}\sigma} \end{pmatrix} \equiv \sum_{\mathbf{k}\sigma} \mathbf{d}^\dagger_{\mathbf{k}\sigma} \mathbf{H}_0 \mathbf{d}_{\mathbf{k}\sigma}. \tag{9.2.2}$$

We take the lattice constant of the square lattice formed from Cu sites as the unit of length. Then, $\zeta_{\mathbf{k}}^{x(y)} = 2it_{dp}\sin\frac{k_{x(y)}}{2}$ and $\zeta_{\mathbf{k}}^{p} = -4t_{pp}\sin\frac{k_x}{2}\sin\frac{k_y}{2}$, where t_{dp} is the transfer energy between a d orbital and a neighboring $p^{x(y)}$ orbital and t_{pp} is that between a p^x orbital and a p^y orbital. Δ_{dp} is the difference of energy levels of d- and p-orbitals. In this study, we take t_{dp} as the unit of energy. The residual parts are described as follows:

$$H_{\mathrm{el-el}} = \frac{U}{N}\sum_{\mathbf{k}\mathbf{k}'}\sum_{\mathbf{q}(\neq 0)} d^{\dagger}_{\mathbf{k}+\mathbf{q}\uparrow} d^{\dagger}_{\mathbf{k}'-\mathbf{q}\downarrow} d_{\mathbf{k}'\downarrow} d_{\mathbf{k}\uparrow}, \tag{9.2.3}$$

$$H_{ph} = \sum_{\mathbf{q}}\sum_{\nu=\{x,y\}} \omega^{\nu}_{\mathbf{q}} b^{\nu\dagger}_{\mathbf{q}} b^{\nu}_{\mathbf{q}}, \tag{9.2.4}$$

and

$$H_{\mathrm{el-ph}} = \frac{1}{N}\sum_{\mathbf{k}\sigma}\sum_{\mathbf{q}}\sum_{\nu=\{x,y\}} g\alpha^{\nu}_{\mathbf{k},\mathbf{q}} p^{\nu\dagger}_{\mathbf{k}+\mathbf{q}\sigma} p^{\nu}_{\mathbf{k}\sigma}(b^{\nu}_{\mathbf{q}} + b^{\nu\dagger}_{-\mathbf{q}}), \tag{9.2.5}$$

where U is the on-site Coulomb repulsion between d orbitals, N is the number of $\mathbf{k}$-space lattice points in the first Brillouin zone (FBZ), and $g\alpha^{\nu}_{\mathbf{k},\mathbf{q}}(\nu = \{x, y\})$ is the electron–phonon matrix element, respectively. We consider that the half-breathing phonon mode [9], in which oxygen ions are vibrating along the x or y directions, is crucial for our problem. Thus, ignoring the other phonon modes, we have the electron–phonon interacting part as Eq. (9.2.5).

Then, we introduced the unperturbed and perturbed Green's functions, which are to be described in 3×3 matrix form. The unperturbed Green's function $\mathbf{G}^{(0)}(\mathbf{k}, z)$ is derived from Eq. (9.2.2) as

$$\mathbf{G}^{(0)}(\mathbf{k}, z) = [z\mathbf{I} - \mathbf{H}_0]^{-1}, \tag{9.2.6}$$

where $\mathbf{I}$ is a 3×3 unit matrix. Using the abbreviation of Fermion Matsubara frequencies $\epsilon_n = \pi T(2n + 1)$ with integer n and temperature T, the perturbed Green's function $\mathbf{G}(\mathbf{k}, z)$ is determined by the Dyson equation:

$$\mathbf{G}(\mathbf{k}, i\epsilon_n)^{-1} = \mathbf{G}^{(0)}(k, i\epsilon_n)^{-1} - \mathbf{\Sigma}(\mathbf{k}, i\epsilon_n), \tag{9.2.7}$$

where $\mathbf{\Sigma}(\mathbf{k}, i\epsilon_n)$ is the self-energy expected to be a diagonal matrix. In order to estimate the d-electron self-energy $\Sigma_d(\mathbf{k}, i\epsilon_n) \equiv \Sigma_{11}(\mathbf{k}, i\epsilon_n)$ in Eq. (9.2.7), we adopt the fluctuation exchange approximation [10] as follows:

$$\Sigma_d(\mathbf{k}, i\epsilon_n) = \frac{T}{N}\sum_{\mathbf{q}m} G_d(\mathbf{k}-\mathbf{q}, i\epsilon_n - i\omega_m)V_{\text{el-el}}(\mathbf{q}, i\omega_m), \tag{9.2.8}$$

where $G_d(\mathbf{k}, i\epsilon_n) \equiv G_{11}(\mathbf{k}, i\epsilon_n)$,

$$V_{\text{el-el}}(\mathbf{q}, i\omega_m) = \frac{3}{2}\frac{U^2\chi(\mathbf{q}, i\omega_m)}{1-U\chi(\mathbf{q}, i\omega_m)} + \frac{1}{2}\frac{U^2\chi(\mathbf{q}, i\omega_m)}{1+U\chi(\mathbf{q}, i\omega_m)} - U^2\chi(\mathbf{q}, i\omega_m), \tag{9.2.9}$$

where $\omega_m = 2m\pi T$ with integer m are Boson Matsubara frequencies, and

$$\chi(\mathbf{q}, i\omega_m) = -\frac{T}{N}\sum_{\mathbf{k}n} G_d(\mathbf{q}+\mathbf{k}, i\omega_m + i\epsilon_n)G_d(\mathbf{k}, i\epsilon_n). \tag{9.2.10}$$

In order to estimate the $p^{x(y)}$-electron self-energy $\Sigma_{x(y)}(\mathbf{k}, i\epsilon_n) \equiv \Sigma_{22(33)}(\mathbf{k}, i\epsilon_n)$ in Eq. (9.2.7), we exploit the Brillouin–Wigner perturbation theory. We adopt the self-consistent one-loop approximation as follows:

$$\Sigma_{x(y)}(\mathbf{k}, i\epsilon_n) = \frac{T}{N}\sum_{\mathbf{q}m} G_{x(y)}(\mathbf{k}-\mathbf{q}, i\epsilon_n - i\omega_m)V^{x(y)}_{\text{el-ph}}(\mathbf{q}, i\omega_m), \tag{9.2.11}$$

where $G_{x(y)}(\mathbf{k}, i\epsilon_n) \equiv G_{22(33)}(\mathbf{k}, i\epsilon_n)$ and $V^{x(y)}_{\text{el-ph}}(\mathbf{q}, i\omega_m)$ is the EPI on $p^{x(y)}$ electron. Our EPI is determined as follows:

$$V^{x(y)}_{\text{el-ph}}(\mathbf{q}, i\omega_m) = \lambda|\alpha^{x(y)}_{\mathbf{q}}|^2\left[\frac{1}{\omega_h + i\omega_m} + \frac{1}{\omega_h - i\omega_m}\right], \tag{9.2.12}$$

where $\lambda = g^2/(2\omega_h)$ and $\alpha^{x(y)}_{\mathbf{q}} = \sin\frac{q_{x(y)}}{2}$. ω_h is the specific phonon energy for the half-breathing mode. As above, we ignore the effects in which EEI and EPI are coupled. Thus, as shown by Eqs. (9.2.8)–(9.2.12), in our formulation, EEI and EPI are completely decoupled. Of course, this assumption is inadequate to analyze the case in which the characteristic energy due to EEI is comparable to the phonon energy. However, as will be seen later, we actually treat the cases with rather high characteristic energy due to EEI. Hence, decoupling EEI with EPI should be justified in our analysis. The electronphonon coupling constant λ does not depend on M_{O} since $\omega_h \propto M_{\text{O}}^{-1/2}$ and $g \propto (M_{\text{O}}\omega_h)^{-1/2}$, where M_{O} is the mass of an oxygen ion. In our model, thus, the isotope effect is reflected on the phonon energy in the electronic self-energies only, but not any changes in the strength.

9.3. Results and Discussion

We need to solve Eqs. (9.2.7)–(9.2.11) in a fully self-consistent manner. During numerical calculations, we divide the FBZ into 128×128 meshes. We prepare $2^{12} = 4096$ Matsubara frequencies for temperature $T \sim 87$ K. As shown later, at this temperature, our calculation can reproduce the important behavior of electrons in normal state. Moreover, to our knowledge, the situation will not be changed if we change T to some extent.

$t_{dp} \sim 1.0$ eV and $t_{pp} \sim 0.55$ eV, which are all common for our calculations. These values are chosen so that we can reproduce the typical Fermi surface of $Bi_2Sr_2CaCu_2O_{8+\delta}$ observed by ARPES [11, 12]. $\Delta_{dp} \sim 1.4$ eV, $U \sim 3.0$ eV, and $\lambda = 0.8$ unless stated. The phonon energy is set as $\omega_h \sim 65(61)$ meV for ^{16}O (^{18}O) material.

We show the numbers of doped holes for our fully self-consistent solutions in Table 9.1. The numbers of doped holes both for ^{16}O and ^{18}O materials are exactly the same to three places of decimals and they correspond to four different hole-doped samples, lightly doped (LD), underdoped (UD), optimally doped (OP), and overdoped (OD), respectively. In Fig. 9.1, we show the color map of the one-particle spectrum at Fermi level $A(\mathbf{k}, 0)$, where

$$A(\mathbf{k}, \varepsilon) \equiv -\frac{1}{\pi}\mathrm{Im}\{\mathrm{Tr}\mathbf{G}(\mathbf{k}, i\epsilon_n)\}_{i\epsilon_n \to \varepsilon}, \tag{9.3.1}$$

in order to indicate the Fermi surfaces for ^{16}O materials. In Eq. (9.3.1), the Padé approximation is exploited for analytic continuation. Furthermore, we calculate the electronic dispersions of the antibonding band $E_d(\mathbf{k})$ along the nodal direction indicated as the cut in Fig. 9.1 for all our doping cases. $E_d(\mathbf{k})$ is determined as the $\mathbf{k}$ point on which $A(\mathbf{k}, \varepsilon)$ has the maximum value

Table 9.1. Number of doped holes. $\delta \equiv n_d^h + n_p^h - 1$.

Isotope	n_d^h	n_p^h	δ	
^{16}O	0.5575	0.4837	0.0412	LD
^{18}O	0.5574	0.4838	0.0412	LD
^{16}O	0.5932	0.5039	0.0971	UD
^{18}O	0.5931	0.5040	0.0972	UD
^{16}O	0.6281	0.5258	0.1540	OP
^{18}O	0.6281	0.5259	0.1540	OP
^{16}O	0.6683	0.5525	0.2209	OD
^{18}O	0.6683	0.5526	0.2209	OD

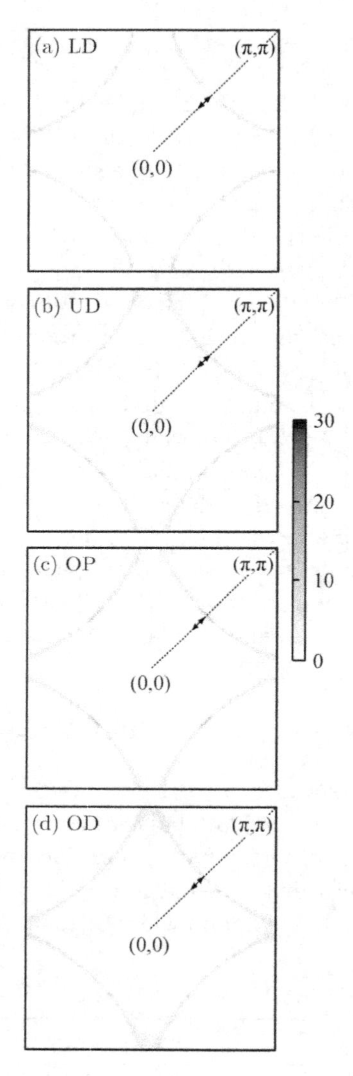

Fig. 9.1. Fermi surfaces for ^{16}O materials indicated in Table 9.1.

at each energy level. Due to this method, the curves of $E_d(\mathbf{k})$ look like a series of line segments, as shown in Fig. 9.2. We compare every $E_d(\mathbf{k})$ of our solution with the one obtained by another fully self-consistent manner, in which the same calculation is performed, except for the EPI. We show

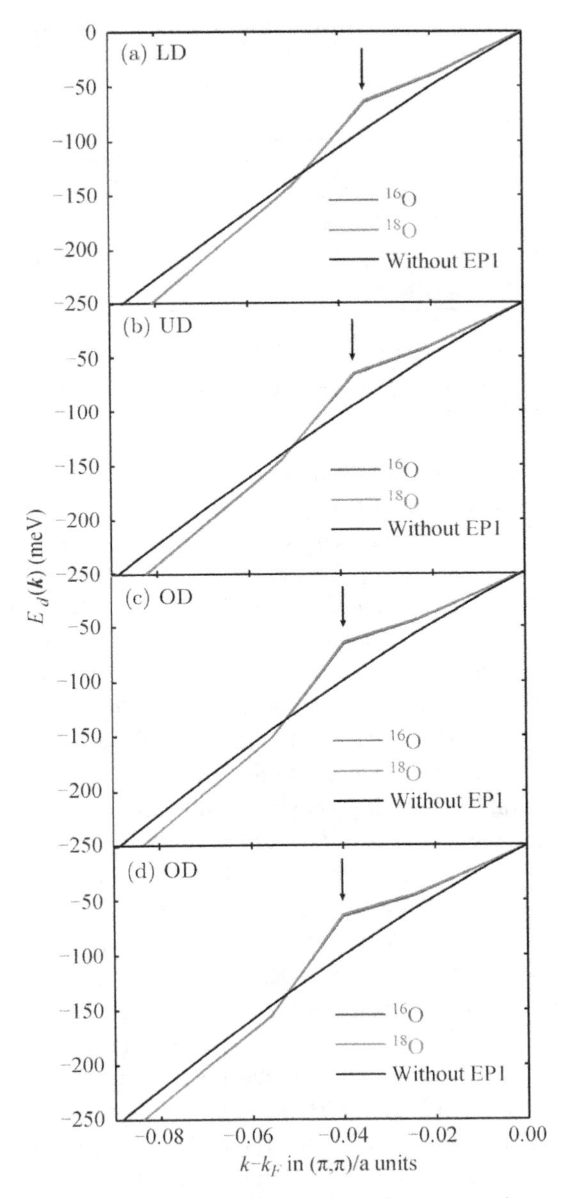

Fig. 9.2. Dispersion kinks along the nodal direction. Momentum is measured from each Fermi surface. Arrows indicate the momenta at which kinks occur.

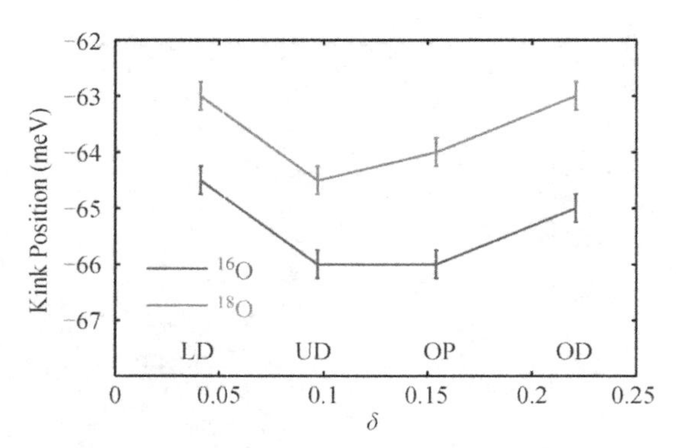

Fig. 9.3. Doping dependence of kink energies. Bars indicate the discretization error during analytic continuation.

our results on these dispersions in Fig. 9.2, where we can easily recognize that the dispersion kinks along the nodal direction appear only when EPI affects the $p^{x(y)}$ electrons. The kink energies were slightly shifted by $^{16}O \rightarrow {}^{18}O$ substitution. In Fig. 9.3, we detail how these kink energies shift depending on hole doping. These theoretically evaluated isotope shifts are at most 2.5 meV, which are much smaller than the ones measured by another group's ARPES experiment [13, 14]. Furthermore, these isotope shifts are almost independent of hole doping while another group insists that they are critically affected [15]. Considering the energy and momentum resolutions in their experiment, it may be hard to detect the subtle isotope shifts and their dependence on hole doping shown in our model.

Let us now look at λ and Δ_{dp} dependences in the dispersion kinks along the nodal direction in detail. Figure 9.4 shows the energy dispersions for $\lambda = 0.8$ (dotted lines) and $\lambda = 1.2$ (solid lines). There is no clear difference in the isotope effect between $\lambda = 1.2$ (2.5 meV) and $\lambda = 0.8$ (2.0 meV), though the dispersion kinks for $\lambda = 1.2$ are distinctly shifted to the high binding energy compared with those for $\lambda = 0.8$. On the other hand, the isotope shift for $\Delta_{dp} = 1.8$ eV (1 meV) is slightly shrank compared to that for $\Delta_{dp} = 1.4$ eV (2.5 meV), though the dispersions depend on the Δ_{dp} considerably, as shown in Fig. 9.5. Hence, our model calculation shows that the isotope shifts are not sensitive to λ and Δ_{dp}. Considering that Δ_{dp} is closely related with EEI in our three band HH model as discussed later, we can be fairly certain that the isotope shifts are determined by the relative strength between EPI and EEI. However, these changes of the isotope shifts are minute, thus, our discussions so far are valid regardless of λ and Δ_{dp}.

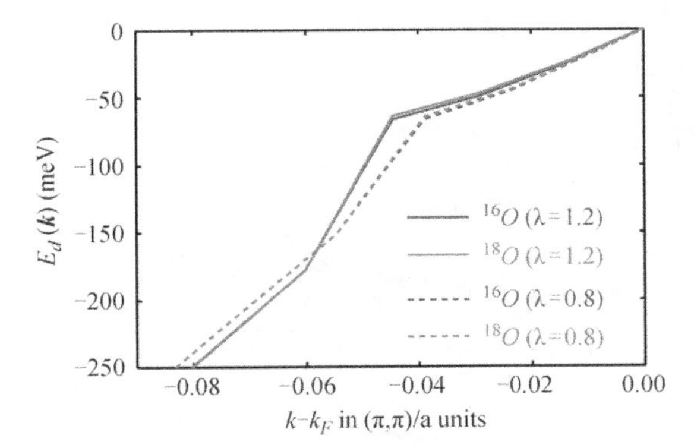

Fig. 9.4. Dispersion kinks along the nodal direction for $\lambda = 1.2$. The number of doped holes for ^{16}O (^{18}O) material is $\delta = 0.1309(0.1310)$. The dashed lines show the ones for OP in Table 9.1.

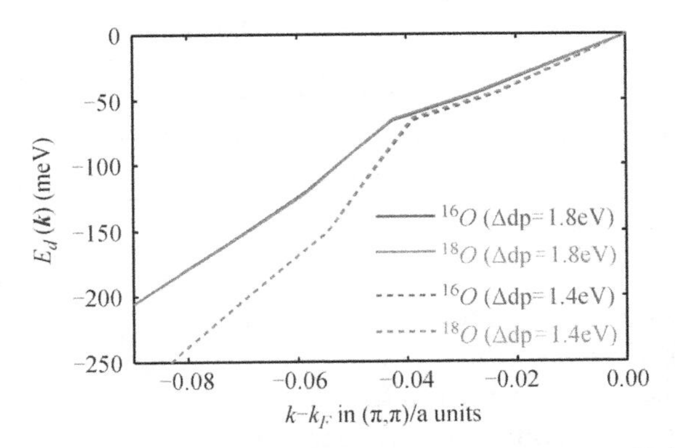

Fig. 9.5. Dispersion kinks along the nodal direction for $\Delta_{dp} = 1.8$ eV. The number of doped holes for ^{16}O(^{18}O) material is $\delta = 0.1621(0.1622)$. The dashed lines show the ones for OP in Table 9.1.

To clarify the EPI effect on the $p^{x(y)}$ electrons described above, we investigate the p-electron self-energy $\Sigma_p(\mathbf{k},\omega) \equiv \Sigma_x(\mathbf{k},\omega) + \Sigma_y(\mathbf{k},\omega)$ along the nodal direction. In Fig. 9.6, we show $\Sigma_p(k,\omega)$ on every six $\mathbf{k}$ points along the nodal direction, located inside Fermi surfaces. The energy where $\Sigma_p(\mathbf{k},\omega)$ is maximal corresponds to the one of the dispersion kink and shifts upward by $^{16}\text{O}\rightarrow^{18}\text{O}$ substitution, as shown in Fig. 9.2. The energy dependence of $\Sigma_p(\mathbf{k},\omega)$ is definitely due to the EPI introduced with the use of Eqs. (9.2.7), (9.2.11), and (9.2.12). Thus, we can conclude that in our

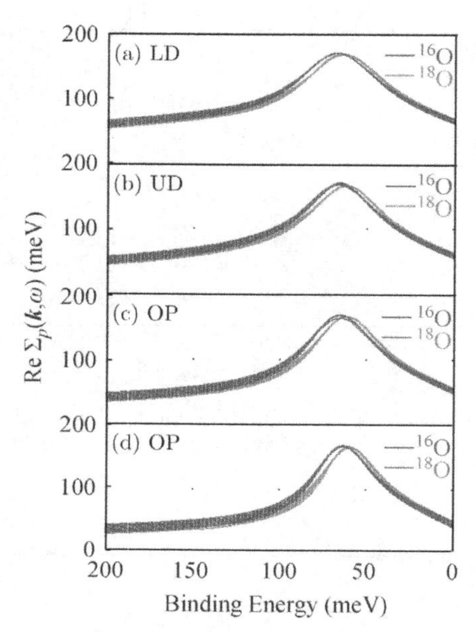

Fig. 9.6. $\Sigma_P(\mathbf{k},\omega)$ on $\mathbf{k} = (\tilde{k}_F - \frac{n}{64})(\pi,\pi)(n = 0,1,\ldots,5)$, where $\tilde{k}_F = \frac{53}{128}$ for (a), $\tilde{k}_F = \frac{57}{128}$ for (b), $\tilde{k}_F = \frac{53}{128}$ for (d), respectively.

solutions for all doping levels from the UD to the OD region, the dispersion kinks along the nodal direction are created *only when the EPI is included.*

Hereafter, we will discuss why the magnetic ingredients hardly bring the dispersion kinks along the nodal direction. Even when EPI does not exist, the electrons in our results are exposed to the strong AF fluctuation originating from the electronic correlation among d electrons. In Fig. 9.7, we show $\Sigma_d(\mathbf{k},\omega)$ on the same six $\mathbf{k}$ points as in Fig. 9.6. It is shown that there is no clear difference in $\Sigma_d(\mathbf{k},\omega)$ between ^{16}O and ^{18}O materials. The energy dependence of $\Sigma_d(\mathbf{k},\omega)$ is definitely due to the electronic correlation introduced with the use of Eqs. (9.2.8)–(9.2.10). We easily recognize that $\Sigma_d(\mathbf{k},\omega)$ uniformly increases with the binding energy and has no maximal value up to 200 meV even for LD case, in which the strong AF fluctuation is expected to be grown. Thus, along the nodal direction, the strong AF fluctuation could cause the renormalization of the Fermi velocity, however, it hardly promotes any anomalous behavior such as kink structure.

Finally, we will discuss how our EEI affects the dispersion for ^{16}O materials. When the on-site Coulomb repulsion U is changed, the Fermi velocity

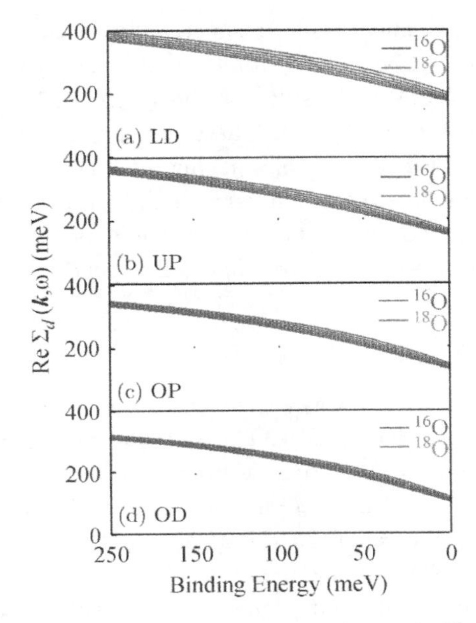

Fig. 9.7. $\Sigma_d(\mathbf{k}, \omega)$ on the same $\mathbf{k}$ points as in Fig. 9.4.

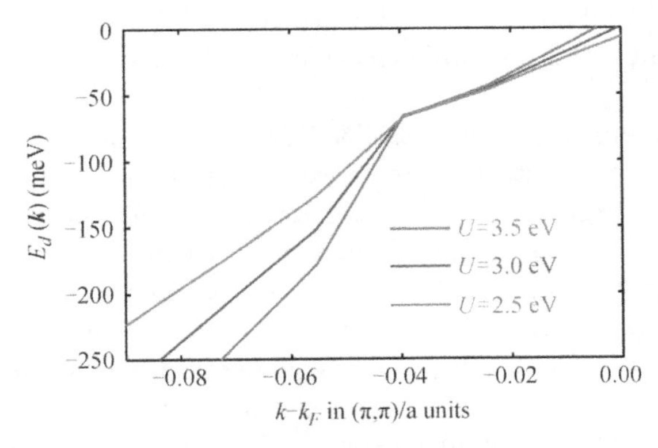

Fig. 9.8. Dispersion kinks along the nodal direction for $U = 2.5, 3.0$, and 3.5 eV. They all correspond to OP. Momentum is measured from the Fermi surface for $U = 3.0$ eV.

is renormalized differently, but this would not change the kink energy so much since the kink energy is determined by the EPI alone. In Fig. 9.8, we lay out our results for three different Us and they all correspond to OP. Their total doped holes δ are slightly different as: $\delta = 0.158, 0.154$, and

0.149 for $U = 2.5, 3.0$, and 3.5 eV, respectively. Hence, the Fermi momentum for $U = 2.5$ eV moves inside (or the Fermi surface shrinks) and the one for $U = 3.5$ eV moves outside (or the Fermi surface enlarges), compared to the one for $U = 3.0$ eV. These changes of the Fermi momenta are small; however, the dispersions at higher energy are quite affected, reflecting the binding energy dependence of $\Sigma_d(\mathbf{k}, \omega)$, as shown in Fig. 9.7. Therefore, the dispersion at higher energy could be changed a lot by the EEI even if the Fermi velocities are almost independent of them.

9.4. Conclusions

By the analysis of our model, we can show that the dispersion kink along the nodal direction occurs due to EPI. The isotope effect upon the electronic dispersion is shown near the kink of energy dispersion, not in the high binding energy portions [13, 14]. Our evaluation of the subtle isotope shifts has been backed by the report of the recent ARPES experiments [16, 17] which show the lack of the unusual isotope effect in the high energy portion [13–15]. Fortunately for us, our scenario was possibly realized in further ARPES experiment [18]. In addition to that, we have investigated how EEI effects on the nodal dispersion. It can hardly affect the kink and the nodal Fermi velocity; however, it can change the dispersion at higher energy. Hence, EPI and EEI play different roles on the nodal energy dispersion, respectively.

Of course, our treatment of EEI is just suited for *weak coupling regime*, and all of our parameter sets employed might be far from the ones for *strong coupling regime*. If we investigate *strong coupling regime* with the use of another approach, the low energy structure corresponding to the superexchange J could appear in the dispersion. However, our results presented here suggest that the structure should appear as a broad peak at higher energy due to the frequency dependence of the strong AF fluctuation, which will grow into J. As we all know, other works have already derive qualitatively similar conclusions on the basis of other models. Some groups adopt $t - J$ models [19–22] and other groups do one-band HH models [23–25]. Furthermore, other groups have succeeded in explaining the ARPES results [26–31]. However, in our 2D three-band HH model, both the electron–electron interaction among the d electrons and the EPI on p electrons are considered according to high-T_c materials. We believe that it is important that quantitatively consistent results with the ARPES experiments can be reproduced from such a model. The advantage will be when our analysis extends to the superconducting state, in which the p electrons play important roles as well as d electrons.

Acknowledgments

The authors are grateful to H. Iwasawa and T. Yanagisawa for their stimulating discussions. The computation in this work was performed on Intel Xeon servers at NeRI in AIST. This work was supported by KAKENHI (19340105).

References

[1] A. Kaminski, M. Randeria, J. C. Campuzano, M. R. Norman, H. Fretwell, J. Mesot, T. Sato, T. Takahashi, and K. Kadowaki, *Phys. Rev. Lett.* **86** (2001), 1070.

[2] A. Lanzara, P. V. Bogdanov, X. J. Zhou, S. A. Kellar, D. L. Feng, E. D. Lu, T. Yoshida, H. Eisaki, A. Fujimori, K. Kishio, J.-I. Shimoyama, T. Noda, S. Uchida, Z. Hussain, and Z.-X. Shen, *Nature (London)* **412** (2001), 510.

[3] X. J. Zhou, T. Yoshida, A. Lanzara, P. V. Bogdanov, S. A. Kellar, K. M. Shen, W. L. Yang, E. Ronning, T. Sasagawa, T. Kakeshita, T. Noda, H. Eisaki, S. Uchida, C. T. Lin, F. Zhou, J. W. Xiong, W. X. Ti, Z. X. Zhao, A. Fujimori, Z. Hussain, and Z.-X. Shen, *Nature (London)* **423** (2003), 398.

[4] Z.-X. Shen, A. Lanzara, S. Ishihara, and N. Nagaosa, *Philos. Mag.* **82** (2002), 1394.

[5] K. Terashima, H. Matsui, D. Hashimoto, T. Sato, T. Takahashi, H. Ding, T. Yamamoto, and K. Kadowaki, *Nat. Phys.* **2** (2006), 27.

[6] V. B. Zabolotnyy, S. V. Borisenko, A. A. Kordyuk, J. Fink, J. Geck, A. Koitzsch, M. Knupfer, B. Büchner, H. Berger, A. Erb, C. T. Lin, B. Keimer, and R. Follath, *Phys. Rev. Lett.* **96** (2006), 037003.

[7] A. A. Kordyuk, S. V. Borisenko, V. B. Zabolotnyy, J. Geck, M. Knupfer, J. Fink, B. Büchner, C. T. Lin, B. Keimer, H. Berger, A. V. Pan, S. Komiya, and Y. Ando, *Phys. Rev. Lett.* **97** (2006), 017002.

[8] J. Lee, K. Fujita, K. McElroy, J. A. Slezak, M. Wang, Y. Aiura, H. Bando, M. Ishikado, T. Matsui, J.-X. Zhu, A. V. Balatsky, H. Eisaki, S. Uchida, and J. C. Davis, *Nature (London)* **442** (2006), 546.

[9] R. J. McQueeney, J. L. Sarrao, P. G. Pagliuso, P. W. Stephens, and R. Osborn, *Phys. Rev. Lett.* **87** (2001), 077001.

[10] N. E. Bickers and D. J. Scalapino, *Ann. Phys. (N.Y.)* **193** (1989), 206.

[11] D. L. Feng, N. P. Armitage, D. H. Lu, A. Damascelli, J. P. Hu, P. Bogdanov, A. Lanzara, F. Ronning, K. M. Shen, H. Eisaki, C. Kim, J.-I. Shimoyama, K. Kishio, and Z.-X. Shen, *Phys. Rev. Lett.* **86** (2001), 5550.
[12] Y.-D. Chuang, A. D. Gromko, A. Fedorov, Y. Aiura, K. Oka, Y. Ando, H. Eisaki, S. I. Uchida, and D. S. Dessau, *Phys. Rev. Lett.* **87** (2001), 117002.
[13] G.-H. Gweon, T. Sasagawa, S. Y. Zhou, J. Graf, H. Takagi, D.-H. Lee, and A. Lanzara, *Nature (London)* **430** (2004), 187.
[14] G.-H. Gweon, S. Y. Zhou, M. C. Watson, T. Sasagawa, H. Takagi, and A. Lanzara, *Phys. Rev. Lett.* **97** (2006), 227001.
[15] G.-H. Gweon, T. Sasagawa, H. Takagi, D.-H. Lee, and A. Lanzara, arXiv: 0708.1027, unpublished.
[16] J. F. Douglas, H. Iwasawa, Z. Sun, A. V. Fedorov, M. Ishikado, T. Saitoh, H. Eisaki, H. Bando, T. Iwase, A. Ino, M. Arita, K. Shimada, H. Namatame, M. Taniguchi, T. Masui, S. Tajima, K. Fujita, S. Uchida, Y. Aiura, and D. S. Dessau, *Nature (London)* **446** (2007), E5.
[17] H. Iwasawa, Y. Aiura, T. Saitoh, H. Eisaki, H. Bando, A. Ino, M. Arita, K. Shimada, H. Namatame, M. Taniguchi, T. Masui, S. Tajima, M. Ishikado, K. Fujita, S. Uchida, J. F. Douglas, Z. Sun, and D. S. Dessau, *Physica C* **463–465** (2007), 52.
[18] H. Iwasawa, J. F. Douglas, K. Sato, T. Masui, Y. Yoshida, Z. Sun, H. Eisaki, H. Bando, A. Ino, M. Arita, K. Shimada, H. Namatame, M. Taniguchi, S. Tajima, S. Uchida, T. Saitoh, D. S. Dessau, and Y. Aiura, unpublished.
[19] O. Rösch and O. Gunnarsson, *Phys. Rev. Lett.* **92** (2004), 146403.
[20] S. Ishihara and N. Nagaosa, *Phys. Rev. B* **69** (2004), 144520.
[21] A. S. Mishchenko and N. Nagaosa, *Phys. Rev. Lett.* **93** (2004), 036402.
[22] A. S. Mishchenko and N. Nagaosa, *Phys. Rev. B* **73** (2006), 092502.
[23] S. Fratini and S. Ciuchi, *Phys. Rev. B* **72** (2005), 235107.
[24] G. Sangiovanni, O. Gunnarsson, E. Koch, C. Castellani, and M. Capone, *Phys. Rev. Lett.* **97** (2006), 046404.
[25] P. Paci, M. Capone, E. Cappelluti, S. Ciuchi, and C. Grimaldi, *Phys. Rev. B* **74** (2006), 205108.
[26] T. Cuk, F. Baumberger, D. H. Lu, N. Ingle, X. J. Zhou, H. Eisaki, N. Kaneko, Z. Hussain, T. P. Devereaux, N. Nagaosa, and Z.-X. Shen, *Phys. Rev. Lett.* **93** (2004), 117003.
[27] T. P. Devereaux, T. Cuk, Z. X. Shen, and N. Nagaosa, *Phys. Rev. Lett.* **93** (2004), 117004.
[28] G. Seibold and M. Grilli, *Phys. Rev. B* **72** (2005), 104519.
[29] E. G. Maksimov, O. V. Dolgov, and M. L. Kulic, *Phys. Rev. B* **72** (2005), 212505.
[30] W. Meevasana, N. J. C. Ingle, D. H. Lu, J. R. Shi, F. Baumberger, K. M. Shen, W. S. Lee, T. Cuk, H. Eisaki, T. P. Devereaux, N. Nagaosa, J. Zaanen, and Z.-X. Shen, *Phys. Rev. Lett.* **96** (2006), 157003.
[31] R. Heid, K.-P. Bohnen, R. Zeyher, and D. Mansker, *Phys. Rev. Lett.* **100** (2008), 137001.

Part III

Polaron and Bipolaron

10
Theory of High-Temperature Superconductivity in Doped Polar Insulators*

A. S. Alexandrov

Department of Physics, Loughborough University, Loughborough LE11 3TU, UK

Many high-temperature superconductors are highly polarizable ionic lattices where the Fröhlich electron–phonon interaction (EPI) with longitudinal optical phonons creates an effective attraction of doped carriers virtually equal to their Coulomb repulsion. The general multipolaron theory is given with both interactions being strong compared with the carrier kinetic energy so that the conventional BCS-Eliashberg approximation is inapplicable. The many-electron system is described by the polaronic $t-J_p$ Hamiltonian with reduced hopping integral, t, allowed double on-site occupancy, large phonon-induced anti-ferromagnetic exchange, $J_p \gg t$, and a high-temperature superconducting state of small superlight bipolarons protected from clustering.

10.1. Introduction

It seems plausible that the true origin of high-temperature superconductivity is found in a proper combination of the finite-range Coulomb repulsion with a significant finite-range EPI as suggested by a growing number of experimental and theoretical studies [1]. In highly polarizable ionic lattices like cuprate superconductors both interactions are quite strong (of the order of 1 eV) compared with the low Fermi energy of doped carriers because of a

*Reprinted with the permission from A. S. Alexandrov. Original published in *Europhys. Lett.* **95** (2011), 27004.

poor screening by non or near-adiabatic carriers [2]. In those conditions the BCS-Eliashberg theory [3] breaks down because of the polaronic collapse of the electron bandwidth [4].

The many-body theory for polarons has been developed for extremely weak and strong EPI. In the weak-coupling limit this problem is reduced to the study of a structure factor of the uniform large polaron gas [5]. For strong coupling the problem is reduced to on-site [6] or inter-site [7, 8] small bipolarons on a lattice. A strong enhancement of T_c was predicted in the crossover region from the BCS-like polaronic to BEC-like bipolaronic superconductivity due to a sharp increase of the density of states in a narrow polaronic band [4], which is missing in the so-called *negative* Hubbard U model. Nevertheless the theory of dense polaronic systems in the intermediate coupling regime remains highly cumbersome, in particular, when EPI competes with strong electron correlations. Corresponding microscopic models with the on-site Hubbard repulsion and the short-range Holstein EPI have been studied using powerful numerical techniques [9, 10].

In most analytical and numerical studies mentioned above and many others both interactions are introduced as input parameters not directly related to the material. Quantitative calculations of the interaction matrix elements can be performed from pseudopotentials using the density functional theory (DFT) [11]. On the other hand, one can express the bare Coulomb repulsion and EPI through material parameters rather than computing them from first principles in many physically important cases [12]. In particular, for a polar coupling to longitudinal optical phonons (the Fröhlich EPI), whichis the major EPI in polar crystals, both the momentum dependence of the matrix element, $M(\mathbf{q})$, and its magnitude are well known, $|M(\mathbf{q})| = \gamma(q))\hbar\omega_0/\sqrt{2N}$ with a dimensionless $\gamma(q) = \sqrt{4\pi e^2/\kappa\Omega\hbar\omega_0 q^2}$, where Ω is a unit cell volume, N is the number of unit cells in a crystal, ω_0 is the optical phonon frequency, and $\kappa = \epsilon_\infty\epsilon_0/(\epsilon_0 - \epsilon_\infty)$. The high-frequency, ϵ_∞ and the static, ϵ_0 dielectric constants are both measurable in a parent polar insulator. As is well known, a two-particle bound state exists even in the weak-coupling regime, $\lambda < 0.5$, due to a quantum (exchange) interaction between two large polarons forming a *large bipolaron* [1] (λ is the familiar EPI constant of the BCS–Eliashberg theory). These weakly coupled large pairs overlap in dense systems, so that their many-particle ground state is a BCS-like superconductor with Cooper pairs (see below).

Here, the analytical multipolaron theory is given in the strong-coupling regime for highly polarizable lattices with $\epsilon_0 \gg 1$.

10.2. Generic Hamiltonian and its Canonical Transformation

The dielectric response function of strongly correlated electrons is *a priori* unknown. Hence, one has to start with a generic Hamiltonian including *unscreened* Coulomb and Fröhlich interactions operating on the same scale since any *ad hoc* assumption on their range and relative magnitude might fail,

$$H = -\sum_{ij}(T_{ij}\delta_{ss'} + \mu\delta_{ij})c_i^\dagger c_j + \frac{1}{2}\sum_{i\neq j}\frac{e^2}{\epsilon_\infty|\mathbf{m}-\mathbf{n}|}\hat{n}_i\hat{n}_j$$
$$+\sum_{\mathbf{q},i}\hbar\omega_0\hat{n}_i[u(\mathbf{m},\mathbf{q})d_{\mathbf{q}} + \text{H.c.}] + H_{ph}. \qquad (10.2.1)$$

Here, $T_{ij} \equiv T(\mathbf{m}-\mathbf{n})$ is the bare hopping integral, μ is the chemical potential, $i = \mathbf{m}, s$ and $j = \mathbf{n}, s'$ include both site $(\mathbf{m}, \mathbf{n})$ and spin (s, s') states, $u(\mathbf{m},\mathbf{q}) = (2N)^{-1/2}\gamma(q)\exp(i\mathbf{q}\cdot\mathbf{m})$, $c_i, d_{\mathbf{q}}$ are electron and phonon operators, respectively, $\hat{n}i = c_i^\dagger c_i$ is a site occupation operator, and $H_{ph} = \sum_{\mathbf{q}}\hbar\omega_0(d_{\mathbf{q}}^\dagger d_{\mathbf{q}} + 1/2)$ is the polar vibration energy.

In highly polarizable lattices with $\epsilon_0 \to \infty$ the familiar Lang-Firsov (LF) [13] canonical transformation e^S is particulary instrumental with $S = -\sum_{\mathbf{q},i}\hat{n}_i[u(\mathbf{m},\mathbf{q})d_{\mathbf{q}} - \text{H.c.}]$. It shifts the ions to new equilibrium positions changing the phonon vacuum, and removes most of *both* interactions from the transformed Hamiltonian, $\tilde{H} = e^S He^{-S}$,

$$\tilde{H} = -\sum_{i,j}(\hat{\sigma}_{ij}\delta_{ss'} + \tilde{\mu}\delta_{ij})c_i^\dagger c_j + H_{ph}, \qquad (10.2.2)$$

where $\hat{\sigma}_{ij} = T(\mathbf{m}-\mathbf{n})\hat{X}_i^\dagger\hat{X}_j$ is the renormalised hopping integral involving the multiphonon transitions described with $\hat{X}_i = \exp[\sum_{\mathbf{q}} u(\mathbf{m},\mathbf{q})d_{\mathbf{q}} - \text{H.c.}]$, and $\tilde{\mu} = \mu + E_p$ is the chemical potential shifted by the polaron level shift,

$$E_p = \frac{2\pi e^2}{\kappa}\int_{\text{BZ}}\frac{d^3q}{(2\pi)^3q^2}. \qquad (10.2.3)$$

Here, the integration goes over the Brillouin zone (BZ) and $E_p = 0.647$ eV in La_2CuO_4 [2]. The electron–phonon coupling constant is dened as $\lambda = 2E_pN(0)$. In the case of 2D carriers with a constant bare density of states, $N(0) = ma^2/2\pi\hbar^2$ per spin, Eq. (10.2.3) places cuprates in the strong-coupling regime, $\lambda \gtrsim 0.5$, if the bare band mass $m > m_e$ (here a is the in-plane lattice constant).

10.3. Weak-Coupling Regime

For comparison, let us first consider the weak-coupling limit, where not only $\lambda < 0.5$ but also the number of phonons dressing the carrier is small, $E_p/\hbar\omega_0 \ll 1$. In this limit one can expand $\hat{X}_i$ in Eq. (10.2.2) in powers of $\gamma(q)$ keeping just single-phonon transitions so that (in the momentum representation)

$$\tilde{H} \approx \sum_{\mathbf{k},s} \xi_{\mathbf{k}} c^{\dagger}_{\mathbf{k},s} c_{\mathbf{k},s} + H_{ph} + \sum_{\mathbf{q},\mathbf{k},s} \tilde{M}(\mathbf{k},\mathbf{q}) c^{\dagger}_{\mathbf{k}+\mathbf{q},s} c_{\mathbf{k},s}(d_{\mathbf{q}} - d^{\dagger}_{-\mathbf{q}}), \quad (10.3.1)$$

where $\xi_{\mathbf{k}} = E(\mathbf{k}) - \tilde{\mu}$, $E(\mathbf{k}) = -\sum_{\mathbf{m}} T(\mathbf{m}) \exp(i\mathbf{m} \cdot \mathbf{k})$ is the bare band dispersion, and $\tilde{M}(\mathbf{k},\mathbf{q}) = \gamma(q)[E(\mathbf{k}+\mathbf{q}) - E(\mathbf{k})]/\sqrt{2N}$ is the transformed EPI matrix element, renormalised by the Coulomb repulsion. There are no other interactions left in the transformed Hamiltonian since the bare Coulomb repulsion is nullied by the Fröhlich EPI.

Applying the BCS–Eliashberg formalism [3] yields the master equation for the superconducting order parameter, $\Delta(\omega_n, \mathbf{k})$,

$$\Delta(\omega_n, \mathbf{k}) = k_B T \sum_{\mathbf{k}', \omega_{n'}} \frac{\tilde{M}(\mathbf{k}, \mathbf{k}-\mathbf{k}')^2 D(\omega_n - \omega_{n'}) \Delta(\omega_{n'}, \mathbf{k}')}{\omega_{n'}^2 + \xi_{\mathbf{k}'}^2 + |\Delta(\omega_{n'}, \mathbf{k}')|^2}, \quad (10.3.2)$$

where $D(\omega_n - \omega_{n'}) = -\hbar\omega_0/[(\omega_n - \omega_{n'})^2 + \hbar^2\omega_0^2]$ is the phonon propagator and $\omega_n = \pi k_B T(2n+1)$ are the Matsubara frequencies ($n = 0, \pm 1, \pm 2, \pm 3, \ldots$). Depending on a particular shape of the band dispersion, Eq. (10.3.2) allows for different symmetries of the order parameter since EPI is not local [14]. Here we conne our analysis to a simple estimate of T_c by assuming a **k**-independent gap function, $\Delta(\omega_n)$. Then factorizing the kernel in Eq. (10.3.2) on the "mass shell", $E(\mathbf{k}') - E(\mathbf{k}) = \omega_{n'} - \omega_n$ and linearizing Eq. (10.3.2) with respect to the gap function one obtains the familiar estimate of the critical temperature, $k_B T_c \approx \hbar\omega_0 \exp[-1/(\lambda - \mu_c^*)]$, where $\mu_c^* = \lambda/(1 + \lambda L)$ is the Coulomb pseudopotential. In our case the weak-coupling BCS superconductivity with $k_B T_c \ll \hbar\omega_0$ exists *exclusively* due to the "Tolmachev–Morel–Anderson" logarithm $L = \ln(\tilde{\mu}/\hbar\omega_0) > 1$, if the EPI is retarded (i.e., $\hbar\omega_0 < \tilde{\mu}$).

10.4. Strong-Coupling Regime

Actually the number of virtual phonons in the polaron cloud is large in oxides and some other polar lattices, $E_p/\hbar\omega_0 \gg 1$ with the characteristic (oxygen) optical phonon frequency $\hbar\omega_0 \lesssim 80$ meV, so that multiphonon vertexes are essential in the expansion of the hopping operator $\hat{\sigma}_{ij}$. To deal with this challenging problem let us single out the coherent hopping in Eq. (10.2.2) averaging $\hat{\sigma}_{ij}$ with respect to the phonon vacuum, and consider the remaining terms as perturbation, $\tilde{H} = H_0 + H_{p-ph}$. Here,

$$H_0 = -\sum_{i,j}(t_{ij}\delta_{ss'} + \tilde{\mu}\delta_{ij})c_i^\dagger c_j + H_{ph} \tag{10.4.1}$$

describes free phonons and polarons coherently propagating in a narrow band with the exponentially diminished hopping integral, $t_{ij} = T(\mathbf{m} - \mathbf{n})\exp[-g^2(\mathbf{m} - \mathbf{n})]$,

$$g^2(\mathbf{m}) = \frac{1}{2N}\sum_{\mathbf{q}} \gamma(q)^2[1 - \cos(\mathbf{q} \cdot \mathbf{m})], \tag{10.4.2}$$

and

$$H_{p-ph} = \sum_{i,j}(t_{ij} - \hat{\sigma}_{ij})\delta_{ss'}c_i^\dagger c_j \tag{10.4.3}$$

is the residual polaron–multiphonon interaction, which is a perturbation at large λ. In the diagrammatic technique the corresponding vertexes have any number of phonon lines as shown in Fig. 10.1 for the second-order in H_{p-ph} polaron self-energy ($\Sigma_p \approx -E_p/2z\lambda^2$) and the phonon self-energy ($\Sigma_{ph} \approx -x\hbar\omega_0/z\lambda^2$) [15], where z is the lattice coordination number and x is the atomic density of carriers. Hence the perturbation expansion in $1/\lambda$ is applied if $\lambda \gg 1/\sqrt{2z}$ [15, 16]. Importantly there is no structural instability in the strong-coupling regime since $|\Sigma_{ph}| \ll \hbar\omega_0$ [15].

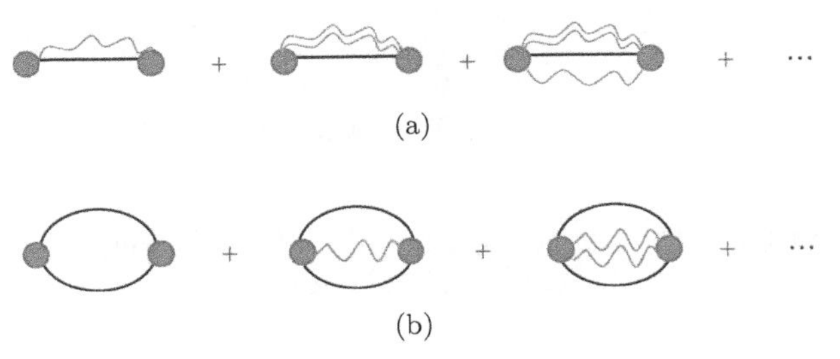

Fig. 10.1. A few diagrams contributing to the second-order in $1/\lambda$ polaron (a) and phonon (b) self-energies with multiphonon vertexes. Straight and wavy lines correspond to the polaron and phonon propagators, respectively.

The LF transformation, Eq. (10.2.2) is exact for any adiabatic ratio $\hbar\omega_0/T(a)$. However, if the perturbation expansion in $1/\lambda$ is restricted by lowest orders, then it significantly overestimates polaron masses in the adiabatic regime, $\hbar\omega_0/T(a) < 1$, for the case of the short-range (Holstein) EPI [1] (here $T(a)$ is the nearest-neighbor bare hopping integral). The polaronic band narrowing factor, $\exp(-g^2)$ becomes very small for this EPI in the strong-coupling regime, which would eliminate any possibility of high-temperature superconductivity and even metallicity of the small Hosltein polarons.

In our case of the long-range (Fröhlich) EPI, Quantum Monte Carlo simulations [17] show that the LF transformation provides numerically accurate polaron masses already in the zero order of the inverse-coupling expansion both in the adiabatic regime as well as in the non-adiabatic one for *any* strength of the Fröhlich EPI. Moreover, such small polarons [17] and small bipolarons [7] are perfectly mobile in the relevant range of the coupling and the adiabatic ratio [1].

The perturbation $H_{p\text{-}ph}$ has no diagonal matrix elements with respect to phonon occupation numbers. Hence it can be removed from the Hamiltonian in the first order using a second canonical transformation $\mathcal{H} = e^{S_2}\tilde{H}e^{-S_2}$ with $(S_2)_{n'n} = \sum_{i,j}\langle n'|(\hat{\sigma}_{ij} - t_{ij})c_i^\dagger c_j|n\rangle/(E_{n'} - E_n)$, where $E_n, E_{n'}$ and $|n\rangle, |n'\rangle$ are the energy levels and the eigenstates of H_0, respectively. Taking into account that the polaron Fermi energy is small compared with the phonon energy at strong coupling and/or sufficiently low doping [2], one can neglect its contribution to $E_{n'} - E_n \approx \hbar\omega_0 \sum_{\mathbf{q}} n'_{\mathbf{q}}$ and project the second-order in $1/\lambda$ Hamiltonian $\mathcal{H}$ onto the phonon vacuum $|0\rangle$ with the following result:

$$\mathcal{H} = -\sum_{i,j}(t_{ij}\delta_{ss'} + \tilde{\mu}\delta_{ij})c_i^\dagger c_i - \sum_{\mathbf{mnm'n'},ss'} V_{\mathbf{mn}}^{\mathbf{m'n'}} c_{\mathbf{m}s}^\dagger c_{\mathbf{n}s} c_{\mathbf{m'}s'}^\dagger c_{\mathbf{n'}s'}, \tag{10.4.4}$$

where

$$V_{\mathbf{mn}}^{\mathbf{m'n'}} = iT_{ij}T_{i'j'}\int_0^\infty dte^{-\delta t}\langle 0|[\hat{X}_i^\dagger(t)\hat{X}_j(t) - e^{-g^2(\mathbf{m-n})}]\hat{X}_{i'}^\dagger\hat{X}_{j'}|0\rangle \tag{10.4.5}$$

and $\hat{X}_i^\dagger(t)$ is the Heisenberg multiphonon operator obtained by replacing d_q in $\hat{X}_i^\dagger$ with $d_q\exp(i\omega_0 t)$. Calculating the integral, Eq. (10.4.5), with $\delta \to +0$ yields

$$V_{\mathbf{mn}}^{\mathbf{m'n'}} = \frac{t_{ij}t_{i'j'}}{\hbar\omega_0}\sum_{k=1}^{\infty}\frac{f(\mathbf{mn},\mathbf{m'n'})^k}{k!k}, \tag{10.4.6}$$

where $f(\mathbf{mn}, \mathbf{m'n'}) = (1/2N)\sum_{\mathbf{q}}\gamma(q)^2[\cos(\mathbf{q}\cdot(\mathbf{m}-\mathbf{n'}))+\cos(\mathbf{q}\cdot(\mathbf{n}-\mathbf{m}))-\cos(\mathbf{q}\cdot(\mathbf{m}-\mathbf{m'}))-\cos(\mathbf{q}\cdot(\mathbf{n}-\mathbf{n'}))]$.

10.5. Polaronic $t-J_p$ Hamiltonian

All matrix elements, Eq. (4.6), of the polaron–polaron interaction are small compared with the polaron kinetic energy except the *exchange* interaction, $J_p(\mathbf{m}-\mathbf{n}) \equiv V_{\mathbf{mn}}^{\mathbf{nm}}$ such that $f(\mathbf{mn}, \mathbf{m'n'}) = 2g^2(\mathbf{m}-\mathbf{n})$. Using $\sum_{k=1}^{\infty} y^k/k!k = -C - \ln(y) + Ei^*(y)$ with $C \approx 0.577$ and $E_i^*(y) \approx e^y/y$ (for large y) one obtains a substantial $J_p(\mathbf{m}) = T^2(\mathbf{m})/2g^2(\mathbf{m})\hbar\omega_0$, which is much larger than the polaron hopping integral, $t/J_p \propto 2\hbar\omega_0 g^2 e^{-g^2}/T(a) \ll 1$ in the strong-coupling limit. Here, t is the nearest-neighbor polaron hopping integrals. Keeping only this exchange we finally arrive with the polaronic "$t-J_p$" Hamiltonian,

$$\mathcal{H} = -\sum_{i,j}(t_{ij}\delta_{ss'} + \tilde{\tilde{\mu}}\delta_{ij})c_i^\dagger c_j + 2\sum_{\mathbf{m}\neq\mathbf{n}} J_p(\mathbf{m}-\mathbf{n})\left(\vec{S}_{\mathbf{m}}\cdot\vec{S}_{\mathbf{n}} + \frac{1}{4}\hat{n}_{\mathbf{m}}\hat{n}_{\mathbf{n}}\right), \tag{10.5.1}$$

where $\vec{S}_{\mathbf{m}} = (1/2)\sum_{s,s'} c_{\mathbf{m}s}^\dagger \vec{\tau}_{ss'} c_{\mathbf{m}s'}$ is the spin-1/2 operator ($\vec{\tau}$ are the Pauli matrices), $\hat{n}_{\mathbf{m}} = \sum_s \hat{n}_i$, and $\tilde{\tilde{\mu}} = \tilde{\mu} + \sum_{\mathbf{m}} J_p(\mathbf{m})$ is the chemical potential further renormalized by H_{p-ph}.

There is a striking difference between this polaronic $t-J_p$ Hamiltonian and the familiar $t-J$ model derived from the repulsive Hubbard U Hamiltonian in the limit $U \gg t$ omitting the so-called three-site hoppings and EPI [18]. The latter model acts in a projected Hilbert space constrained to no double occupancy. Within this standard $t-J$ model the bare transfer amplitude of electrons (t) sets the energy scale for incoherent transport, while the Heisenberg interaction ($J \propto t^2/U$) allows for spin flips leading to coherent hole motion with an effective bandwidth determined by $J \ll t$. Using the Gutzwiller-type approximation to remove the constraint results in an unconstrained $t-J$ model also containing aband narrowing, but purely electronic rather than phononic origin [19]. On the contrary in our polaronic $t-J_p$ Hamiltonian, Eq. (10.5.1), there is no constraint on the double on-site occupancy since the Coulomb repulsion is negated by the Fröhlich EPI. The polaronic hopping integral t leads to the coherent (bi)polaron band and the anti-ferromagnetic exchange of purely phononic origin J_p bounds polarons into small superlight inter-site bipolarons. Last but not least the difference is in the "+" sign in the last term of Eq. (10.5.1) proportional to $\hat{n}_{\mathbf{m}}\hat{n}_{\mathbf{n}}$, which protects the ground superconducting state from the bipolaron clustering, in contrast with the "−" sign in the similar term of the standard $t-J$ model, where the phase separation is expected at sufficiently large J [20].

The cancellation of the bare Coulomb repulsion by the Fröhlich EPI is accurate up to a $1/\epsilon_0$ correction. This correction produces a long-range residual repulsion of (bi)polarons in the transformed Hamiltonian, Eq. (10.5.1), which is small as soon as $\epsilon_0 \gg e^2/aJ_p$. The residual repulsion results in some screening of the Coulomb interactions responsible for the doping dependence of the bipolaron binding energy and of the (bi)polaron mass [2]. In layered polar insulators the static dielectric constant could be anisotropic, which together with local field corrections might result in different EPI matrix elements for in-plane and out-of-plane polarised optical phonons, respectively. The difference is not considered here.

10.6. Projection onto Bipolaronic Hamiltonian and High-T_c

The polaronic $t-J_p$ Hamiltonian, Eq. (10.5.1), is analytically solvable in the limit of sufficiently low atomic density of carriers, $x \ll 1$. Neglecting the first term in $\mathcal{H}$, which is the polaron kinetic energy proportional to $t \ll J_p$, one can readily diagonalise the remaining spin-exchange part of the Hamiltonian. Its ground state is an ensemble of inter-site singlet bipolarons with the binding energy $\Delta_b = J_p$ localised on nearest neighbor sites. Such small bipolarons repel each other and single polarons via a short-range repulsion of about J_p.

The kinetic energy operator in Eq. (10.5.1) connects singlet configurations in the first and higher orders with respect to the polaronic hopping integrals. Taking into account only the lowest-energy degenerate singlet configurations and discarding all other configurations one can project the $t-J_p$ Hamiltonian onto the inter-site bipolaronic Hamiltonian using the bipolaron annihilation operators $B_{\mathbf{m}} = 2^{-1/2}(c_{\mathbf{m}\uparrow}c_{\mathbf{m}+\mathbf{a}\downarrow} - c_{\mathbf{m}\downarrow}c_{\mathbf{m}+\mathbf{a}\uparrow})$, where $\mathbf{a}$ connects nearest neighbors [7]. These operators are similar to the bond-order operators introduced later by Newns and Tsuei [21], which are weakly coupled in their model with the single-plane lattice vibrations via a nonlinear (two-phonon) EPI (Eq. (10.4.4) in Ref. [21]). Actually it is well known that

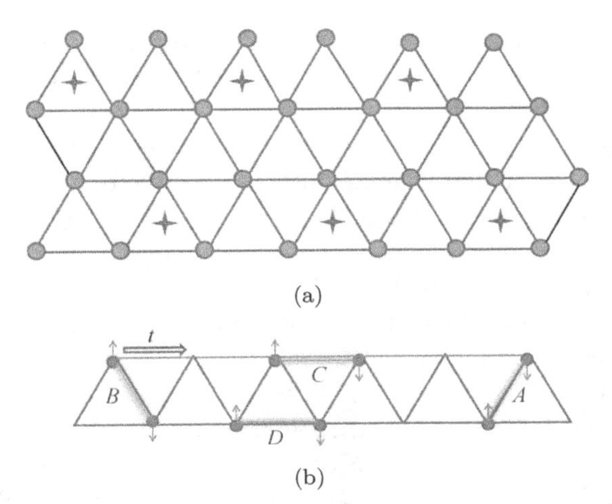

Fig. 10.2. A plane of the 3D polar lattice of anions (circles) and cations (crosses) (a) with doped carriers on anions bound by the polaronic exchange into four degenerate singlet bipolarons A, B, C and D (b).

the nonlinear anharmonic corrections to EPI are small compared with the linear Fröhlich interaction in real 3D solids, which makes the 2D model of Ref. [21] of cuprate superconductors unfeasible.

Our strong-coupling projection is illustrated using a polar lattice, sketched in Fig. 10.2(a), of anion-cation triangular planes (the in-plane lattice constant is a and the nearest-neighbor hopping distance is $a/2$) separated by the out-of-plane lattice constant c. For a zig-zag ladder-fragment of the lattice, Fig. 10.2(b), the projected bipolaronic Hamiltonian in the nearest-neighbor hopping approximation is

$$H_b = -t \sum_n B_n^\dagger A_n + D_n^\dagger B_n + D_n^\dagger A_n + C_n^\dagger B_n + A_{n+1}^\dagger C_n + A_{n+1}^\dagger B_n + B_{n-1}^\dagger A_n + C_{n-1}^\dagger A_n + \text{H.c.}, \tag{10.6.1}$$

where A, B, C, D are annihilation operators of the four degenerate singlets. Fourier transforming H_b yields four bipolaronic bands,

$$E_{1,2}(K) = -t[\cos(Ka/4) \pm \sqrt{1 + 4\sin(Ka/8)^4}],$$
$$E_{3,4}(K) = t[\cos(Ka/4) \pm \sqrt{1 + 4\cos(Ka/8)^4}]$$

with the center-of-mass momentum $\hbar K$. Expanding in powers of K one obtains the effective mass of these small singlets, $m^{**} = 10m^*$, where

$m^* = 2\hbar^2/5t(a/2)^2$ is the polaron mass. A similar Hamiltonian can be derived also for a square lattice, if next-nearest-neighbor hopping integrals are taken into account [22].

Small bipolarons are hard-core bosons with the short-range repulsion of the radius $r = a/2$ and a huge anisotropy of their effective mass since their inter-plane hopping is possible only in the second order of t [23]. The occurrence of superconductivity in bipolaronic strong-coupling systems is not controlled by the pairing strength, but by the phase coherence among the electron pairs [6]. While in two-dimensions Bose condensation does not occur in either the ideal or the interacting system, there is a phase transition to a superfluid state at $T_c = 2\pi n_b\hbar^2/k_B m^{**} \ln[\ln(1/n_b r^2)]$ in the dilute Bose gas [24, 25] (here, n_b is the boson density per unit area). Using Eqs. (10.2.3), (10.4.2) we obtain $E_p \approx 0.4 E_c$ and $g^2 \approx 0.18 E_p/\hbar\omega_0$, allowing for a quantitative estimate of T_c (here $E_c = 2e^2 c/\pi\kappa a^2$). With typical values of $a = 0.4\,\mathrm{nm}$, $c = 1.2\,\mathrm{nm}$, $\kappa = 5$, the bare band mass $m = m_e$, $\hbar\omega_0 = 80\,\mathrm{meV}$ and the moderate atomic density of polarons, $x = 0.1$ (avoiding an overlap of bipolarons) one obtains $E_p \approx 0.55\mathrm{eV}$, $g^2 \approx 1.24$, and $T_c \approx 205\,\mathrm{K}$. Importantly, the projection procedure of reducing Eq. (10.5.1) to Eq. (10.6.1) is well justified since the ratio $t/J_p \approx 0.1$ is small and $k_B T_c \ll J_p$, so that only the lowest singlet configurations can be included while discarding the others.

In conclusion, it seems very likely that a peculiar cancellation of the long-range Coulomb repulsion by the long-range Fröhlich EPI can help much in producing high-temperature superconductivity in doped polar insulators such as cuprates and other oxides, for instance BaKBiO. The polaronic $t{-}J_p$ Hamiltonian, Eq. (10.5.1) derived here from the bare long-range Coulomb interactions could provide a novel avenue for analytical and computational studies of superconductivity in complex ionic lattices since the repulsive Hubbard U model and its strong-coupling $t{-}J$ projection do not explain high T_c [26].

Acknowledgments

The author thanks A. Bratkovsky, J. Bonča, J. Devreese, H. Fehske, V. Kabanov, D. Mihailović, P. Prelovsek, J. Samson and G. Sica for helpful discussions. This work was partially supported by the Royal Society (London).

References

[1] A. S. Alexandrov and J. T. Devreese, *Advances in Polaron Physics* (Springer, Berlin, 2009).
[2] A. S. Alexandrov and A. M. Bratkovsky, *Phys. Rev. Lett.* **105** (2010), 226408.
[3] G. M. Eliashberg, *Zh. Eksp. Teor. Fiz.* **39** (1960), 1437 (*Sov. Phys. JETP* **12** (1960), 1000).
[4] A. S. Alexandrov, *Zh. Fiz. Khim.* **57** (1983), 273 (*Russ. J. Phys. Chem.* **57** (1983), 167).
[5] L. F. Lemmens, F. Brosens, and J. T. Devreese, *Phys. Status Solidi* (b) **82** (1977), 439.
[6] A. S. Alexandrov and J. Ranninger, *Phys. Rev. B* **23** (1981), 1796; **24** (1981), 1164.
[7] A. S. Alexandrov, *Physica C (Amsterdam)* **182** (1991) 327; *Phys. Rev. B* **53** (1996), 2863.
[8] S. Aubry, *J. Phys. IV, Colloq.* **03**, No. C2 (1993), C2-349; J. Bonča and S. A. Trugman, *Phys. Rev. B* **64** (2001), 094507; A. S. Alexandrov and P. E. Kornilovitch, *J. Phys.: Condens. Matter* **14** (2002), 5337; A. Macridin, G. A. Sawatzky, and M. Jarrell, *Phys. Rev. B* **69** (2004), 245111; J. P. Hague, P. E. Kornilovitch, J. Samson, and A. S. Alexandrov, *Phys. Rev. Lett.* **98** (2007), 037002.
[9] H. Fehske and S. A. Trugman, *Polarons in Advanced Materials* (Springer, Dordrecht, 2007), pp. 393–461; A. S. Mishchenko and N. Nagaosa, *Polarons in Advanced Materials* (Springer, Dordrecht, 2007), pp. 503–544.
[10] L. Vidmar, J. Bonca, S. Maekawa, and T. Tohyama, *Phys. Rev. Lett.* **103** (2009), 186401.
[11] T. Bauer and C. Falter, *Phys. Rev. B* **80** (2009), 094525.
[12] G. D. Mahan, *Many-Particle Physics* (Plenum, New York, 1990).
[13] I. G. Lang and Y. A. Firsov, *Zh. Eksp. Teor. Fiz.* **43** (1962) 1843 (*Sov. Phys. JETP* **16** (1962), 1301).
[14] A. S. Alexandrov, *Phys. Rev. B* **77** (2008), 094502.
[15] A. S. Alexandrov, *Phys. Rev. B* **46** (1992), 2838.
[16] D. M. Eagles, *Phys. Rev.* **145** (1966), 645; A. A. Gogolin, *Phys. Stat. Solidi B* **109** (1982), 95.
[17] A. S. Alexandrov and P. E. Kornilovitch, *Phys. Rev. Lett.* **82** (1999), 807.
[18] J. E. Hirsch, *Phys. Rev. Lett.* **54** (1985), 1317; J. Spalek, *Phys. Rev. B* **37** (1988), 533; C. Gros, R. Joynt, and T. M. Rice, *Phys. Rev. B* **36** (1987), 381.
[19] F. C. Zhang, C. Gros, T. M. Rice, and H. Shiba, *Supercond. Sci. Technol.* **1** (1988), 36.
[20] V. J. Emery, S. A. Kivelson, and H. Q. Lin, *Phys. Rev. Lett.* **64** (1990), 475.
[21] D. M. Newns and C. C. Tsuei. *Nat. Phys.* **3** (2007), 184.
[22] A. S. Alexandrov and P. E. Kornilovitch, *J. Phys.: Condens. Matter.* **14** (2002), 5337.

[23] A. S. Alexandrov, V. V. Kabanov, and N. F. Mott, *Phys. Rev. Lett.* **77** (1996), 4796.
[24] N. Popov, *Theor. Math. Phys.* **11** (1972), 565.
[25] D. S. Fisher and P. C. Hohenberg, *Phys. Rev. B* **37** (1988), 4936.
[26] A. S. Alexandrov and V. V. Kabanov, *Phys. Rev. Lett.* **106** (2011), 136403.

11
High-Temperature Superconductivity: The Explanation*

A. S. Alexandrov

Department of Physics, Loughborough University,
Loughborough LE11 3TU, UK

Soon after the discovery of the first high-temperature superconductor by Georg Bednorz and Alex Müller in 1986, the late Sir Nevill Mott in answering his own question 'Is there an explanation?' (*Nature* **327** (1987), 185) expressed the view that the Bose–Einstein condensation (BEC) of small bipolarons, predicted by us in 1981, could be the one. Several authors then contemplated BEC of real-space tightly bound pairs, but with a purely electronic mechanism of pairing rather than with an electron–phonon interaction (EPI). However, a number of other researchers criticized the bipolaron (or any real-space pairing) scenario as incompatible with some angle-resolved photoemission spectra, with experimentally determined effective masses of carriers and unconventional symmetry of the superconducting order parameter in cuprates. Since then, the controversial issue of whether EPI is crucial for high-temperature superconductivity or is weak and inessential has been one of the most challenging problems of contemporary condensed matter physics. Here I outline some developments in the bipolaron theory suggesting that the true origin of high-temperature superconductivity is found in a proper combination of strong electron–electron correlations with a significant finite-range (Fröhlich) EPI, and that the theory is fully compatible with key experiments.

*Reprinted with the permission from A. S. Alexandrov. Original published in *Phys. Scr.* **83** (2011) 038301.

11.1. Real Space and Cooper Pairs

There is still little consensus on the origin of high-temperature superconductivity in cuprates [1] and other related compounds. The only consensus that exists is that charge carriers are bound into pairs with an integer spin. Pairing of two fermionic particles has been evidenced in cuprate superconductors [2] from the quantization of magnetic flux in units of the flux quantum $\phi_0 = h/2e$.

A long time ago, London suggested that the remarkable superfluid properties of ^{4}He were intimately linked to the Bose–Einstein condensation (BEC) of the entire assembly of Bose particles [3]. The crucial demonstration that superfluidity was linked to the Bose particles and the BEC came after experiments on liquid ^{3}He, whose atoms were fermions, which failed to show the characteristic superfluid transition within a reasonable wide-temperature interval around the critical temperature for the onset of superfluidity in ^{4}He. In sharp contrast, ^{3}He becomes a superfluid only below a very low temperature of some 0.0026K. Here, we have a superfluid formed from *pairs* of two ^{3}He fermions below this temperature.

The three orders of magnitude difference between the critical superfluidity temperatures of ^{4}He and ^{3}He kindles the view that the BEC might represent the 'smoking gun' of high-temperature superconductivity [4]. Unfortunately electrons are fermions. Therefore, it is not surprising at all that the first proposal for high-temperature superconductivity, made by Ogg Jr in 1946 [5], was the pairing of individual electrons. If two electrons are chemically coupled together, the resulting combination is a boson with the total spin $S = 0$ or $S = 1$. Thus, an ensemble of such two-electron entities can, in principle, be condensed into the Bose–Einstein superconducting condensate. This idea was further developed as a natural explanation of superconductivity by Schafroth [6] and by Blatt and Butler in 1955 [7].

However, with one or two exceptions [8], the Ogg–Schafroth picture was condemned and practically forgotten because it neither accounted quantitatively for the critical parameters of "old" (i.e., low T_c) superconductors, nor did it explain the microscopic nature of the attractive force that could overcome the natural Coulomb repulsion between two electrons that constitute a Bose pair. The same model that yields a rather precise estimate of the critical temperature of ^{4}He leads to an utterly unrealistic result for superconductors, namely $T_c = 10^4$K with the atomic density of electron pairs of about 10^{22} per cm^3 and with the effective mass of each boson twice the electron mass, $m^{**} = 2m_e$.

The failure of this "bosonic" picture of individual electron pairs became fully transparent when Bardeen *et al.* [9] proposed that two electrons in a superconductor were indeed correlated in real space, but on a very large (practically macroscopic) coherence length ξ of about 10^4 times the average inter-electron spacing. Bardeen–Cooper–Schrieffer (BCS) theory was derived from an early demonstration by Fröhlich [10] that conduction electrons in states near the Fermi energy attract each other on account of their weak interaction with vibrating ions of a crystal lattice. Cooper then showed that any two electrons were paired in the momentum space due to their quantum interaction (i.e the Pauli exclusion principle) with all other electrons in the Fermi surface. These Cooper pairs strongly overlap in the real space, in sharp contrast with the model of non-overlapping real-space pairs discussed in Refs. [5–7]. Highly successful for metals and alloys with a low T_c, the BCS theory led some theorists (e.g. Ref. [11]) to the conclusion that there should be no superconductivity above 30K. Whereas the Ogg–Schafroth phenomenology predicted unrealistically high values of T_c, the BCS theory left perhaps only a limited hope for the discovery of new materials that could superconduct at liquid-nitrogen or higher temperatures.

11.2. Strong-Coupling Superconductivity Beyond the BCS–Migdal–Eliashberg Approximation

It became clear now that the Ogg–Schafroth and BCS descriptions are actually two opposite extremes of the electron–phonon interaction (EPI). On a phenomenological level, Eagles [12] proposed pairing without superconductivity in some temperature range solving simultaneous equations for the BCS gap and for the Fermi energy when an electron–electron attraction becomes greater than some critical value, and that superconductivity sets in at a lower temperature, of the order of the BEC temperature of the pairs in some low-carrier-density compounds like $SrTiO_3$. Lateron, extending the BCS theory to the strong interaction between electrons and ion vibrations, a Bose *liquid* of tightly bound electron pairs surrounded by the lattice deformation (i.e., of *small bipolarons*) was predicted [13, 14]. A further prediction was that high-temperature superconductivity should exist in the crossover region of the EPI strength from the BCS-like to bipolaronic superconductivity [15] (Fig. 11.1).

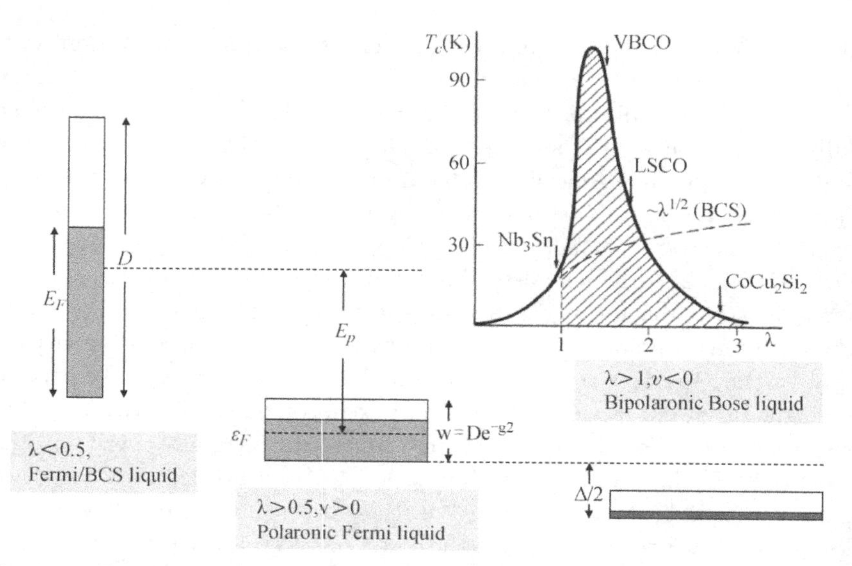

Fig. 11.1. Breakdown of the BCS–ME approximation due to the polaron bandwidth collapse at some critical coupling λ of the order of 1 (here, v is the polaron–polaron interaction). The inset shows the dependence of T_c versus λ in the polaron–bipolaron crossover region compared with the BCS–ME dependence (dotted line). (Reproduced from Ref. [93]. © American Physical Society, 1988.)

Compared with the early Ogg–Schafroth view, two fermions (now small polarons) are bound into a small bipolaron by the lattice deformation. Such bipolaronic states, at first sight, have a mass too large to be mobile. Actually, earlier studies [16, 17] considered small bipolarons as entirely localized objects. However, it has been shown analytically [13, 15, 18, 19] and using different numerical techniques [20, 21] that small bipolarons are itinerant quasi-particles existing in the Bloch states at temperatures below the characteristic phonon frequency which coherently tunnel through the lattice with a reasonable effective mass in particular if the EPI is finite-range. As a result, the superconducting critical temperature, proportional to the inverse mass of a bipolaron, is reduced in comparison with the "ultra-hot" local-pair Ogg–Schafroth superconductivity, but turns out to be much higher than the BCS prediction (Fig. 11.1). A strong enhancement of T_c in the crossover region from BCS-like polaronic to BEC-like bipolaronic superconductivity is entirely due to a sharp increase of the density of states in a narrow polaronic band [15], which is missing in the so-called *negative* Hubbard U model [22, 23]. Quite remarkably Bednorz and Müller noted in their Nobel Prize lecture that in their ground-breaking search for high-T_c superconductivity, they were stimulated and guided by the polaron model. Their expectation was that if 'an electron and a surrounding lattice distortion with a high effective mass can travel through the lattice as a whole, and a strong electron–lattice coupling exists, an insulator could be turned into a high-temperature superconductor' [24].

After we showed [15] — unexpectedly for many researchers — that the BCS–Migdal–Eliashberg (BCS–ME) theory breaks down at the EPI coupling $\lambda \gtrsim 0.5$ for any adiabatic ratio $h\omega_0/E_F$, multipolaron physics gained particular attention [25]. The parameter $\lambda\hbar\omega_0/E_F$, which is supposed to be small in the BCS–ME theory [26, 27], becomes in fact large at $\lambda \gtrsim 0.5$ since the electron bandwidth is narrowed and the Fermi energy, E_F, is renormalized down exponentially below the characteristic phonon energy, $\hbar\omega_0$ (Fig. 11.1) [4]. Nevertheless, as noted in the unbiased comment by Jorge Hirsch [28], in order to explain the increasingly higher T_cs found in supposedly 'conventional' materials, values of the electron–phonon coupling constant λ larger than 1 have been used in the conventional BCS–ME formalism. This formalism completely ignores the polaronic collapse of the bandwidth, but regrettably continues to be used by some researchers irrespective of whether λ is small or large.

11.3. Key Pairing Interaction and Unconventional Symmetry of the Order Parameter

In general, the pairing mechanism of carriers could be not only "phononic" as in the BCS theory or its strong-coupling bipolaronic extension [4] but also "excitonic", "plasmonic", "magnetic", "kinetic" or due to purely repulsive Coulomb interaction combined with an unconventional pairing symmetry of the order parameter [4]. Actually, following the original proposal by Anderson, many authors ([29] and references therein) assumed that the electron–electron interaction in novel superconductors was very strong but repulsive and provided high T_c without phonons via e.g. superexchange, spin fluctuations, excitons or any other non-phononic mechanism. A motivation for this concept can be found in the earlier work by Kohn and Luttinger (KL) [30], who showed that the Cooper pairing of fermions with a weak hard-core repulsion was possible in a finite orbital momentum state. However, the same work showed that T_c of hard-core repulsive fermions was well below the mK scale, and more importantly the KL pairing with moderate values of angular momenta (p or d) was impossible for *charged* fermions with the realistic finite-range Coulomb repulsion [31, 32] in disagreement with some recent claims [33]. Also advanced simulations with a (projected) BCS-type trial wave function [34], using the sign-problem-free Gaussian–Basis Monte Carlo algorithm (GBMC), showed that the simplest repulsive Hubbard model did not account for high-temperature superconductivity in the intermediate and strong-coupling regimes either.

On the other hand, some density functional theory (DFT) calculations [35, 36] found small EPI insufficient to explain high critical temperatures within the BCS–ME framework, while other first-principles studies found large EPI in cuprates [37] and in recently discovered iron-based compounds [38]. It is commonplace that DFT underestimates the role of the Coulomb correlations and non-adiabatic effects, predicting an anisotropy of electron-response functions much smaller than that experimentally observed in the layered high-T_c superconductors. Adiabatic DFT calculations could not explain the optical infrared c-axis spectra and the corresponding electron–phonon coupling in the metallic state of the cuprates. On the other hand, these spectra are well described within the non-adiabatic response approach of [37]. There is a strong nonlocal polar EPI along the c-axis in the cuprates together with optical conductivity as in an ionic insulator even in the well-doped "metallic" state [37]. The inclusion of a short-range repulsion (Hubbard U) via the LDA+U algorithm [39] also significantly enhances the EPI strength due to a poor screening of some particular phonons. Substantial isotope effects on the carrier mass and a number of other independent observations (see e.g. [40] and references therein) unambiguously show that lattice vibrations play a significant although unconventional role in high-temperature superconductors. Overall, it seems plausible that the true origin of high-temperature superconductivity should be found in a proper combination of strong electron–electron correlations with a significant EPI [41].

We have recently calculated the EPI strength, the phonon-induced electron–electron attraction, and the carrier mass renormalization in layered superconductors at different dopings using a continuum approximation for the renormalized carrier energy spectrum and the RPA dielectric response function [42].

If, for instance, we start with a parent insulator as La_2CuO_4, the magnitude of the Fröhlich EPI is unambiguously estimated using the static, ϵ_s, and high-frequency, ϵ_∞, dielectric constants [18, 43]. To assess its strength, one can apply an expression for the polaron binding energy (polaronic level shift) E_p, which depends only on the measured ϵ_s and ϵ_∞:

$$E_p = \frac{e^2}{2\epsilon_0 \kappa} \int_{\mathrm{BZ}} \frac{d^3q}{(2\pi)^3 q^2}. \tag{11.3.1}$$

Here, the integration is over the Brillouin zone (BZ), $\epsilon_0 \approx 8.85 \times 10^{-12}\,\mathrm{Fm}^{-1}$ is the vacuum permittivity and $\kappa = \epsilon_s \epsilon_\infty / (\epsilon_s - \epsilon_\infty)$. In the parent insulator, the Fröhlich interaction alone provides the binding energy of two holes, $2E_{\mathrm{p}}$, an order of magnitude larger than any magnetic interaction ($E_p = 0.647\mathrm{eV}$ in La_2CuO_4 [43]). Actually, Eq. (11.3.1) underestimates the polaron binding energy, since the deformation potential and/or molecular-type (e.g. [44]) EPIs are not included.

It was argued earlier [18] that the interaction with c-axis polarized phonons in cuprates would also remain strong at finite doping due to a poor screening of high-frequency electric forces as confirmed in some pump–probe [45, 46] and photoemission [47, 48] experiments. However, a quantitative analysis of the doping-dependent EPI remained elusive because the dynamic dielectric response function, $\epsilon(\omega, \mathbf{q})$, was unknown. Recent observations of the quantum magnetic oscillations in some underdoped [49] and overdoped [50] cuprate superconductors are opening up the possibility for a quantitative assessment of EPI in these and related doped ionic lattices with the quasi-2D carrier energy spectrum. The oscillations revealed cylindrical Fermi surfaces, enhanced effective masses of carriers (ranging from $2m_{\mathrm{e}}$ to $6m_{\mathrm{e}}$) and the astonishingly low Fermi energy, E_F, which appears to be well below 40 meV in underdoped Y–Ba–Cu–O [49] and less than or about 400 meV in heavily overdoped Tl2201 [50]. Photoemission spectroscopies ([47, 48] and references therein) do not show small Fermi-surface pockets and there are alternative interpretations of slow magnetic oscillations in underdoped cuprates unrelated to Landau quantization, for example [51]. However, a poorly screened strong EPI [42] is not sensitive to particular band structures and Fermi surfaces, but originates in the low Fermi energy, which is supported by other independent experiments [52]. Such low Fermi energies make the ME adiabatic approach to EPI inapplicable in these compounds. Indeed, the ME non-crossing approximation breaks down at $\lambda\hbar\omega_0/E_F > 1$ when the crossing diagrams become important. The characteristic oxygen vibration energy is about $\hbar\omega_0 = 80$ meV in oxides; therefore the ME theory cannot be applied even for a weak EPI with the coupling constant $\lambda < 0.5$. In the strong coupling regime, $\lambda \gtrsim 0.5$, the effective parameter $\lambda\hbar\omega_0/E_F$ becomes large irrespective of the adiabatic ratio, $\hbar\omega_0/E_F$, because the Fermi energy shrinks exponentially due to the polaron narrowing of the band [15]. Since carriers in cuprates are in the non-adiabatic (underdoped) or near-adiabatic (overdoped) regimes, $E_F \lesssim \hbar\omega_0$, their energy spectrum renormalizes by EPI and the polaron–polaron interactions can be found with the familiar small-polaron canonical transformation at *any coupling* λ [53].

With doping the attraction and the polaron mass drop [42]. Nevertheless, on-site and inter-site attractions induced by EPI remain well above the superexchange (magnetic) interaction J (about 100 meV) at any doping since the non-adiabatic carriers cannot fully screen high-frequency electric fields. The polaron mass [42] agrees quite well with the experimental masses [49, 50]. Decreasing the phonon frequency lowers the attraction and increases the polaron mass in underdoped compounds with little effect on both quantities at overdoping. Hence, the Fröhlich EPI with high-frequency optical phonons turns out to be the key pairing interaction in underdoped cuprates and remains the essential player at overdoping. What is more surprising is that EPI is clearly beyond the BCS–ME approximation since its magnitude is larger than or comparable with the Fermi energy and the carriers are in the non-adiabatic or near-adiabatic regimes.

Together with the deformation potential and Jahn–Teller EPIs, the Fröhlich EPI overcomes the direct Coulomb repulsion at distances comparable with the lattice constant even without any retardation [15]. Since EPI is not local in the non-adiabatic electron system with poor screening, it can provide the d-wave symmetry of the pairing state [54]. Remarkably, the internal symmetry of an individual bipolaron in the underdoped regime can be different from the symmetry of the BEC if the pairing takes place with non-zero center-of-mass momentum [55]. All these conditions point to a crossover from bipolaronic to polaronic superconductivity [15] in cuprates with doping.

11.4. Pseudogap, Superconducting Gap, Angle-Resolved Photoemission Spectroscopy and Tunnelling Spectra of Cuprate Superconductors

A detailed microscopic theory capable of describing unusual angle-resolved photoemission spectroscopy (ARPES) and tunnelling data has so far remained elusive. Soon after the discovery of high-T_c superconductivity [1], a number of tunnelling, photoemission, optical, nuclear spin relaxation and electron-energy-loss spectroscopies revealed an anomalously large gap in cuprate superconductors existing well above the superconducting critical temperature, T_c. The gap, now known as the pseudogap,was originally assigned [56] to the binding energy of real-space preformed hole pairs — small bipolarons — bound by a strong EPI. Since then, many alternative explanations of the pseudogap have been proposed.

The present-day scanning tunnelling spectroscopy (STS) [57–59], intrinsic tunnelling spectroscopy [60] and ARPES [61, 62] have offered a tremendous advance in the understanding of the pseudogap phenomenon in cuprates and some related compounds. Both extrinsic (see [57, 59] and references therein) and intrinsic [60] tunnelling as well as high-resolution ARPES [61] have found another energy scale, reminiscent of a BCS-like "superconducting" gap that opens at T_c accompanied by the appearance of Bogoliubov-like quasi-particles [61] around the node. Earlier experiments with a time-resolved pump–probe demonstrated two distinct gaps, one a temperature-independent pseudogap and the other a BCS-like gap [63]. Another remarkable observation is the spatial nanoscale inhomogeneity of the pseudogap observed with STS [57–59] which is presumably related to an unavoidable disorder in doped cuprates. Essentially, the doping and magnetic field dependence of the superconducting gap compared with the pseudogap and their different real-space profiles have prompted the opinion that the pseudogap is detrimental to superconductivity [60].

Without a detailed microscopic theory that could describe highly unusual tunnelling and ARPES spectra, the relationship between the pseudogap and the superconducting gap has remained a mystery [61]. Recently, we have developed the bipolaron theory of ARPES [64] and tunnelling [65, 66] by taking into account real-space pairing, coherence effects in a single-particle excitation spectrum and disorder. Our theory accounts for major peculiarities in extrinsic and intrinsic tunnelling in cuprate superconductors and in ARPES.

Real-space pairs, whatever the pairing interaction, can be described as a charged Bose liquid on a lattice if the carrier density is relatively small avoiding their overlap [4]. The superfluid state of such a liquid is the true BEC, rather than a coherent state of overlapping Cooper pairs. Single-particle excitations of the liquid are thermally excited single polarons propagating in a doped insulator band or are localized by impurities. Different from the BCS case, their *negative* chemical potential, μ, is found outside the band by about half of the bipolaron binding energy, Δ_p, both in the superconducting and normal states [4]. Here, in the superconducting state ($T < T_c$), following [67] one takes into account that polarons interact with the condensate via the same potential that binds the carriers. As in the BCS case, the single-quasi-particle energy spectrum, ϵ_ν, is found using the Bogoliubov transformation, $\epsilon_\nu = [\xi_\nu^2 + \Delta_{c\nu}^2]^{1/2}$. However, this spectrum is different from the BCS quasi-particles because the chemical potential is negative with respect to the bottom of the single-particle band, $\mu = -\Delta_p$. A single-particle gap, Δ, is defined as the minimum of ϵ_ν. Without disorder, for a point-like pairing potential with the s-wave coherent gap, $\Delta_{\mathbf{ck}} \approx \Delta_c$, one has [67] $\Delta(T) = [\Delta_p^2 + \Delta_c(T)^2]^{1/2}$. The full gap varies with temperature from $\Delta(0) = [\Delta_p^2 + \Delta_c(0)^2]^{1/2}$ at zero temperature down to the temperature-independent $\Delta = \Delta_p$ above T_c, which qualitatively describes some earlier and more recent [60] observations including the Andreev reflection in cuprates (see [67] and references therein).

To calculate ARPES and the tunnelling conductance, we adopted the first-principle 'LDA+GTB' band structure [68] amended with impurity band-tails [64] (Fig. 11.2). It explains the charge-transfer gap, $E_{\rm ct}$, sharp "quasi-particle" peaks near $(\pi/2, \pi/2)$ of the Brillouin zone and a high-energy 'waterfall' observed by ARPES in underdoped cuprate superconductors [64]. The chemical potential is found in the single-particle

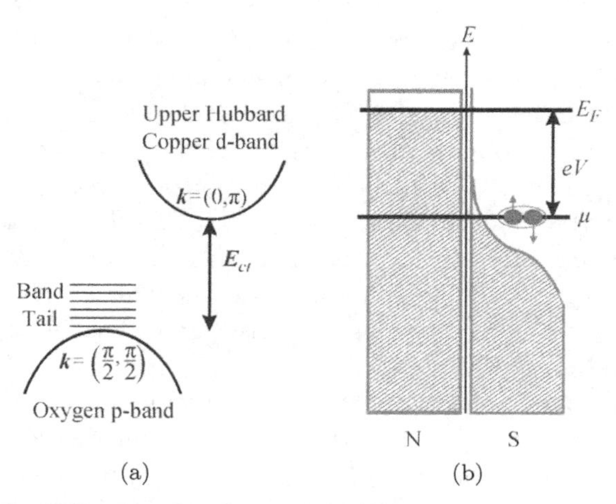

Fig. 11.2. LDA+GTB energy band structure of the cuprates, with the impurity localized states [64] shown as horizontal lines in (a); NS model densities of states (b), showing the band-tail in the bosonic superconductor. (Reproduced from Ref. [65]. © American Physical Society, 2010.)

band-tail within the charge-transfer gap at the bipolaron mobility edge (Fig. 11.2(b)), in agreement with the S-N-S tunnelling experiments [69]. Such a band structure explains an insulating-like low-temperature normal-state resistivity as well as many other unusual normal-state properties of underdoped cuprates [4].

The bipolaron theory captures key unusual signatures of the experimental tunnelling conductance in cuprates, such as the low-energy coherent gap, the high-energy pseudogap and the asymmetry [65]. In the case of atomically resolved STS, one should use a *local* band-tail DOS $\rho(E, \mathbf{r})$, which depends on different points of the scan area $\mathbf{r}$ due to a nonuniform dopant distribution, rather than an averaged DOS or spectral functions measured with ARPES. As a result the pseudogap shows nanoscale inhomogeneity, while the low-energy coherent gap is spatially uniform, as observed in [59]. Increasing the doping level tends to diminish the bipolaron binding energy, Δ_p, since the pairing potential becomes weaker due to a partial screening of EPI with low-frequency phonons [42, 70]. However, the coherent gap, Δ_c, which is the product of the pairing potential and the square root of the carrier density [67], can remain almost a constant or even increase with doping, as also observed in Ref. [59].

11.5. Concluding Remarks on Lattice (bi)polarons in High-Temperature Superconductors

A growing number of observations tell us that high-T_c cuprate superconductors [1] are not the conventional BCS superconductors [9], but represent a realization of strong-coupling non-adiabatic polaronic and bipolaronic superconductivity [4]. The fundamental origin of such a strong departure of superconducting cuprates from conventional BCS behaviour stems from the poorly screened Fröhlich EPI of the order of 1 eV, routinely neglected in the Hubbard U and t–J models [29]. This interaction with optical phonons is poorly screened because the charge carriers are found in the non-adiabatic or near adiabatic regime with their plasmon frequency below or near the characteristic frequency of optical phonons. Since screening is poor, the magnetic interaction remains small compared with the Fröhlich EPI at any doping of cuprates. Consequently, to generate an adequate theory of high-temperature superconductivity, finite-range Coulomb repulsion and the Fröhlich EPI must be treated on an equal footing. When both the interactions are strong compared with the kinetic energy of carriers, our theory predicts the low-energy state in the form of mobile inter-site bipolarons at underdoping and mobile small polarons at overdoping [4].

There is abundant independent evidence in favour of (bi)polarons [25] and 3D BEC in cuprate superconductors [71, 72]. The substantial isotope effect on the carrier mass [73–76] predicted for (bi)polaronic conductors in [77] is the most compelling evidence for (bi)polaronic carriers in cuprate superconductors. High-resolution ARPES [47, 48, 78] provides another piece of evidence for the strong EPI in cuprates and related doped layered compounds apparently with c-axis-polarized optical phonons. These, as well as tunnelling spectroscopies of cuprates [79, 80] and recent pump–probe experiments [45, 46], unambiguously show that the Fröhlich EPI is important in highly polarizable ionic lattices.

A parameter-free estimate of the Fermi energy using the magnetic-field penetration depth [52] and the magnetic quantum oscillations [49] found its very low value, $\epsilon_F \lesssim 50$ meV, supporting the real-space pairing in underdoped cuprate superconductors.

Magnetotransport and thermal magnetotransport data strongly support preformed bosons in cuprates. In particular, many high-magnetic-field studies revealed a non-BCS upward curvature of the upper critical field $H_{c2}(T)$ (see [81] for a review of experimental data), predicted for the BEC of charged bosons in the magnetic field [82]. The Lorenz number, $L = \kappa_e/T\sigma$, differs significantly from the Sommerfeld value L_e of the standard Fermi-liquid theory if carriers are double-charged bosons [83]. Here κ_e and σ are electron thermal and electrical conductivities, respectively. Alexandrov and Mott [83] predicted a rather low Lorenz number for bipolarons, $L \approx 0.15L_e$, due to the double elementary charge of bipolarons and also due to their nearly classical distribution function above T_c. Direct measurements of the Lorenz number using the thermal Hall effect [84] produced a value of L just above T_c, almost the same as predicted by the bipolaron model.

Single polarons, *localized* within an impurity band-tail, coexist with bipolarons in the charge-transfer doped Mott–Hubbard insulator. They account for sharp "quasi-particle" peaks near $(\pi/2, \pi/2)$ of the Brillouin zone and high-energy "waterfall" effects observed with ARPES in cuprate superconductors [64]. This 'band-tail' model also accounts for two energy scales in ARPES and in the extrinsic and intrinsic tunnelling, their temperature and doping dependence, and for the asymmetry and inhomogeneity of extrinsic tunnelling spectra of cuprates [65]. On the other hand, essentially different doping and magnetic field dependence of the superconducting gap compared with the pseudogap and their different real-space profiles have prompted the opinion [60] that the pseudogap is not connected with the so-called "preformed" Cooper pairs advocated by Emery and Kivelson [85] as an alternative to the BEC of real-space pairs. The unusual normal state diamagnetism uncovered by torque magnetometry has also been convincingly explained as the normal state (Landau) diamagnetism of charged bosons [86].

Overall the real-space pairing seems to be a remarkable feature of cuprates no matter what the microscopic pairing mechanism is. A lattice disorder introduces additional complexity to the problem since an interference of impurity potential with the lattice distortion, which accompanies the polaron movement, contributes to polaron and bipolaron localization [87]. Self-organized discrete dopant networks [88] lead to multiscale complexity for key materials as well. However, the detailed microscopic physics of the bosonic many-body state seems to be irrelevant for fitting their electrodynamic properties [89].

As was emphasized in a number of early qualitative [71] (1990s) and more recent numerical studies [90, 91] of strongly correlated electrons with a significant EPI, the anti-ferromagnetic spin fluctuations facilitate a doping-induced lattice polaron formation profoundly lowering the required strength of the bare EPI, but playing virtually no role in pairing compared with the EPI [42, 92]. It is quite surprising that despite clear evidence for lattice polarons in cuprate and related superconductors, there are still opinions (e.g. [29]) suggesting that EPI is inessential or that polaron formation does not help, but hinders the pairing instability, very much in contrast to the notion advanced by this and other authors that in fact lattice polarons explain high-T_c superconductivity.

References

[1] J. G. Bednorz and K. A. Müller, *Z. Phys. B* **64** (1986), 189.
[2] C. E. Gough, M. S. Colclough, E. M. Forgan, R. G. Jordan, M. Keene, C. M. Muirhead, A. I. M. Rae, N. Thomas, J. S. Abell, and S. Sutton, *Nature* **326** (1987), 855.
[3] F. London, *Phys. Rev.* **54** (1938), 947.
[4] A. S. Alexandrov, (2003) Institute of Physics Publishing.
[5] R. A. Ogg Jr., *Phys. Rev.* **69** (1946), 243.
[6] M. R. Schafroth, *Phys. Rev.* **100** (1955), 463.

[7] J. M. Blatt and S. T. Butler, *Phys. Rev.* **100** (1955), 476.
[8] P. P. Edwards, C. N. R. Rao, N. Kumar, and A. S. Alexandrov, *Chem. Phys. Chem.* **7** (2006), 2015.
[9] J. Bardeen, L. N. Cooper, and J. R. Schrieffer, *Phys. Rev.* **108** (1957), 1175–204.
[10] H. Fröhlich, *Phys. Rev.* **79** (1950), 845.
[11] M. L. Cohen and P. W, Anderson, *Superconductivity in d-and f-Band Metals* (AIP, New York, 1972), pp 17–27.
[12] D. M. Eagles, *Phys. Rev.* **186** (1969), 456.
[13] A. S. Alexandrov and J. Ranninger, *Phys. Rev. B* **23** (1981), 1796.
[14] A. S. Alexandrov and J. Ranninger, *Phys. Rev. B* **24** (1981), 1164.
[15] A. S. Alexandrov, *Zh. Fiz. Khim.* **57** (1983), 273; A. S. Alexandrov, *Russ. J. Phys. Chem.* **57** (1983), 167.
[16] P. W. Anderson, *Phys. Rev. Lett.* **34** (1975), 953.
[17] B. K. Chakraverty, *J. Physique Lett.* **40** (1979), L-99.
[18] A. S. Alexandrov, *Phys. Rev. B* **53** (1996), 2863.
[19] S. Aubry, bipolarons *Polarons in Advanced Materials* (Springer, Dordrecht, 2007), pp 311–71.
[20] J. P. Hague, P. E. Kornilovitch, J. H. Samson, and A. S. Alexandrov. *Phys. Rev. Lett.* **98** (2007), 037002.
[21] J. Bonča and S. A. Trugman, *Phys. Rev. B* **64** (2001), 094507.
[22] R. Micnas, J. Ranninger, and S. Robaszkiewicz, *Rev. Mod. Phys.* **62** (1990), 113.
[23] Q. J. Chen, J. Stajic, S. Tan, and K. Levin, *Phys. Rep.* **412** (2005), 1–88.
[24] J. G. Bednorz and K. A. Müller, *Angew. Chem. Int. Edn. Engl.* **27** (1988), 735.
[25] A. S. Alexandrov and J. T. Devreese, *Advances in Polaron Physics* (Springer, Heidelberg, 2009).
[26] A. B. Migdal, *Sov. Phys. — JETP* **7** (1958), 996–1001.
[27] G. M. Eliashberg, *Sov. Phys. — JETP* **11** (1960), 696–702.
[28] J. E. Hirsh, *Phys. Scr.* **80** (2009), 035702.
[29] P. W. Anderson, P. A. Lee, M. Randeria, T. M. Rice, N. Tiverdi, and F. C. Zhang, *J. Phys.: Condens. Matter* **16** (2004), R755.
[30] W. Kohn and J. M. Luttinger, *Phys. Rev. Lett.* **15** (1965), 524.
[31] J. M. Luttinger, *Phys. Rev.* **150** (1966), 202.
[32] A. S. Alexandrov and A. A. Golubov, *Phys. Rev. B* **45** (1992), 4769.
[33] S. Raghu, S. A. Kivelson, and D. J. Scalapino, *Phys. Rev. B* **81** (2010), 224505.
[34] T. Aimi and M. Imada, *J. Phys. Soc. Japan* **76** (2007), 113708.
[35] F. Giustino, M. L. Cohen, and S. G. Louie, *Nature* **452** (2008), 975.
[36] R. Heid, K. P. Bohnen, R. Zeyher, and D. Manske, *Phys. Rev. Lett.* **100** (2008), 137001.
[37] T. Bauer and C. Falter, *Phys. Rev. B* **80** (2009), 094525.
[38] F. Yndurain and J. M. Soler, *Phys. Rev. B* **79** (2009), 134506.
[39] P. Zhang, S. G. Louie, and M. L. Cohen, *Phys. Rev. Lett.* **98** (2007), 067005.

[40] A. S. Alexandrov and G. M. Zhao, *Phys. Rev. B* **80** (2009), 136501.
[41] A. S. Alexandrov, C. Di Castro, I. Mazin, and D. Mihailovic, *Adv. Condens. Matter Phys.* **2010** (2010), 206012.
[42] A. S. Alexandrov and A. M. Bratkovsky, *Phys. Rev. Lett.* **105** (2010), 226408.
[43] A. S. Alexandrov and A. M. Bratkovsky, *J. Phys.: Condens. Matter* **11** (1999), L531.
[44] K. A. Müller, *J. Phys.: Condens. Matter* **19** (2007), 251002.
[45] Z. Radović, N. Božović and I. Božović, *Phys. Rev. B* **77** (2008), 092508.
[46] C. Gadermaier *et al.*, *Phys. Rev. Lett.* **105** (2010), 257001.
[47] W. Meevasana *et al.*, *Phys. Rev. Lett.* **96** (2006), 157003.
[48] W. Meevasana *et al.*, *New J. Phys.* **12** (2010), 023004.
[49] N. Doiron-Leyraud *et al.*, *Nature* **447** (2007), 565.
[50] B. Vignolle *et al.*, *Nature* **455** (2008), 952.
[51] A. S. Alexandrov, *J. Phys.: Condens. Matter* **20** (2008), 192202.
[52] A. S. Alexandrov, *Physica C* **363** (2001), 231.
[53] A. S. Alexandrov and P. E. Kornilovitch, *Phys. Rev. Lett.* **82** (1999), 807.
[54] A. S. Alexandrov, *Phys. Rev. B* **77** (2008), 094502.
[55] A. S. Alexandrov, *J. Supercond.: Inc. Novel Magn.* **17** (2004), 53.
[56] A. S. Alexandrov and D. K. Ray, *Phil. Mag. Lett.* **63** (1991), 295.
[57] K. K. Gomes *et al.*, *Nature* **447** (2007), 569.
[58] J. Lee *et al.*, *Science* **325** (2009), 1099.
[59] T. Kato, T. Maruyama, S. Okitsu, and H. Sakata, *J. Phys. Soc. Japan* **77** (2008), 054710.
[60] V. M. Krasnov, *Phys. Rev. B* **79** (2009), 214510.
[61] W. S. Lee *et al.*, *Nature* **450** (2007), 81.
[62] R. H. He *et al.*, *Nat. Phys.* **5** (2009), 119.
[63] J. Demsar, B. Podobnik, V. V. Kabanov, Th. Wolf, and D. Mihailovic, *Phys. Rev. Lett.* **82** (1999), 4918.
[64] A. S. Alexandrov and K. Reynolds, *Phys. Rev. B* **76** (2007), 132506.
[65] A. S. Alexandrov and J. Beanland, *Phys. Rev. Lett.* **104** (2010), 026401.
[66] J. Beanland and A. S. Alexandrov, *J. Phys.: Condens. Matter* **22** (2010), 403202.
[67] A. S. Alexandrov and A. F. Andreev, *Europhys. Lett.* **54** (2001), 373.
[68] M. M. Korshunov, V. A. Gavrichkov, S. G. Ovchinnikov, I. A. Nekrasov, Z. V. Pchelkina, and V. I. Anisimov, *Phys. Rev. B* **72** (2005), 165104.
[69] I. Bozovic, G. Logvenov, M. A. J. Verhoeven, P. Caputo, E. Goldobin, and T. H. Geballe, *Nature* **422** (2003), 873.
[70] A. S. Alexandrov, V. V. Kabanov, and N. F. Mott, *Phys. Rev. Lett.* **77** (1996), 4796.
[71] A. S. Alexandrov and N. F. Mott, *Int. J. Mod. Phys. B* **8** (1994), 2075.
[72] A. S. Alexandrov, *J. Phys.: Condens. Matter* **19** (2007), 125216.
[73] G. M. Zhao and D. E. Morris, *Phys. Rev. B* **51** (1995), 16487.
[74] G. M. Zhao, M. B. Hunt, H. Keller, and K. A. Müller, *Nature* **385** (1997), 236.
[75] R. Khasanov *et al.*, *Phys. Rev. Lett.* **92** (2004), 057602.

[76] A. Bussmann-Holder and H. Keller, *Polarons in Advanced Materials* (Springer, Dordrecht, 2007), pp. 599–621.
[77] A. S. Alexandrov, *Phys. Rev. B* **46** (1992), 14932.
[78] A. Lanzara et al., *Nature* **412** (2001), 510.
[79] G. M. Zhao, *Phys. Rev. Lett.* **103** (2009), 236403.
[80] H. Shim, P. Chaudhari, G. I. Logvenov, and I. van Bozovic, *Phys. Rev. Lett.* **101** (2008), 247004.
[81] V. N. Zavaritsky, V. V. Kabanov, and A. S. Alexandrov, *Europhys. Lett.* **60** (1998), 127.
[82] A. S. Alexandrov, *Phys. Rev. B* **48** (1993), 10571.
[83] A. S. Alexandrov and N. F. Mott, *Phys. Rev. Lett.* **71** (1993), 1075.
[84] Y. Zhang, N. P. Ong, Z. A. Xu, K. Krishana, R. Gagnon, and L. Taillefer, *Phys. Rev. Lett.* **84** (2000), 2219.
[85] V. J. Emery and S. A. Kivelson, *Nature* **374** (1995), 434.
[86] A. S. Alexandrov, *Phys. Rev. Lett.* **96** (2006), 147003.
[87] J. C. Phillips, A. R. Bishop, and A. Saxena, *Rep. Prog. Phys.* **66** (2003), 2111.
[88] J. C. Phillips, *Adv. Condens. Matter Phys.* **2010** (2010), 250891.
[89] A. S. Alexandrov, *J. Phys.: Condens. Matter.* **22** (2010), 426004.
[90] H. Fehske and S. A. Trugman, *Polarons in Advanced Materials* (Springer, Dordrecht, 2007), pp 393–461.
[91] A. S. Mishchenko and N. Nagaosa, Spectroscopic properties of polarons in polarons in strongly correlated systems by exact diagrammatic Monte Carlo method, *Polarons in Advanced Materials* (Springer, Dordrecht, 2007), pp. 503–544.
[92] L. Vidmar, J. Bonča, S. Maekawa, and T. Tohyama, *Phys. Rev. Lett.* **103** (2009), 186401.
[93] A. S. Alexandrov, *Phys. Rev. B* **38** (1988), 925.

12
Polaronic Effect and Its Impact on T_c for Novel Layered Superconducting Systems*

V. Kresin

Lawrence Berkeley National Laboratory, University of California, 70A-3307, 1 Cyclotron Road, Berkeley, CA 94720, USA

The crystal lattice of a complex compound may contain a subsystem of ions with each one possessing two close equilibrium positions (double-well structure). For example, the oxygen ions in the cuprates form such a subsystem. In such a situation, it is impossible to separate electronic and local vibrational motions. This leads to a large increase in the effective strength of the electron–lattice interaction, which is beneficial for pairing.

12.1. Introduction

This paper is concerned with the lattice dynamics and the impact of electron–lattice interaction on the properties of novel superconductors, especially the high-T_c cuprates. More specifically, we address the question: why the presence of polaronic states is beneficial for superconductivity? As is known, polaronic effect provided the main motivation for the original search for high T_c in cuprates [1].

Consider a complex compound where one ionic subsystem is characterized by two close equilibrium positions (double-well potential). Such strongly anharmonic potential leads to a peculiar non-adiabatic polaronic

*Reprinted with the permission from V. Kresin. Original published in *J. Supercond. Nov. Magn.* **23**: (2010) 179–182.

effect; indeed, in this case the electronic and local lattice degrees of freedom turn out to be unseparable (see below). In fact, oxygen ions in the cuprates do form such a subsystem; the double-well structure has been observed experimentally in Ref. [2] and studied theoretically by Wolf and the author in Ref. [3], see also the reviews [4, 5]. One can demonstrate — and this is the goal of the present paper — that the presence of such a structure leads to a noticeable increase in the effective strength of electron–lattice coupling relative to the usual case of a single potential minimum for the ionic coordinate.

An impact of nonlinear lattice dynamics, including anharmonicity, on the pairing has been studied in a number of interesting papers (see, e.g., [6–9]); direct interaction of electrons and non-harmonic lattice has been studied. Here, our focus is different. The thing is that the pairing interaction is described by the matrix element connecting an initial and virtual states. One can show that even in the absence of direct interaction with the double-well structure, the resulting coupling constant will increase. This is due to the increased phase space for virtual transitions or, in other words, due to an increased number of these transitions. Below we describe the corresponding formalism.

12.2. Double-Well Structure; Diabatic Representation

According to the adiabatic theory, the equation

$$\begin{aligned} \hat{H}_{\vec{r}}\psi_n(\vec{r},\vec{R}) &= \varepsilon_n(\vec{R})\psi_n(\vec{r},\vec{R}); \\ \hat{H}_{\vec{r}} &= \hat{T}_{\vec{r}} + V(\vec{r},\vec{R}) \end{aligned} \tag{12.2.1}$$

determines the electronic terms $\varepsilon_n(\vec{R})$ which depend parametrically on ionic positions. Here, $\{\vec{r},\vec{R}\}$ are the electronic and ionic coordinates, $\hat{T}_{\vec{r}}$ is the operator of kinetic energy of electrons, $V(\vec{r},\vec{R})$ is the total potential energy, and $\psi(\vec{r},\vec{R})$ is the electronic wave function. The electronic term forms the potential for the ionic motion.

Consider now the special case when the electronic term for some ionic subsystem (e.g., for oxygen ions in the cuprates) contains two close minima positions. As an example, we focus on the case when the double-well potential corresponds to some direction; for layered systems this direction is perpendicular to the layers ($OZ\|c$). As for the dependence of the ionic motion

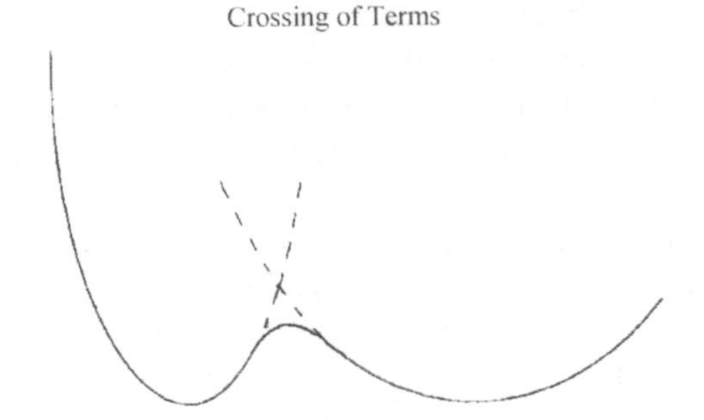

Fig. 12.1. Transformation to the diabatic representation. Solid line: initial term (double-well structure); *broken lines*: crossing terms.

on X, Y, it is described by usual harmonic dynamics with single minima equilibrium positions. Therefore, $\varepsilon_n(\vec{R}) \equiv \varepsilon_n(\vec{\rho}, Z)$ with the double-well structure in the Z direction, where $\vec{\rho}$ denotes the in-plane ionic position.

We employ the tight-binding approximation. As a first step, we should write down the local wave function. In our case, the ion is affected by the double-well potential (in Z direction; Fig. 12.1). At this stage it is very convenient to employ the diabatic representation [10–12]. As a result of the transformation, the double-well potential is replaced by two crossing harmonic energy terms (Fig. 12.1). Each of these terms contains also its vibrational manifold. Therefore, the local state is described by two groups of terms ("a" and "b"), so that

$$\Psi_{\text{loc}} = c_a \Psi_a(\vec{r}, \vec{R}) + c_b \Psi_b(\vec{r}, \vec{R}), \tag{12.2.2}$$

where $\Psi_{a(b)}(\vec{r}, \vec{R}) = \psi_{a(b)}(\vec{r}, \vec{R}_0^{a(b)}) \chi_{a(b)}(z - z_0^{a(b)})$.

Note that $\Psi_a = \psi_s \chi_{s\nu}; \Psi_b = \psi_{s'} \chi_{s'\nu'}$. Here, $\{s, s'\}$ and $\{\nu, \nu'\}$ are the electronic and vibrational quantum numbers. If both terms are equivalent, then $c_a^2 = c_b^2 = 0.5$. Because of the inter-term tunneling, the energy levels which correspond to isolated terms are split into symmetric and antisymmetric levels; the scale of the splitting is of order ε_{ab}, where (see [12])

$$\begin{aligned} \varepsilon_{ab} &= \int d\vec{R} \chi_a \hat{H}_{\vec{r};ab} \chi_b, \\ \hat{H}_{\vec{r};ab} &= \int d\vec{r} \psi_b^* \hat{H}_{\vec{r}} \psi_a. \end{aligned} \tag{12.2.3}$$

It is essential that, contrary to the usual adiabatic picture, the operator $\hat{H}_{\vec{r}}$ defined by (12.2.1) has non-diagonal terms in the diabatic representation.

One can see directly from (12.2.2) that the wave function Ψ_{loc} cannot be written as a product of electronic and ionic wave functions. In other words, the electronic and local ionic motions cannot be separated like in the usual adiabatic theory. Such a polaronic effect is a strong non-adiabatic phenomenon.

In the tight-binding approximation each local term, including those formed by the transformation (Fig. 12.1), is broadened into the energy band. Note at first that the local electronic function (see (12.2.2)) $\psi_i(\vec{r}, \vec{R}_0^i)(i = a, b)$ can be written in the form $\psi_i(\vec{r} - \vec{R}_0^n), \vec{R}_0^n$, which corresponds to the crossing point for the nth ion ($i = a, b$). Indeed, since the electronic wave function has a scale of the length of the bond that greatly exceeds the amplitude of vibrations and the distance "δ" between the minima, one can neglect the difference between $\vec{R}_0^n$ and $\vec{R}_0^i$.

The total wave function contains also the vibrational part. At first, let us separate the local vibrational mode that corresponds initially to the double-well structure, or, in the diabatic approximation, to two crossing terms (see above). As for the dependence of ionic motion on X and Y, it can be described by a set of normal modes. Then the total wave function has a form:

$$\begin{aligned} \Psi(\vec{r}, \vec{R}) &= u_{\vec{k}}(\vec{r}, \vec{R}) e^{i\vec{k}\vec{r}} \Phi_{\text{vib}}, \\ u_{\vec{k}}(\vec{r}, \vec{R}) &= \sum_n \Psi_{\text{loc}}^n(\vec{r}, \vec{R}) e^{i\vec{k}(\vec{R}_0^n - \vec{r})}. \end{aligned} \tag{12.2.4}$$

Ψ_{loc} is determined from (12.2.2). The local vibrational wave functions are centered at z_0^i and at $z_0^i - \delta$. If the terms are similar, then $\chi^b(z) = \chi^a(z - \delta)$.

One should stress a key difference between the ionic motions described by the function

$$\Phi_{\text{vib}}(R_x, R_y) = \prod \varphi(\Omega_m) \tag{12.2.5}$$

(Ω_m are the frequencies of the usual normal modes) and the function $\chi(R_z)$. There is no in-plane ionic hopping between various ionic sites. However, the oxygen can tunnel between the two minima, and this is reflected in a noticeable overlap of vibrational functions χ_i corresponding to the two crossing terms (Fig. 12.1).

12.3. Superconducting State

Let us focus on the electron–lattice interaction and its impact on pairing. We consider the case when the main contribution to the interaction is coming from the interaction with usual normal (harmonic) modes Q_m (one can easily describe a more general case). Then the interaction Hamiltonian can be written in the usual form:

$$\hat{H}' = (\partial V/\partial Q_m)\delta Q_m. \tag{12.3.1}$$

Summation over m is implied. The equation for the pairing amplitude has a form

$$\Delta(\omega_n) = T\sum_{\omega_{n'}}\int \varsigma d\xi N_F D(\omega_n - \omega_{n'}; \tilde{\Omega})F^+(\omega_{n'}). \tag{12.3.2}$$

Here, $D = \tilde{\Omega}^2[\tilde{\Omega}^2 + (\omega_n - \omega_{n'})^2]^{-1}$ is the phonon propagator, $\omega_n = (2n + 1)\pi T$; $\tilde{\Omega}$ is the characteristic phonon frequency, $F^+(\omega_n) = \Delta(\omega_n)[\omega_n^2 + \xi_{m\nu}^2 + \Delta^2(\omega_n)]^{-1}$ is the Gor'kov's pairing function [13] (see, e.g., Ref. [14]), $\xi_{n\nu}$ is the excitation energy relative to the Fermi energy, N_F is the density of states. It is essential that $\xi_{n\nu}$ is the energy of virtual transitions which contribute to pairing. Note that the integrand contains the matrix element ($\zeta = g^2$) which has a form

$$g = \int \Psi^{*f}\hat{H}'\Psi^i d\vec{r}d\vec{R}, \tag{12.3.3}$$

where $\hat{H}'$ is defined by (12.3.1), i and f denote the initial and virtual states, and the total function has a form (12.2.4).

Let us introduce some characteristic frequency $\tilde{\Omega}$ and a single mode $\tilde{Q}$ which provides a main contribution to the pairing (cf. (12.3.2)). Then the product $\prod \varphi\nu_m(Q_m)$ can be replaced by the function $\varphi_\nu(\tilde{Q})$. Correspondingly, $H' = (\partial V/\partial\tilde{Q})_0\delta\tilde{Q}$. In other words, we model the phonon spectrum as consisting of two modes. One of them, harmonic mode $\tilde{Q}$, provides a major contribution to the coupling interaction; and the second mode, which is characterized by double well structure (Fig. 12.1), provides an additional number of the virtual states. This model can be easily generalized. One can see from (12.3.3) that the average value of the coupling matrix element is a sum

$$g_{\mathrm{av}} = g_0 + g_1, \tag{12.3.4}$$

where

$$g_0 = \int d\vec{r} d\tilde{Q} \sum_n \psi_s(\vec{r} - \vec{R}_0^n) \times \psi_{s'}(\vec{r} - \vec{R}_0^n)(\partial V/\partial \tilde{Q})_0 \delta \tilde{Q}_{\mathrm{lav}},$$

$$g_1 \approx g_0 \bar{F}^{ab}, \tag{12.3.4$'$}$$

where

$$F^{ab} = 2 \sum_{s,\nu_1} \int \chi_{s\nu}(z - z_0^a) \chi_{s_1\nu'}(z - z_0^b). \tag{12.3.4$''$}$$

As usual (see, e.g., Ref. [15]), the displacement $\delta\tilde{Q}$ can be expanded in the Fourier series.

It is very essential that the Franck–Condon factor $F^{ab} \neq 0$. Indeed, (12.3.2) assumes summation over virtual states, that is integration over ξ' and summation over ν'. The vibrational wave functions for the same term are orthogonal. However, this is not the case for different crossing terms.

The additional contribution g_1 corresponds to virtual transitions which are accompanied by the change of electronic terms upon ionic tunneling between the two minima. In other words, g_1 corresponds to transitions to the states with various ν' (for given ν, see (12.2.2)). It is important to note that the contribution comes only from the terms with the same symmetry, that is from the terms with the same sign of the coefficient "b" (see (12.2.2)).

Therefore, there is an additional contribution (relative to the usual case), caused by the transitions between the local vibrational levels. Such an increase in phase space for virtual transitions leads to an increase in the value of the coupling matrix element g.

Consider in more detail the Franck–Condon factor F^{ab}. Assume for concreteness that the initial state corresponds to $\nu_a = 0$. Equation (12.3.2) contains a sum $\sum_{\nu'} F^{ab}_{\nu'0}$ which represents the contributions of virtual transitions to the manifold of the coherence states. In the diabatic representation (see above), a single non-harmonic term is replaced by two crossing terms, and each of them can be treated in harmonic approximation. The Franck–Condon factor is equal (see, e.g., Ref. [16]): $F^{ab} = (\bar{\beta}^{\nu}/\nu!) \exp(-\beta)$, where $\beta = (\delta/\alpha)^2$ (α is the amplitude of vibrations). One can see that $F_{\nu'0}$ contains a small factor $\exp[-(\delta/\alpha)^2]$ which describes the probability of the ionic tunneling between two minima. But this smallness is compensated by summation over all virtual transitions. If, for example, $(\delta/\alpha) = 2$, then $g_1 \approx 0.7g_0$. If $g_0 \approx 1$, then $g = g_0 + g_1 \approx 1.7$. The value of T_c is determined by the coupling constant $\lambda \propto \zeta \propto g^2$, and we obtain $\lambda \approx 3$. Therefore, the polaronic effect leads to a large increase in the effective strength of the electron–lattice coupling. As was stressed above, this enhancement arises due to the increase in phase space for virtual transitions which promote the pairing.

In summary, the electron–lattice interaction was studied for the case of a complex system when the electronic term (potential energy surface) for some ionic degree of freedom contains two close equilibrium positions (double-well structure). An example of such a system is offered by the oxygen ions in cuprates. Using the diabatic representation, one can describe this anharmonic case as motion in the potential formed by two crossing harmonic terms. The presence of such terms results in the formation of coherent ionic states with a concomitant increase in the number of virtual transitions. Importantly, these very transitions are the key ingredient entering the equation for the pairing order parameter. Since the vibrational wave functions for different terms are not orthogonal, the inter-term transitions are not forbidden. The increased phase space for virtual transitions brings about an increase in the coupling constant and thereby in the value of T_c.

Acknowledgments

The author is grateful to L. Gor'kov for interesting discussion. The research was supported by DARPA.

Open Access

References

[1] G. Bednorz and K. Mueller, *Z. Phys. B* **64** (1986), 189.

[2] D. Haskel, E. Stern, D. Hinks, D. Mitchell, and J. Jorgenson, *Phys. Rev. B* **56** (1997), 521.
[3] V. Kresin and S. Wolf, *Phys. Rev. B* **49** (1994), 3652.
[4] V. Kresin, Y. Ovchinnikov, and S. Wolf, *Phys. Rep.* **431** (2006), 231.
[5] V. Kresin and S. Wolf, *Rev. Mod. Phys.* **81** (2009), 481.
[6] J. Hui and P. Allen, *J. Phys. F* **4** (1974), L42.
[7] J. Hardy and J. Flocken, *Phys. Rev. Lett.* **60** (1988), 2191.
[8] A. Bussmann-Holder and A. Bishop, *Phys. Rev. B* **56** (1997), 5297.
[9] A. Bussmann-Holder and A. Bishop, *J. Supercond. Nov. Magn.* **10** (1997), 1557.
[10] T. O'Malley, *Phys. Rev.* **252** (1967), 98.
[11] V. Kresin and W. Lester, *Chem. Phys.* **90** (1984), 935.
[12] V. Kresin. *J. Chem. Phys.* **128** (2008), 094706.
[13] L. Gor'kov, *J. Exp. Theor. Phys.* **7** (1958), 505.
[14] A. Abrikosov, L. Gor'kov, and I. Dzyaloshinski, *Methods of Quantum Field Theory in Statistical Physics.* (Dover, New York, 1963).
[15] A. Anselm, *Introduction to Semiconductor Theory* (Prentice Hall, New York, 1982).
[16] L. Landau and E. Lifshits, *Quantum Mechanics*, Sec. 41 (Pergamon, New York, 1976).

Part IV

Other Evidences

13 Fine Structure in the Tunneling Spectra of Electron-Doped Cuprates: No Coupling to the Magnetic Resonance Mode*

G. M. Zhao

Department of Physics and Astronomy, California State University, Los Angeles, California 90032, USA
Department of Physics, Faculty of Science, Ningbo University, Ningbo, People's Republic of China

We reanalyze high-resolution scanning tunneling spectra of the electron-doped cuprate $Pr_{0.88}LaCe_{0.12}CuO_4$ (T_c = 24 K). We find that the spectral fine structure below 35 meV is consistent with strong coupling to a bosonic mode at about 16 meV, in quantitative agreement with early tunneling spectra of $Nd_{1.85}Ce_{0.15}CuO_4$. Since the energy of the bosonic mode is significantly higher than that (9.5–11 meV) of the magnetic resonancelike mode observed by inelastic neutron scattering, the coupling feature at about 16 meV cannot arise from strong coupling to the magnetic mode. The present work thus demonstrates that the magnetic resonancelike mode cannot be the origin of high-temperature superconductivity in electron-doped cuprates.

13.1. Introduction

The microscopic pairing mechanism for high-temperature superconductivity in cuprates remains elusive despite tremendous efforts for over 20 years.

*Reprinted with the permission from G. M. Zhao. Original Published in *Phys. Rev. Lett.* **103** (2009), 236403.

The most central issue is the origin of the bosonic modes mediating the electron pairing. Most workers believe that magnetic resonance modes, which have been observed in various hole-doped double-layer cuprate systems, predominantly mediate the electron pairing. Recent observation of a magnetic resonancelike mode at 9.5–11.0 meV in two electron-doped cuprates [1, 2] seems to suggest that the magnetic resonance is a universal property of all cuprate systems and thus essential to the pairing mechanism of high-temperature superconductivity. However, this speculated magnetic pairing mechanism is seriously undermined by recent optical experiments [3] which showed that the electron–boson spectral function $\alpha^2(\omega)F(\omega)$ is independent of magnetic field, in contradiction with the theoretical prediction based on the magnetic pairing mechanism (see Fig. 9 of Ref. [3]). In contrast, extensive studies of various unconventional oxygen-isotope effects in hole-doped cup-rates have clearly shown strong electron–phonon interactions and the existence of polarons [4–12]. Neutron scattering [13], angle-resolved photoemission (ARPES) [14, 15], and Raman scattering [16] experiments have also demonstrated strong electron–phonon coupling. Further, ARPES data [17] and tunneling spectra [18–22] have consistently provided direct evidence for strong coupling to multiple-phonon modes in hole-doped cuprates. Therefore, electron–phonon coupling in hole-doped cuprates should play an important role in the pairing mechanism.

On the other hand, the role of electron–phonon coupling in the pairing mechanism of electron-doped cuprates has not been clearly demonstrated. Early tunneling spectra in $Nd_{1.85}Ce_{0.15}CuO_4$ (NCCO) suggested predominantly phonon-mediated pairing [23] while the oxygen-isotope exponent α_O in $Pr_{1.85}Ce_{0.15}CuO_4$ was found to be 0.08 ± 0.01 (Ref. [24]), which is significantly below 0.5, expected for the phonon-mediated mechanism. Moreover, surface-sensitive ARPES experiments showed very weak electron–phonon coupling [25], which may support an alternative mechanism where the 10 meV magnetic resonance mode is mainly responsible for the pairing. If this were the case, the strong coupling to the 10 meV magnetic excitation would show up in single-particle tunneling microscopy that has a much higher energy resolution than ARPES. However, the early tunneling spectra [23] do not seem to show this coupling feature at about 10 meV. One might argue that the absence of this coupling feature in the early data could be due to a low experimental resolution and poor sample quality. Therefore, it is essential to obtain reproducible high-resolution single-particle tunneling spectra and analyze the spectra in a correct way to unambiguously address this issue.

Here, we reanalyze high-resolution single-particle tunneling spectra of the electron-doped cuprate $Pr_{0.88}LaCe_{0.12}CuO_4$ (T_c = 24 K) [26]. The d^2I/dV^2 spectra reveal one dip and two peak features below V = 35 mV, where I is the tunneling current and V is the bias voltage. We find that these fine features are consistent with strong coupling to a bosonic mode at about 16 meV, in quantitative agreement with early tunneling spectra [23] of $Nd_{1.85}Ce_{0.15}CuO_4$. Since the energy of the bosonic mode is significantly higher than that (9.5–11 meV) of the magnetic resonancelike mode observed by inelastic neutron scattering [1, 2], this coupling feature cannot arise from strong coupling to the magnetic mode. The present work thus demonstrates that the magnetic resonancelike mode cannot be the origin of high-temperature superconductivity in electron-doped cuprates.

For conventional superconductors, the energies of the phonon modes coupled to electrons can be precisely determined from the second derivative tunneling spectra d^2I/dV^2. Measured from the isotropic s-wave superconducting gap Δ, the energy positions of the dips (minima) in d^2I/dV^2 correspond to those of the peaks in the electronphonon spectral function $\alpha^2(\omega)F(\omega)$ (Refs. [27, 28]). In a recent paper [29], attempting to show an important role of phonons in the electron pairing, the authors assign the energy (52 meV) of a peak position in d^2I/dV^2 spectra of $Bi_2Sr_2CaCu_2O_{8+\delta}$ to the energy of a phonon mode. Such an assignment is incorrect because the energies of phonon modes are equal to the energies of dip positions rather than peak positions in d^2I/dV^2 (see Fig. 13.1 paper and also Refs. [21, 22, 27, 28]). The same mistake occurs in a more recent article [26] where the authors also assign the energy (10.5 meV) of a peak position in d^2I/dV^2 spectra of $Pr_{0.88}LaCe_{0.12}CuO_4$ (PLCCO) to the energy of a bosonic mode. Since this mistakenly assigned mode energy (10.5 meV) is very close to the energy (9.5–11 meV) of the magnetic resonancelike mode measured by inelastic neutron scattering [1, 2], the authors [26] conclude that the magnetic resonance mode mediates electron pairing in electron-doped cuprates.

Figure 13.1 shows the normalized second derivative $(d^2I/dV^2)_S/(dI/dV)_N$ for the conventional superconductor Pb along with the electron–phonon spectral function $\alpha^2(\omega)F(\omega)$ (where S represents the superconducting state and N the normal state). The figure is reproduced from Ref. [27]. It is apparent that the dip features in $(d^2I/dV^2)_S$ match precisely with

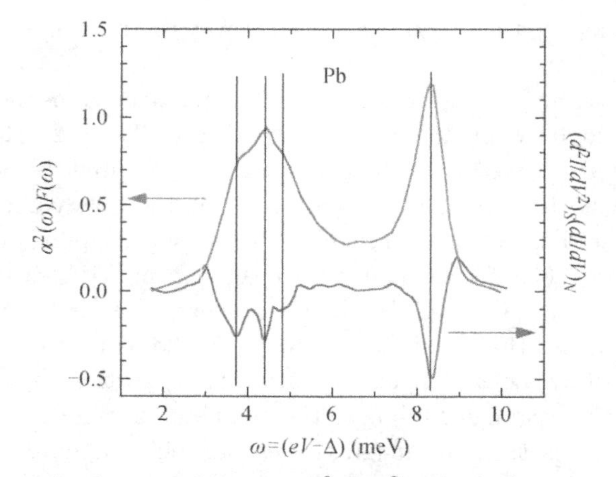

Fig. 13.1. Normalized second derivative $(d^2I/dV^2)_S/(dI/dV)_N$ for the conventional superconductor Pb along with the electron–phonon spectral function $\alpha^2(\omega)F(\omega)$ (where S represents the superconducting state and N the normal state). The figure is reproduced from Ref. [27]. It is apparent that the dip features in $(d^2I/dV^2)_S$ match precisely with the peak features in $\alpha^2(\omega)F(\omega)$ (see the vertical solid lines).

the peak features in $\alpha^2(\omega)F(\omega)$ (see the vertical solid lines). Therefore, the energy positions of bosonic modes strongly coupled to electrons correspond to the energy positions of the dip features (rather than the peak features) in $(d^2I/dV^2)_S$. As a matter of fact, this is true not only for Pb but also for any boson-mediated superconductor, as demonstrated both experimentally and theoretically [28].

In Fig. 13.2(a), we show the tunneling conductance $(dI/dV)_S$ in the superconducting state for the electron-doped $Pr_{0.88}LaCe_{0.12}CuO_4$ crystal. The data are digitized from Ref. [26]. The normal-state tunneling conductance $(dI/dV)_N$ is approximated by a straight line, which is obtained by conservation of states; i.e., the superconducting spectral deviation above the line is balanced by the deviation below. This straight-line approximation for the normal-state conductance was also used in Ref. [26]. Then, we obtain the normalized conductance $(dI/dV)_S/(dI/dV)_N$, which is shown in Fig. 13.2(b).

For an anisotropic gap function $\Delta(\theta)$, the directional dependence of the differential tunneling conductance is given by [30]

$$\frac{dI}{dV} \propto \int_0^{2\pi} p(\theta-\theta_0)\mathrm{Re}\left[\frac{eV-i\Gamma}{\sqrt{(eV-i\Gamma)^2-\Delta^2(\theta)}}\right]N(\theta)d\theta, \tag{13.1.1}$$

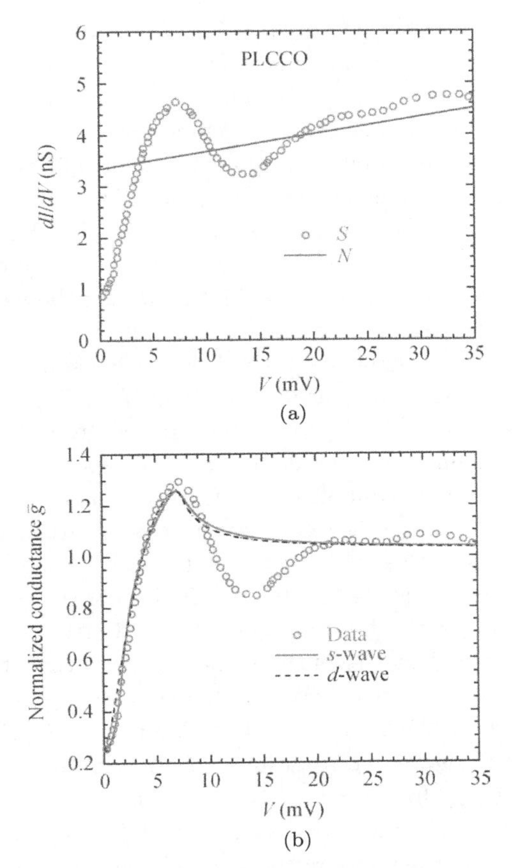

Fig. 13.2. (a) Tunneling conductance $(dI/dV)_S$ in the superconducting state for the electron-doped $Pr_{0.88}LaCe_{0.12}CuO_4$ (PLCCO) crystal (○). The data are digitized from Ref. [26]. The normal-state tunneling conductance $(dI/dV)_N$ is approximated by a straight line, which is obtained with conservation of states. (b) Normalized conductance $\bar{g} = (dI/dV)_S/(dI/dV)_N$. The solid line is the numerically calculated curve in terms of an anisotropic s-wave gap symmetry and the dashed line is the numerically calculated curve in terms of d-wave gap symmetry.

where $N(\theta)$ represents the anisotropy of the band dispersion, Γ is the lifetime broadening parameter of an electron, $p(\theta - \theta_0)$ is the angle dependence of the tunneling probability and equal to $\exp[-\beta \sin^2(\theta - \theta_0)]$, and θ_0 is the angle of the tunneling barrier direction measured from the Cu-O bonding direction. For simplicity, we assume a cylindrical Fermi surface so that both $N(\theta)$ and β are independent of the angle. The solid line is the numerically calculated curve using $\Gamma = 0.73\,\text{meV}$, $\beta = 6$, $\theta_0 = 0.18\pi$, and an anisotropic s-wave gap function: $\Delta = 4.4(1.0 - 0.6\sin 4\theta)$ meV. The dashed line is the numerically calculated curve using $\Gamma = 0.40\,\text{meV}$, $\beta = 7, \theta_0 = \pi/4$, and a simple d-wave gap function: $\Delta = 7.2\cos 2\theta\,\text{meV}$. It is interesting that the calculated curves for the anisotropic s-wave and d-wave gaps are all in good agreement with the data. We further find that the isotropic s-wave gap is inconsistent with the data. Therefore, the tunneling spectrum alone rules out isotropic s-wave gap symmetry on the top surface of the crystal but cannot make the distinction between d-wave and anisotropic s-wave gap symmetry.

In Fig. 13.3(a), we show the second derivative spectrum $(d^2I/dV^2)_S$ together with the first derivative spectrum $(dI/dV)_S$ for the $Pr_{0.88}LaCe_{0.12}CuO_4$ crystal. The figure is reproduced from Ref. [26]. In the $(d^2I/dV^2)_S$ spectrum, there are two peak features at $E_{p1} = 17.8$ meV and $E_{p2} = 29.0$ meV and one dip feature at $E_{d1} = 23.4$ meV,as indicated in the figure. It is clear that the dip feature is just halfway between the two peak features. The energy position Δ_R of the peak in the $(dI/dV)_S$ spectrum is about 7.0 meV. Following the result of Fig. 13.1 for Pb, the energy of the bosonic mode coupled strongly to electrons is $\Omega_1 = E_{d1} - \Delta_R = 16.4$ meV, which is slightly lower than the energy of a very strong coupling feature observed in hole-doped cuprates (e.g., 20 meV in $Bi_2Sr_2CaCu_2O_{8+\delta}$ [20, 21, 31] and 18 meV in $La_{2-x}Sr_xCuO_4$ [22]).

Figure 13.3(b) shows a histogram of the occurrences of Δ_R and the energies E_{p1} and E_{p2} for a map of tunneling spectra on a 64 Å × 64 Å area of the sample. The data are taken from Ref. [26]. According to the result in Fig. 13.3(a), the midpoint between E_{p1} and E_{p2} should mark E_{d1}. Then, the difference between E_{d1} and Δ_R is found to be 15.9 meV, that is, $\Omega_1 = 15.9\,\text{meV}$. More bosonic modes would be revealed if these spectra were extended to higher energies.

Figure 13.4 shows electron–boson spectral functions for two $Nd_{1.85}Ce_{0.15}CuO_4$ samples along with $-(2\Delta_R)d\bar{g}/d\omega$ for PLCCO (where $\omega = eV - \Delta_R$). The electron–boson spectral functions are reproduced from Ref. [23] and $-(2\Delta_R)d\bar{g}/d\omega$ is numerically calculated from Fig. 13.2(b) after the data are smoothened. The energies of the lowest bosonic modes in the spectral functions of the two NCCO samples are 15.2 and 17.2 meV, respectively. A simple average of the mode energies of the two NCCO samples is 16.2 meV, which is in quantitative agreement with the energy position (16.3 meV) of the peak in $-(2\Delta_R)d\bar{g}/d\omega$ (or $-d^2I/d\omega^2$) and with the

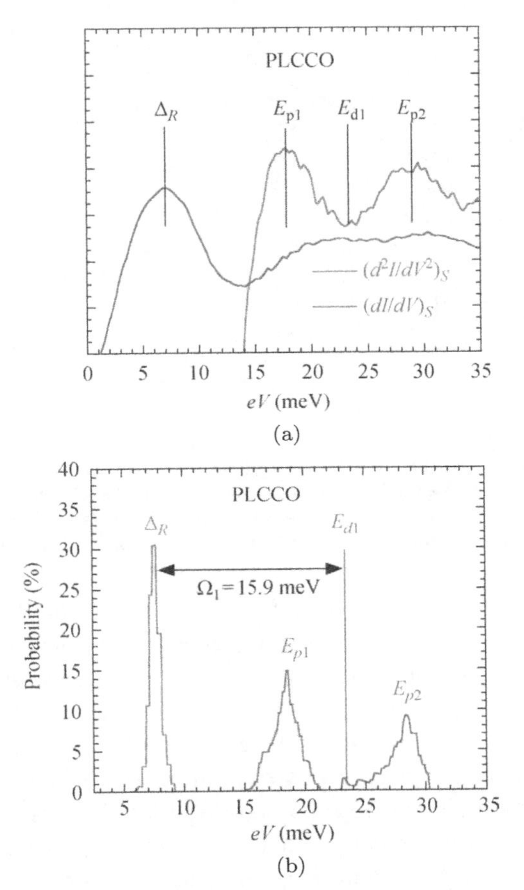

Fig. 13.3. (a) The second derivative spectrum $(d^2I/dV^2)_S$ together with the first derivative spectrum $(dI/dV)_S$ for the $Pr_{0.88}LaCe_{0.12}CuO_4$ crystal. The figure is reproduced from Ref. [26]. (b) A histogram of the occurrences of Δ_R and the energies E_{p1} and E_{p2} for a map of tunneling spectra on a 64 Å × 64 Å area of the sample. The data are taken from Ref. [26].

average mode energy (15.9 meV) deduced from Fig. 13.3. This quantitative agreement clearly indicates that the strong coupling feature at about 16 meV in the tunneling spectra of electron-doped cuprates is intrinsic.

Having established the bosonic mode energy and statistics, we now discuss the nature of this mode. The measured mode energy of about 16 meV rules out its connection to the magnetic resonance mode which has an energy of 9.5 meV in NCCO (Ref. [2]) and 11 meV in PLCCO

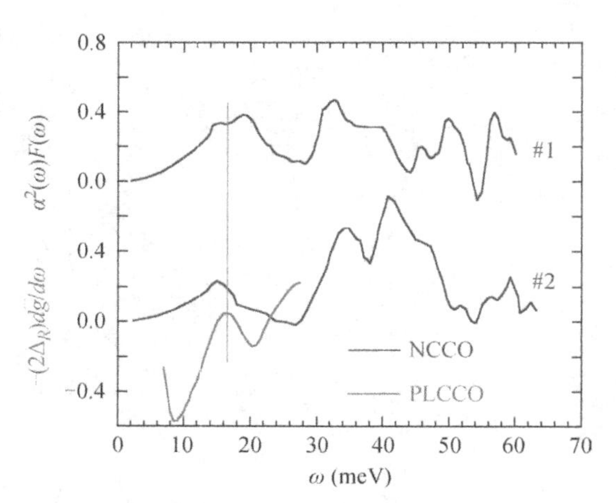

Fig. 13.4. Electron–boson spectral functions (top two curves) for two $Nd_{1.85}Ce_{0.15}CuO_4$ samples along with $-(2\Delta_R)d\bar{g}/d\omega$ (bottom curve) for PLCCO (where $\omega = eV - \Delta_R$). The electron–boson spectral functions are reproduced from Ref. [23] and $-(2\Delta_R)d\bar{g}/d\omega$ is numerically calculated from Fig. 13.2(b) after the data are smoothened.

(Ref. [1]). Alternatively, this 16 meV bosonic mode should be associated with unscreened c-axis polar phonons (transverse optical phonons). Both ARPES data [32] of hole-doped $Bi_2Sr_2CuO_6$ and a theoretical study [33] indicate that the transverse optical (TO) phonons are strongly coupled to electrons due to the unscreened long-range interaction along the c axis and play a predominant role in electron pairing. This long-range electron–phonon interaction should be present in any layered system but is often ignored when one theoretically calculates electron–phonon coupling. For electron-doped $Pr_{1.85}Ce_{0.15}CuO_4$, the lowest TO modes with energies of 15.6 meV (E_u symmetry) and 17.0 meV (A_{2u}) were identified by infrared reflectivity measurements [34]. Since these two TO modes have energies very close to the energy of the bosonic mode seen in the tunneling spectra, it is natural that the 16 meV coupling feature is associated with strong coupling of these TO phonon modes to electrons.

In summary, we have reanalyzed the high-resolution tunneling spectra of the electron-doped $Pr_{0.88}LaCe_{0.12}CuO_4$. We find that the spectral fine structure below 35 meV is consistent with strong coupling to a bosonic mode at about 16 meV, in quantitative agreement with early tunneling spectra [23] of $Nd_{1.85}Ce_{0.15}CuO_4$. Since the energy of the bosonic mode is significantly higher than that (9.5–11 meV) of the magnetic resonance-like mode observed by inelastic neutron scattering, the coupling feature at about 16 meV cannot arise from strong coupling to the magnetic mode. The present work thus demonstrates that the magnetic resonancelike mode cannot be the origin of high-temperature superconductivity in electron-doped cuprates.

References

[1] S. D. Wilson, P. C. Dai, S. L. Li, S. X. Chi, H. J. Kang, and J. W. Lynn, *Nature (London)* **442** (2006), 59.
[2] J. Zhao, P. C. Dai, S. L. Li, P. G. Freeman, Y. Onose, and Y. Tokura, *Phys. Rev. Lett.* **99** (2007), 017001.
[3] Y. S. Lee, K. Segawa, Z. Q. Li, W. J. Padilla, M. Dumm, S. V. Dordevic, C. C. Homes, Yoichi Ando, and D. N. Basov, *Phys. Rev. B* **72** (2005), 054529.
[4] G. M. Zhao, K. K. Singh, and D. E. Morris, *Phys. Rev. B* **50** (1994), 4112.
[5] G. M. Zhao and D. E. Morris, *Phys. Rev. B* **51** (1995), 16487.
[6] G. M. Zhao, K. K. Singh, A. P. B. Sinha, and D. E. Morris, *Phys. Rev. B* **52** (1995), 6840.
[7] G. M. Zhao, M. B. Hunt, H. Keller, and K. A. Müller, *Nature (London)* **385** (1997), 236.
[8] G. M. Zhao, K. Conder, H. Keller, and K. A. Müller, *J. Phys. Condens. Matter* **10** (1998), 9055.
[9] G. M. Zhao, H. Keller, and K. Conder, *J. Phys. Condens. Matter* **13** (2001), R569.
[10] G. M. Zhao, *Philos. Mag. B* **81** (2001), 1335.
[11] G. M. Zhao, V. Kirtikar, and D. E. Morris, *Phys. Rev. B* **63** (2001), 220506.
[12] R. Khasanov *et al.*, *Phys. Rev. Lett.* **92** (2004), 057602.
[13] R. J. McQueeney, Y. Petrov, T. Egami, M. Yethiraj, G. Shirane, and Y. Endoh, *Phys. Rev. Lett.* **82** (1999), 628.
[14] A. Lanzara *et al.*, *Nature (London)* **412** (2001), 510.
[15] O. Rösch *et al.*, *Phys. Rev. Lett.* **95** (2005), 227002.
[16] O. V. Misochko, E. Ya. Sherman, N. Umesaki, K. Sakai, and S. Nakashima, *Phys. Rev. B* **59** (1999), 11495.
[17] X. J. Zhou *et al.*, *Phys. Rev. Lett.* **95** (2005), 117001.
[18] S. I. Vedeneev, P. Samuely, S. V. Meshkov, G. M. Eliashberg, A. G. M. Jansen, and P. Wyder, *Physica (Amsterdam)* **198C** (1992), 47.
[19] D. Shimada, Y. Shiina, A. Mottate, Y. Ohyagi, and N. Tsuda, *Phys. Rev. B* **51** (1995), 16495.
[20] R. S. Gonnelli, G. A. Ummarino, and V. A. Stepanov, *Physica (Amsterdam)* **275C** (1997), 162.
[21] G. M. Zhao, *Phys. Rev. B* **75** (2007), 214507.
[22] H. Shim, P. Chaudhari, G. Logvenov, and I. Bozovic, *Phys. Rev. Lett.* **101** (2008), 247004.

[23] Q. Huang, J. F. Zasadzinski, N. Tralshawala, K. E. Gray, D. G. Hinks, J. L. Peng, and R. L. Greene, *Nature (London)* **347** (1990), 369.
[24] G. M. Zhao and D. E. Morris, *Phys. Rev. B* **50** (1994), 3454.
[25] N. P. Armitage *et al.*, *Phys. Rev. B* **68** (2003), 064517.
[26] F. C. Niestemski, S. Kunwar, S. Zhou, S. L. Li, H. Ding, Z. Q. Wang, P. C. Dai, and V. Madhavan, *Nature (London)* **450** (2007), 1058.
[27] W. L. McMillan and J. M. Rowell, *Phys. Rev. Lett.* **14** (1965), 108.
[28] J. P. Carbotte, *Rev. Mod. Phys.* **62** (1990), 1027; W. L. McMillan and J. M. Rowell, *Superconductivity* (Marcel Dekker, New York, 1969), Vol. 1, p. 561; E. L. Wolf, *Principles of Electron Tunneling Spectroscopy* (Oxford University Press, New York, 1985); D. Shimada, N. Tsuda, U. Paltzer, and F. W. de Wette, Physica (Amsterdam) **298C** (1998), 195.
[29] J.-H. Lee, K. Fujita, K. McElroy, J. A. Slezak, M. Wang, Y. Aiura, H. Bando, M. Ishikado, T. Masui, J.-X. Zhu, A. V. Balatsky, H. Eisaki, S. Uchida, and J. C. Davis, *Nature (London)* **442** (2006), 546.
[30] K. Suzuki *et al.*, *Phys. Rev. Lett.* **83** (1999), 616.
[31] G. M. Zhao, *Phys. Rev. B* **71** (2005), 104517.
[32] W. Meevasana *et al.*, *Phys. Rev. Lett.* **96** (2006), 157003.
[33] W. Meevasana, T. P. Devereaux, N. Nagaosa, Z.-X. Shen, and J. Zaanen, *Phys. Rev. B* **74** (2006), 174524; T. Bauer and C. Falter, *Phys. Rev. B* **80** (2009), 094525.
[34] M. K. Crawford *et al.*, *Solid State Commun.* **73** (1990), 507.

14

Identification of the Bulk Pairing Symmetry in High-Temperature Superconductors: Evidence for an Extended *s* wave with Eight Line Nodes*

G. M. Zhao

Physik-Institut der Universität Zürich, CH-8057 Zürich, Switzerland

We identify the intrinsic bulk pairing symmetry for both electron- and hole-doped cuprates from the existing bulk-and nearly bulk-sensitive experimental results such as magnetic penetration depth, Raman scattering, single-particle tunneling, Andreev reflection, nonlinear Meissner effect, neutron scattering, thermal conductivity, specific heat, and angle-resolved photoemission spectroscopy. These experiments consistently show that the dominant bulk pairing symmetry in hole-doped cuprates is of extended *s* wave with eight line nodes and of anisotropic *s* wave in electron-doped cuprates. The proposed pairing symmetries do not contradict some surface- and phase-sensitive experiments that show a predominant *d*-wave pairing symmetry at the degraded surfaces. We also quantitatively explain the phase-sensitive experiments along the *c* axis for both $Bi_2Sr_2CaCu_2O_{8+y}$ and $YBa_2Cu_3O_{7-y}$.

14.1. Introduction

An unambiguous determination of the symmetry of the order parameter (pair wave function) in cuprates is crucial to the understanding of the

*Reprinted with the permission from G. M. Zhao. Original published in *Phys. Rev. B* **64** (2001), 024503.

pairing mechanism of high-temperature superconductivity. In recent years, many experiments have been designed to test the order-parameter (OP) symmetry in the cuprate superconductors. However, contradictory conclusions have been drawn from different experimental techniques [1–16], which can be classified as being bulk sensitive and surface sensitive. For example, the magnetic penetration depth measurements and polarized Ramanscattering experiments are bulk sensitive. Angle-resolved photoemission spectroscopy (ARPES) is essentially a surface-sensitive technique. However, the ARPES data for $Bi_2Sr_2CaCu_2O_{8+y}$ (BSCCO) should nearly reflect the bulk properties since the cleaved top surface contains an inactive Bi-O layer, and the superconducting coherent length along the c axis is very short. The single-particle tunneling experiments can probe the bulk electronic density of states when the mean free path is far larger than the thickness of the degraded surface layer [17]. Therefore, the single-particle tunneling experiments along the CuO_2 planes are almost bulk sensitive due to a large in-plane mean free path (>100 Å). In contrast, the phase-sensitive experiments based on the Josephson tunneling are rather surface sensitive (since pair tunneling is limited by the coherence length, which is rather short in cuprates), so that they might not probe the intrinsic bulk superconducting state if the surfaces are strongly degraded. In this case, the observed product of the critical current times the junction normal-state resistance (I_cR_N) will be very small compared with the Ambegaokar-Baratoff limit. Then the OP symmetry at surfaces may be different from the one in the bulk [18]. Therefore, the surface- and phase-sensitive experiments do not necessarily provide an acid test for the intrinsic bulk OP symmetry.

Here, we identify the intrinsic bulk pairing symmetry for both electron- and hole-doped cuprates from the existing bulk-and nearly bulk-sensitive experimental results such as magnetic penetration depth, Raman scattering, single-particle tunneling, Andreev reflection, nonlinear Meissner effect, neutron scattering, thermal conductivity, specific heat, and ARPES. These experiments consistently show that the dominant bulk pairing symmetry in hole-doped cuprates is of extended s wave with eight line nodes and of anisotropic s wave in electron-doped cuprates. The proposed pairing symmetries do not contradict some surface- and phase-sensitive experiments that show a d-wave pairing symmetry at the degraded surfaces. The extended s-wave pairing symmetry deduced from the bulk-sensitive experiments is also in quantitative agreement with the well-designed phase-sensitive experiments along the c axis for both $Bi_2Sr_2CaCu_2O_{8+y}$ and $YBa_2Cu_3O_{7-y}$ (YBCO).

14.2. The Pairing Symmetry in Hole-Doped Cuprates

14.2.1. *The Pairing Symmetry in $Bi_2Sr_2CaCu_2O_{8+y}$*

We first examine the high-resolution ARPES data obtained for $Bi_2Sr_2CaCu_2O_{8+y}$ crystals [14, 16]. From the ARPES data, one can determine the angle dependence of the super-conducting gap with a resolution as high as ± 2 meV [16]. Due to the complication arising from a possible superlattice contribution in the X quadrant, we only use the data obtained for the Y quadrant to extract the gap function. In Fig. 14.1, we show the angle dependence of the superconducting gap $\Delta(\theta)$ in the Y quadrant for slightly overdoped and heavily overdoped BSCCO single crystals. The data were taken from Refs. [14] and [16]. Here θ is the angle measured from the Cu-O bonding direction. For the slightly overdoped sample (Fig. 14.1(a)), the gap Δ_D at $\theta = 45°$ (diagonal direction) is very small (3.5 ± 2.5 meV), and the gap symmetry could be consistent with a d-wave symmetry, i.e., $\Delta(\theta) = \Delta \cos 2\theta$. On the other hand, the gap along the diagonal direction ($\Gamma - Y$) for the heavily overdoped sample (Fig. 14.1(b)) is not small (9 ± 2 meV), which is obviously inconsistent with the d-wave pairing symmetry. A similar evolution of the gap function with the doping has been observed by the bulk-sensitive polarized Raman scattering [7], which also shows that the difference in the magnitudes of the gaps along the Cu-O bonding direction and the diagonals becomes smaller and smaller towards overdoping.

The question is: what functional form of $\Delta(\theta)$ can fit the angle dependence of the gap shown in Fig. 14.1? In general, the gap can be expressed as $\Delta(\theta) = \Delta_s + \Delta_d \cos 2\theta + \Delta_g \cos 4\theta + \cdots$. In the case of $\Delta_d \backsimeq 0$, one has

$$\Delta(\theta) = \Delta(\cos 4\theta + s), \tag{14.2.1}$$

where s is the parameter reflecting the isotropic s-wave component. This gap function has eight line nodes for $s < 1$, while there are no nodes for $s > 1$. The gap function (Eq. (14.2.1)) is also called extended s wave (denoted by s^* wave). The polarized Raman data for an optimally doped $HgBa_2CaCu_2O_{6+y}$ are in good agreement with the s^*-wave gap function [6].

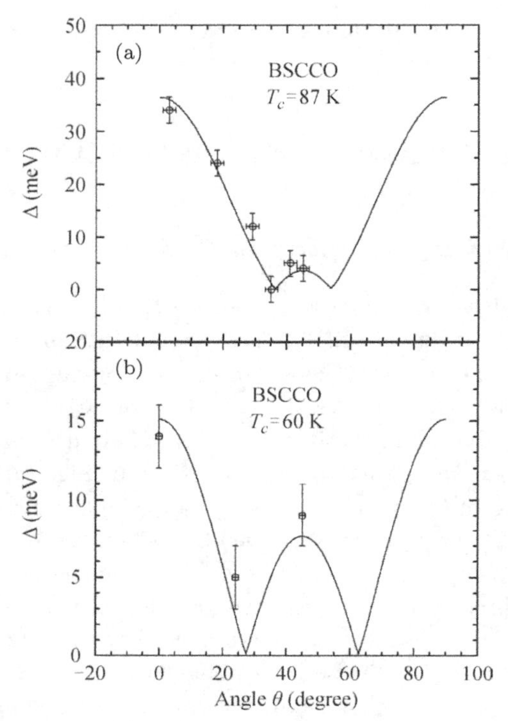

Fig. 14.1. The angle dependence of the superconducting gap $\Delta(\theta)$ in the Y quadrant for $Bi_2Sr_2CaCu_2O_{8+y}$ (BSCCO) crystals: (a) slightly overdoped sample with $T_c = 87$ K; (b) heavily overdoped sample with $T_c = 60$ K. The magnitudes of the gap were extracted from ARPES data (Refs. [14, 16]). Here, θ is the angle measured from the Cu-O bonding direction.

If we take the absolute value of $\Delta(\theta)$, then

$$|\Delta(\theta)| = |\Delta(\cos 4\theta + s)|. \tag{14.2.2}$$

We fit the data of Fig. 14.1 by Eq. (14.2.2). It is remarkable that the fits are rather good. This indicates that the ARPES data may be consistent with the extended s-wave symmetry. The ARPES specified maximum gap Δ_M at $\theta = 0$ for the slightly overdoped sample is 36 ± 3 meV, which is much larger than the value ($\sim$28 meV) determined from break junction spectra [21]. On the other hand, the ARPES determined Δ_M value (15 ± 2 meV) for the heavily overdoped sample with $T_c = 60$ K is very close to the value (18 ± 2 meV) inferred from a break junction spectrum of a similar crystal with $T_c = 62$ K [22]. The discrepancy in the former case may be due to the fact that the doping level in the top layer with the ARPES probes could be slightly lower than in the bulk (i.e., the top CuO_2 layer might be slightly underdoped). Thus, the ARPES experiments on the BSCCO single crystals are nearly bulk sensitive, in contrast to the ARPES experiments on other cuprates, which are essentially surface sensitive.

If the proposed gap functions (Eqs. (14.2.1) and (14.2.2)) are indeed relevant, they should be also consistent with other bulk-sensitive experimental results such as the in-plane magnetic penetration depth $\lambda_{ab}(T)$. Since there are eight line nodes in the proposed gap function, the change of the in-plane penetration depth at low temperatures should be proportional to T. Following the procedure in Ref. [23], we can readily show that the slope

$$d\lambda_{ab}(T)/dT = [\lambda_{ab}(0)\ln 2/\Delta_M]\sqrt{(1+s)/(1-s)} \tag{14.2.3}$$

Compared with the d-wave symmetry, the magnitude of the slope $d\lambda_{ab}(T)/dT$ is enhanced by a factor of $\sqrt{(1+s)/(1-s)}$. In terms of Δ_M and Δ_D, we find that $s = (\Delta_M - \Delta_D)/(\Delta_M + \Delta_D)$ and $\Delta = (\Delta_M + \Delta_D)/2$. Then, Eq. (14.2.3) can be rewritten as

$$d\lambda_{ab}(T)/dT = \lambda_{ab}(0)\ln 2/\sqrt{\Delta_M \Delta_D}. \tag{14.2.4}$$

It is interesting to see that $d\lambda_{ab}(T)/[\lambda_{ab}(0)dT]$ is inversely proportional to $\sqrt{\Delta_M \Delta_D}$, namely, the geometric average of Δ_M and Δ_D.

The single-particle tunneling spectroscopy can probe the superconducting density of states with fine energy resolution and considerable directionality. For an isotropic s-wave superconductor, the characteristic dI/dV vs V curve in the SIN (where S represents a superconductor, and I and N are the insulating and normal-metal layers, respectively) tunneling junctions exhibits a steplike peak at a voltage $V_p = \Delta/e$. For an anisotropic gap function $\Delta(\theta)$, the directional dependence of the tunneling differential conduction is given by [24]

$$\frac{dI}{dV}(V,\theta_o) \propto \int_0^{2\pi} p(\theta-\theta_0)\times R\left[\frac{eV - i\Gamma}{\sqrt{(eV-i\Gamma)^2 - \Delta^2(\theta)}}\right] N(\theta)d\theta. \tag{14.2.5}$$

Here, $N(\theta)$ represents the anisotropy of the band dispersion, Γ is the lifetime broadening parameter of an electron; $p(\theta - \theta_0)$ is the angle dependence of the tunneling probability that decays exponentially as $p(\theta - \theta_0) = \exp[-\beta \sin^2(\theta - \theta_0)]$ (θ_0 is the angle of the tunneling barrier direction), and the parameter β decreases with decreasing barrier resistance R_N. For simplicity, we assume a cylindrical Fermi surface, so that both $N(\theta)$ and

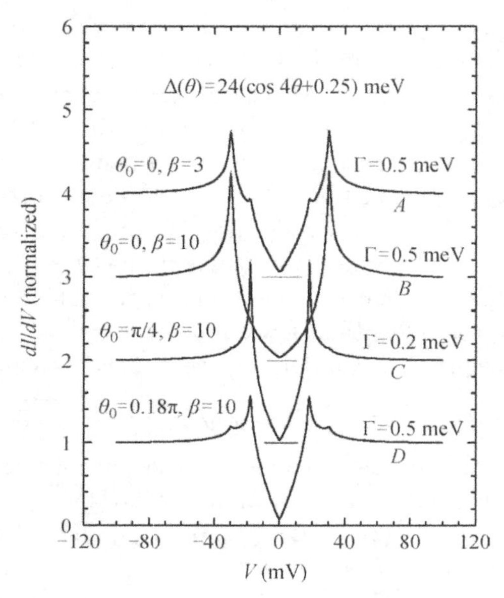

Fig. 14.2. Numerically calculated curves of the renormalized dI/dV for a gap function of $\Delta(\theta) = \Delta(\cos 4\theta + s)$ with $\Delta = 24$ meV and $s = 0.25$. The four curves correspond to different values of the parameters Γ, β, and θ_0, which are indicated in the figure. The curves A, B, and C are vertically shifted up by 3, 2, and 1, respectively.

β are independent of the angle. This will not change the basic features of the dI/dV curve. In Fig. 14.2, we show the numerically calculated results of the renormalized dI/dV for a gap function of $\Delta(\theta) = \Delta(\cos 4\theta + s)$ with $\Delta = 24$ meV and $s = 0.25$. One can readily show that the maximum gap is $\Delta_M = (1 + s)\Delta = 30$ meV at $\theta = 0$, and the gap along the diagonal directions is $\Delta_D = (1 - s)\Delta = 18$ meV. From Fig. 14.2, one can see that either two or four peak features appear clearly in the dI/dV curves, depending on the tunneling barrier direction and/or the β value. For a small β value (corresponding to a small barrier resistance), four peak features are well defined (see curve A). The peak positions are located at $eV = \pm\Delta_M$ and $\pm\Delta_D$. Therefore, from the peak positions, we can determine Δ_M and Δ_D.

In Fig. 14.3, we plot the normalized dI/dV curve at 14 K for an SIS (where S = BSCCO and I is the insulating layer) break junction on a BSCCO crystal that is slightly overdoped ($T_c = 90$ K) [25]. The junction has a very low barrier resistance ($\sim$200Ω, Ref. [25]), indicating a small

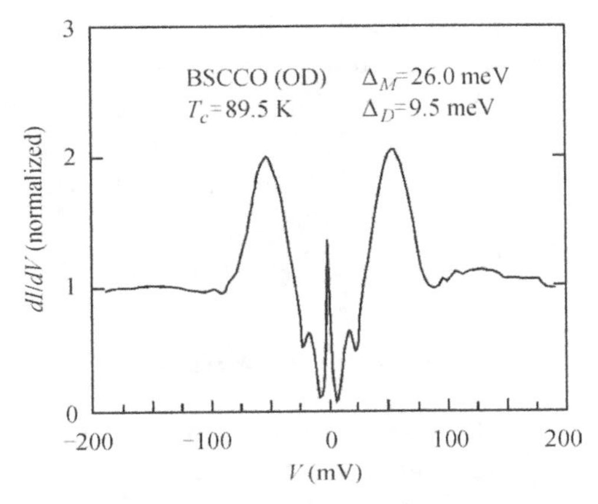

Fig. 14.3. Normalized dI/dV curves at 14 K for the SIS break junctions on a slightly overdoped BSCCO crystal. The spectra were taken from Ref. [25].

β value. It is remarkable that there are four well-defined peak features in the spectrum that resemble curve A in Fig. 14.2. The pronounced zero-bias peak arises from Josephson tunneling [21, 25]. When the barrier resistance is above $2k\Omega$, the inner gap features disappear [25], in agreement with curve B in Fig. 14.2. We would like to mention that, for SIS break junctions, the peak positions are located at $eV = \pm 2\Delta_M$ and $\pm 2\Delta_D$. From the spectra, we obtain $\Delta_M = 26 \pm 0.5$ meV, and $\Delta_D = 9.5 \pm 0.5$ meV. The Δ_M value obtained from the break junction spectrum is the same as that found from the c-axis intrinsic tunneling junctions made of the insulating Bi–O layers [26]. From the Δ_M and Δ_D values, we deduce a gap function $\Delta(\theta) = \Delta(\cos 4\theta + s)$ with $\Delta = 17.75$ meV and $s = 0.46$. With this gap function and $\lambda_{ab}(0) = 2690 \pm 150$ Å [27], we calculate from Eq. (14.2.4) that $d\lambda_{ab}(T)/dT = 10.2 \pm 0.6$ Å/K, in excellent agreement with the measured values (10.2 ± 0.2 Å/K) [2, 3]. Similarly, the earlier break junction spectra for an overdoped BSCCO with $T_c = 86$ K also indicate double gap features at $\Delta_M = 24 \pm 2$ meV and at $\Delta_D = 12 \pm 1$ meV (Ref. [28]). The tunneling spectra are in good agreement with ARPES data for an over-doped BSCCO with $T_c = 83$ K [19]. The ARPES experiment clearly showed that $\Delta_M = 20 \pm 2$ meV, and $\Delta_D = 12 \pm 2$ meV [19]. Moreover, the inner gap features also appear in SIS break junction spectra of a heavily overdoped crystal with $T_c = 62$ K, corresponding to $\Delta_D = 7.5 - 9.0$ meV (Refs. [22, 29]). The magnitude of Δ_D is in excellent agreement with that found from the ARPES experiment (see Fig. 14.1(b)).

We would like to point out that the values of Δ_M determined from the Raman spectrum of B_{1g} symmetry may be overestimated due to the fact that the extended van Hove singularity is slightly below the Fermi level. In this case, the spectra would show double peaks at Raman shifts of $2\Delta_M$ and $2\sqrt{\Delta_M^2 + \xi_{vH}^2}$, where ξ_{vH} is the energy position of the van Hove singularity below the Fermi level. When $\xi_{vH} \ll \Delta_M$, one can only see a single broad peak slightly below $2\sqrt{\Delta_M^2 + \xi_{vH}^2}$.

14.2.2. *The Pairing Symmetry in $YBa_2Cu_3O_{7-y}$*

Evidence for an extended *s*-wave pairing symmetry in $YBa_2Cu_3O_{7-y}$ (YBCO) also comes from single-particle tunneling spectra. Figure 4 shows the scanning tunneling spectrum for a slightly overdoped $YBa_2Cu_3O_{7-y}$ crystal [30]. Four peak features appear in this spectrum that are similar to curve D in Fig. 14.2. From the peak positions, we obtain $\Delta_M = 30 \pm 2$ meV, and $\Delta_D = 19 \pm 1$ meV. The size of $\Delta_M \simeq 30$ meV is consistent with a break junction spectrum [17], and a scanning tunneling spectrum along the *a*-axis direction [8]. A gap feature with $\Delta_D = 19$ meV was also seen in a scanning tunneling spectrum [8] that is very similar to curve C in Fig. 14.2.

Now we discuss the Andreev reflection. Since there is a sign change about its nodal directions in our extended *s*-wave order parameter, the Andreev-bound surface states can be formed. This will lead to a zero-bias conduction peak if tunneling is nearly along one of the nodal directions, and the bare Fermi velocities between the cuprates and normal metals (e.g., Ag and Au) are well matched. For hole-doped cuprates, the bare Fermi velocity v_F^b strongly depends on the angle θ, that is, v_F^b is small along the bonding direction and large along the diagonal directions. This implies that the observation of the Andreev reflection is difficult for tunneling along the bonding direction since the value of v_F^b along this direction is small compared with that of Au or Ag. Due to the strong anisotropy of v_F^b in cuprates, the Andreev reflection mainly probes the gap feature at $eV = \Delta_D$. If tunneling is along one of the diagonal directions, and the angle between the nodal and diagonal directions is far larger than the half tunneling angle (depending on β), one can see an *s*-wave-like gap approximately equal to Δ_D in the Andreev reflection spectra. Indeed an *s*-wave like gap feature at $eV \simeq 20$ meV has been observed in the Andreev reflection spectra of several YBCO crystals with $T_c = 90$ K [31]. We would like to mention that, in general, the double gap features should also appear in the

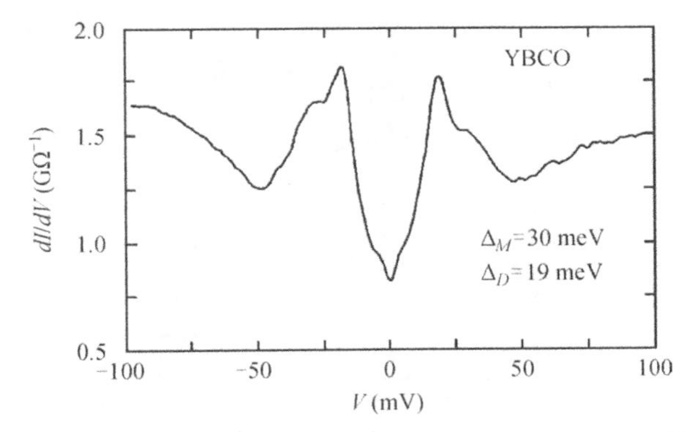

Fig. 14.4. Scanning tunneling spectrum for a slightly overdoped $YBa_2Cu_3O_7$ (YBCO) crystal. The spectrum was taken from Ref. 30.

Andreev reflection spectra when the β value is small and v_F^b does not have a significant an-isotropy.

The tunneling data of YBCO (Fig. 14.4) are thus consistent with a gap function $\Delta(\theta) = \Delta(\cos 4\theta + s)$ with $\Delta = 24.5$ meV and $s = 0.225$. This gap function is in quantitative agreement with the a-axis $\lambda_a(T)$ data (which reflect magnetic screening in CuO_2 planes) for a fully oxygenated YBCO crystal [32]. From Eq. (14.2.4), we calculate $d\lambda_a(T)/dT = 4.0$ Å/K using $\lambda_a(0) = 1600$ Å (Ref. [32]), $\Delta_D = 19$ meV, and $\Delta_M = 30$ meV. We will get the same value of $d\lambda_a(T)/dT$ if we use $\Delta_D = 21$ meV, and $\Delta_M = 27$ meV. For a d-wave gap function $\Delta(\theta) = \Delta_M \cos 2\theta$ with $\Delta_M = 30\,\text{meV}$, the calculated $d\lambda_a(T)/dT = 3.2$ Å/K. The measured value of $d\lambda_a(T)/dT$ is 4 Å/K [32]. It is evident that the extended s-wave gap function is in much better agreement with experiment than the d-wave gap function.

Now we calculate the temperature dependence of $\lambda_{ab}^2(0)/\lambda_{ab}^2(T)$ for the s^*-wave gap function. For a cylindrical Fermi surface [2]

$$\frac{\lambda_{ab}^2(0)}{\lambda_{ab}^2(T)} = 1 + (1/\pi)\int_0^{2\pi}\int_0^{\infty} d\theta d\epsilon \frac{\partial f}{\partial E}. \tag{14.2.6}$$

Here $E = \sqrt{\epsilon^2 + \Delta^2(\theta, T)}$, f is the Fermi-Dirac distribution function, $\Delta(\theta, T) = \Delta(T)(\cos 4\theta + s)$, and $\Delta(T) = \Delta \tanh(2.2\sqrt{T/T_c - 1})$ (Ref. [33]). In Fig. 14.5, we compare the experimental data for YBCO (Ref. [32]) (open circles) and the numerically calculated result (solid line) for the above deduced gap function $\Delta(\theta) = 24.5(\cos 4\theta + 0.225)$ meV. It is remarkable that the data are in quantitative agreement with the calculated result

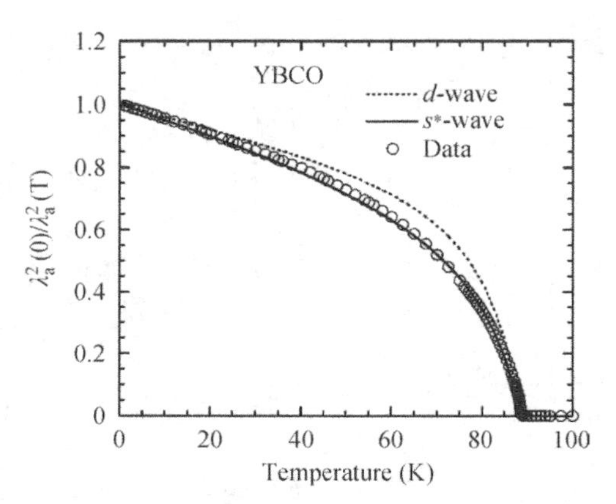

Fig. 14.5. Temperature dependence of the a axis $\lambda_a^2(0)/\lambda_a^2(T)$ for a very high-quality $YBa_2Cu_3O_7$ (YBCO) crystal with $T_c = 88.7$ K. The data were taken from Ref. 32. The solid line is the calculated curve for the s^*-wave gap function deduced from the tunneling spectrum in Fig. 14.4. The dash line is the calculated curve for a d-wave gap function with $\Delta_M = 30\,\text{meV}$.

without any fitting parameters. The dashed line is the calculated result for a d-wave gap function $\Delta(\theta) = \Delta_M \cos 2\theta$ with $\Delta_M = 30$ meV. It is clear that the agreement between the data and the calculated curve is poor for the d-wave symmetry. It is worthy to note that the temperature dependence of $\lambda_{ab}^2(0)/\lambda_{ab}^2(T)$ is mainly determined by the gap function, so the shape of the Fermi surface has little effect on $\lambda_{ab}^2(0)/\lambda_{ab}^2(T)$.

The gap function of YBCO deduced from the tunneling and the $\lambda_a(T)$ data is also consistent with the measured transverse magnetization m_T in the Meissner state [4], as plotted in Fig. 14.6. This bulk-sensitive experiment shows a very small sine fourfold component of the transverse magnetization, that is at least four times smaller than the predicted value from the d wave symmetry. This indicates that the dominant pairing symmetry is not the d-wave. Using the formulas reported in Ref. [34], we can calculate the sine components of the transverse magnetization for the s^*-wave gap function deduced above. We find that the sine fourfold component for the s^*-wave OP is a factor of 8.9 smaller than for the pure d-wave OP. The predicted sine-Fourier amplitude at period $2\pi/4$ is indicated by a horizontal solid line in Fig. 14.6. The calculated amplitudes at $2\pi/2, 2\pi/3, 2\pi/4$, and $2\pi/5$ are similar while the ones at other periods are much smaller. It is clear that the predicted amplitudes at all the periods are below the noise level, which is about 5×10^{-10} emu [4]. Therefore, the very small nonlinear

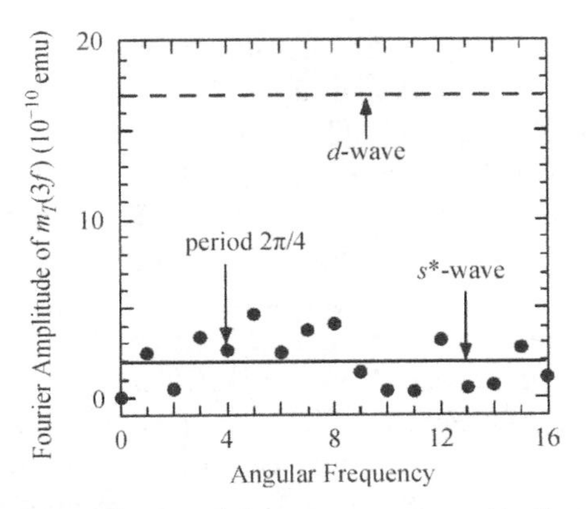

Fig. 14.6. Sine-Fourier amplitudes of the transverse magnetization m_T in the Meissner state for a high-quality $YBa_2Cu_3O_7$ (YBCO) crystal. The data were taken from Ref. [4]. The solid line is the predicted sine-Fourier amplitude at $2\pi/4$ for the s^*-wave gap function deduced from the tunneling spectrum in Fig. 14.4 and the a axis $\lambda_a(T)$ data in Fig. 14.5. The calculated amplitudes at $2\pi/2, 2\pi/3, 2\pi/4,$ and $2\pi/5$ are similar while the ones at other periods are much smaller. The dash line is the predicted sine-Fourier amplitude at $2\pi/4$ for a d-wave gap function.

Meissner effect observed in the overdoped YBCO is in agreement with the s^*-wave OP or with a nodeless OP (Ref. [4]) rather than with the d-wave OP. A nodeless OP symmetry is in contradiction to the observed linear T dependence of the thermal conductivity down to a very low temperature (50 mK) [35].

In addition, we further show that the s^*-wave gap function is in quantitative agreement with the low-temperature thermal conductivity, specific heat, and surface Andreevbound states. By replacing Δ_M with $\sqrt{\Delta_M \Delta_D}$ in the equations for the low-temperature electronic thermal conductivity κ_{el} and specific heat C_{el} for the d-wave gap function in the clean limit [35], we obtain the following equations for an s^*-wave gap function:

$$\frac{\kappa_{el}}{T} = \frac{k_B^2 v_F k_F}{6d\sqrt{\Delta_M \Delta_D}}, \tag{14.2.7}$$

and

$$\frac{C_{el}}{T^2} = \frac{9\zeta(3) k_B^3 k_F}{\pi \hbar v_F d\sqrt{\Delta_M \Delta_D}}. \tag{14.2.8}$$

Here, v_F and k_F are the Fermi velocity and momentum along the nodal directions, respectively, d is the average interlayer distance, and $\zeta(3) = 1.20$. One should note that impurity scattering tends to suppress the values of both κ_{el}/T and C_{el}/T^2. The Fermi velocity along the nodal directions has recently been obtained for YBCO from the studies of surface Andreev-bound states [36]. The deduced Fermi velocity v_F is $(1.2 \pm 0.2) \times 10^5$ m/s, which is a factor of 2 smaller than the measured Fermi velocity along the diagonal directions from the ARPES data of BSCCO [37]. This suggests that the nodal directions might be far away from the diagonal directions. For the s^*-wave gap function deduced above for overdoped YBCO, the nodal directions are about 19° away from the diagonal directions (i.e., at $\theta = 26°$). Indeed, from the ARPES data of BSCCO [37], one can clearly see that the Fermi velocity at $\theta = 26°$ is smaller than that at $\theta = 45°$ by a factor of about 2. Substituting $v_F = 1.2 \times 10^5$ m/s, $k_F = 0.7$ Å^{-1} [37], $d = 5.85$ Å, $\Delta_M = 30$ meV and $\Delta_D = 19$ meV into Eqs. (14.2.7) and (14.2.8), we obtain $\kappa_{el}/T = 0.12$ mW/K^{2}cm and $C_{el}/T^2 = 0.24$ mJ/moleK^3. The calculated values are in excellent agreement with the measured values: $\kappa_{el}/T = 0.14 \pm 0.03$ mW/K^{2}cm (Ref. [35]) and $C_{el}/T^2 = 0.20 \pm 0.05$ mJ/moleK^3 (Ref. [38]).

Moreover, thermal conductivity of YBCO as a function of the angle of an in-plane magnetic field relative to the crystal axes has been studied both theoretically and experimentally [39, 40]. A theoretical calculation for the angular dependence of the magnetothermal conductivity [39] shows that an extended s-wave gap produces a more symmetric angular variation than a d-wave gap. It appears that both sets of experimental data [39, 40] are more consistent with an extended s-wave gap than a d-wave gap.

14.2.3. *The Pairing Symmetry in $La_{2-x}Sr_xCuO_4$*

The polarized Raman scattering data [41] for nearly optimally doped $La_{2-x}Sr_xCuO_4$ (LSCO) with $T_c = 37$ K yield $2\Delta_M/k_BT_c = 7.7$. From the measured value of $d\lambda_{ab}(T)/[\lambda_{ab}(0)dT]$ for the optimally doped LSCO [42], one can readily calculate $2\sqrt{\Delta_M\Delta_D}/k_BT_c = 4.2$ using Eq. (14.2.4). Then we get $2\Delta_D/k_BT_c = 2.3$, i.e., $\Delta_D = 3.8$ meV. This value is in good agreement with the Andreev reflection spectrum of optimally doped LSCO [43], which shows the s-wave-like gap feature at $eV \simeq 3.5$ meV. Therefore, three independent bulk-sensitive experiments on optimally doped LSCO consistently suggest a gap function: $\Delta(\theta) = 8.1(\cos 4\theta + 0.53)$ meV with $\Delta_D = 3.8$ meV and $\Delta_M = 12.5$ meV.

Now we can quantitatively explain the neutron-scattering experiment on an optimally doped LSCO single crystal [44]. The experiment shows that low-energy magnetic excitations are peaked at the quartet of wavevectors $(0.5 \pm 0.135, 0.5)$ and $(0.5, 0.5 \pm 0.135)$ in the normal state, and a spin gap with energy of about 6.7 meV appears in the low-temperature superconducting state. The magnitude of the spin gap should be equal to twice the superconducting gap along the incommensurate wavevectors (i.e., at $\theta = 39°$) [45]. From the gap function deduced above, we calculate $2\Delta(39°) = 6.2\,\text{meV}$, in remarkably good agreement with experiment. Moreover, it was also found [44] that the spin gap at $\theta = 45°$ is 6 ± 2 meV, which is consistent with $2\Delta_D = 7.6$ meV within experimental uncertainty. Obviously, the *d*-wave gap function is incompatible with the large spin gap observed along the diagonal direction. The neutron data might also be consistent with an isotropic spin gap, as suggested by Lake *et al.* [44]. However, the isotropic spin gap is incompatible with the T^3 dependence of the spin-lattice relaxation rate observed in hole-doped cuprates. Only with the s^*-wave gap function for LSCO can one quantitatively explain the neutron experiment, Raman scattering, magnetic penetration depth, Andreev reflection, and magnetic resonances.

14.3. The Pairing Symmetry in Electron-Doped Cuprates

The recent measurements of $\lambda_{ab}(T)$ in an electron-doped $Pr_{1.85}Ce_{0.15}CuO_{4-y}$ (PCCO) reveal contradictory results [46, 47]. In a high-quality PCCO thin film with the lowest residual resistivity and the highest T_c, the temperature dependence of $[\lambda_{ab}(T) - \lambda_{ab}(0)]/\lambda_{ab}(0)$ is consistent with an *s*-wave pairing symmetry with a reduced energy gap $2\Delta(0)/k_BT_c = 2.9$ [46]. On the other hand, the low-temperature $\lambda_{ab}(T)$ in less ideal PCCO single crystals exhibits a power-law temperature dependence, as expected from a dirty *d*-wave superconductor [47].

We show that these apparently conflicting data might well be reconciled by a deeper understanding of how microstructure affects screening. It is well known, for example, that the screening length in the weakly coupled Josephson array of grains is dominated by the magnitude and temperature dependence of the Josephson coupling current between array elements [48]. Thus, tunnel coupling across grain boundaries and/or planar defects (weak links), rather than the BCS response of the grains themselves, mainly determines the magnetic screening length, surface resistance, and critical current (see the review article in Ref. [49]). The extrinsic effect due to the weak links can lead to a linear T dependence in the effective $\lambda_{ab}(T)$ at low temperatures and to a large residual surface resistance [50]. Similarly, Hebard *et al.* [51] showed that the current-induced nucleation of vortex–anti-vortex pairs at defects can make an additional extrinsic contribution to the screening length, i.e., a pinning penetration depth $\lambda^{p}_{ab}(T)$. Within this scenario, the $\lambda^{p}_{ab}(T)$ in zero magnetic field is given by [51]

$$\lambda^{p}_{ab}(t) = \lambda^{p}_{ab}(0)/(1 - t^2), \tag{14.3.1}$$

where $t = T/T_c$, $\lambda^{p}_{ab}(0) = [\Phi_0/H_c(0)]\sqrt{2N_d/\pi}$, Φ_0 is the flux quantum, N_d is the areal density of uniformly distributed defects, and $H_c(0)$ is the zero-temperature critical field. In the presence of the external dc field H, the expression for $\lambda^{p}_{ab}(0, H)$ has to be modified [50]. The total screening length is $\lambda_{ab}(t) = \sqrt{[\lambda^{L}_{ab}(t)]^2 + [\lambda^{p}_{ab}(t)]^2}$, where $\lambda^{L}_{ab}(t)$ is the intrinsic London penetration depth [51]. Assuming an s-wave pairing symmetry, we readily show that the $\lambda_{ab}(T)$ at low temperatures [below $0.2\Delta(0)/k_B$] is given by

$$\lambda_{ab}(T) = \lambda_{ab}(0) + \frac{[\lambda^{L}_{ab}(0)^2]}{\lambda_{ab}(0)}\sqrt{\pi\Delta(0)/2k_BT}$$
$$\times \exp[-\Delta(0)/k_BT] + \frac{\lambda^{2}_{ab}(0) - [\lambda^{L}_{ab}(0)]^2}{\lambda_{ab}(0)T_c^2}T^2. \tag{14.3.2}$$

It is clear that the T^2 dependence of $\lambda_{ab}(T)$ at low temperatures in zero field can be completely caused by the extrinsic effect, that is, the nucleation of vortex–anti-vortex pairs at defects. If N_d is negligible, $\lambda_{ab}(0) = \lambda^{L}_{ab}(0)$, and the second term in Eq. (14.3.2) is absent. Then we recover the BCS expression [52],

$$\lambda_{ab}(T) = \lambda_{ab}(0) + \lambda_{ab}(0)\sqrt{\pi\Delta(0)/2k_BT}\exp[-\Delta(0)k_BT]. \tag{14.3.3}$$

In Fig. 14.7(a), we plot temperature dependence of $\lambda_{ab}(T)$ below 6 K for a PCCO single crystal (the data are from Ref. [47]). The zero-temperature in-plane penetration depth $\lambda_{ab}(0)$ was measured to be 2500 Å [47]. This crystal shows T_c^{onset} at 22 K (defined by the onset of diamagnetism) and T_c^{mid} at 19 K [defined as the inflection point on $\lambda_{ab}(T)$] [47]. A wide superconducting transition in this crystal manifests a rather low quality of the crystal.

We fit the data by Eq. (14.3.2) with two fitting parameters $\Delta(0)$ and $\lambda^{L}_{ab}(0)$ and with a fixed $T_c = 20.5$ K (the average of T_c^{onset} and T_c^{mid}). The solid line is the fitted curve by Eq. (14.3.2). It is remarkable that the fit is very good. This can be seen more clearly in Fig. 14.7(b) where

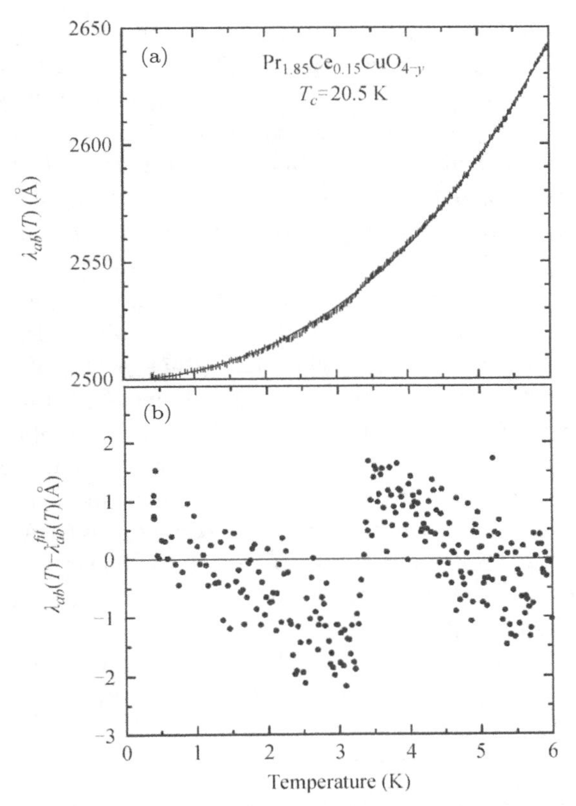

Fig. 14.7. (a) Temperature dependence of $\lambda_{ab}(T)$ below 6 K for a PCCO single crystal. The solid line is the fitted curve by Eq. (14.3.2) with $2\Delta(0)/k_B T_c = 2.9$ and $\lambda^L_{ab}(0) =$ 1643 Å. The value of $\lambda^L_{ab}(0)$ was found to be 1600 ± 100 Å from the optical data Ref. [53]. (b) The difference between the data and the fitted curve. The data are from Ref. [47].

the difference between the data and the fitted curve is plotted. There is a negligible systematic error (the deviation is less than the magnitude of the data scattering). From the fit, we find $\Delta(0)/k_B = 29.6 \pm 0.1$ K, and $\lambda^L_{ab}(0) = 1643$ Å. The deduced $\lambda^L_{ab}(0)$ is in excellent agreement with the value (1600 ± 100 Å) obtained from the optical data [53]. The magnitude of $2\Delta(0)/k_B T_c = 2.9$ is also the same as the one deduced from a high-quality film where the T^2 term is absent [46]. The value of $\Delta(0)$ justifies the fit to the data below 6 K, namely, $0.2\Delta(0)/k_B$. Therefore, the $\lambda_{ab}(T)$ data for the crystal are in quantitative agreement with an anisotropic s-wave pairing symmetry with no nodes.

From the values of $\lambda_{ab}^L(0)$ and $\lambda_{ab}(0)$, we calculate $\lambda_{ab}^p(0) = 1884$ Å. Using the relation $\lambda_{ab}^p(0) = [\Phi_0/H_c(0)]\sqrt{2N_d/\pi}$ and $H_c(0) = 2$ kOe [54], we estimate $N_d = 5.2 \times 10^{10}/\mathrm{cm}^2$, corresponding to one defect over 1333 Cu sites. This implies that a small density of defects can produce a quite large $\lambda_{ab}^p(0)$ that contributes a substantial T^2 term in $\lambda_{ab}(T)$.

In order to rule out the possibility that the data can be also consistent with a d-wave symmetry in the dirty limit, we plot the data as $1 - \lambda_{ab}^2(0)/\lambda_{ab}^2(T)$ vs T^2 in Fig. 14.8. It is apparent that the quantity $1 - \lambda_{ab}^2(0)/\lambda_{ab}^2(T)$ is proportional to T^2 below about 5 K. For a dirty d-wave superconductor, a crossover from T^2 to T dependence should be seen at a temperature $T^* \simeq \lambda_{ab}(0) \ln 2/[\Delta_M(0) d\lambda_{ab}/dT^2]$, where $\Delta_M(0)$ is the maximum gap at zero temperature [55]. Using $\lambda_{ab}(0) = 2500$ Å [47], $d\lambda_{ab}/dT^2 = 3.7$ Å/K^2 [47], and $\Delta_M(0) = 2.5T_c$ [56], one finds $T^* \simeq 9$ K. There is no such crossover at any temperatures up to 11 K (see Fig. 14.8). Only a possible crossover from the T^2 to a higher power-law dependence is seen at about 5 K. Therefore, the data cannot agree with the d-wave pairing symmetry. Furthermore, the absence of the linear T term in $\lambda_{ab}(T)$ indicates that the extrinsic contribution to $\lambda_{ab}(T)$ due to weak links [50] is negligible in this crystal.

In Fig. 14.9, we show $[\lambda_{ab}(T) - \lambda_{ab}(0)]/\lambda_{ab}(0)$ as a function of temperature for a high-quality PCCO thin film (the data are from Ref. [46]).

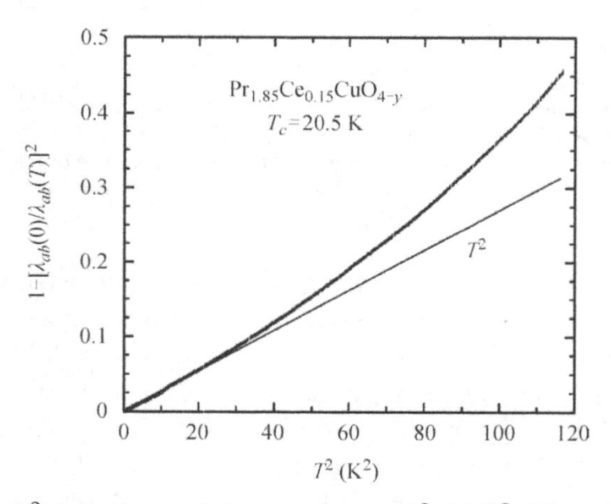

Fig. 14.8. The T^2 dependence of the quantity $1 - \lambda_{ab}^2(0)/\lambda_{ab}^2(T)$ over 0.4–10.8 K for the same PCCO crystal as the one in Fig. 14.7. The crossover from the T^2 to a higher power-law dependence starts at about 5 K. There is no crossover from the T^2 to the T dependence at $T^* \simeq 9$ K.

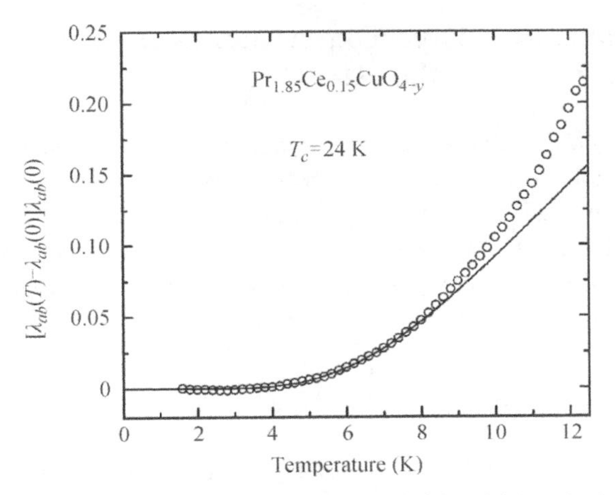

Fig. 14.9. Temperature dependence of $[\lambda_{ab}(T)-\lambda_{ab}(0)]/\lambda_{ab}(0)$ for a high-quality PCCO thin film with the lowest residual resistivity and the highest T_c. The solid line is the fitted curve by Eq. (14.3.3) with $2\Delta(0)/k_BT_c = 2.7$. The data are from Ref. 46.

The film has the lowest residual resistivity ($< 50\mu\Omega$ cm) and the highest T_c (24 K) reported for the PCCO system [46]. This indicates a high quality of the film, which was grown using molecular beam epitaxy. The optimal quality of the film may be due to the fact that a homogeneous oxygen reduction can be easily achieved in thin films. Since the data at low temperatures are quite flat, it appears that there is neither a T^2 nor T contribution. We thus fit the data below 6.5 K by Eq. (14.3.3) with one fitting parameter $\Delta(0)$. The best fit gives $\Delta(0)/k_B = 31.9\pm0.1$ K, which justifies the fit to the data below 6.5 K [$\sim 0.2\Delta(0)/k_B$]. This leads to $2\Delta(0)/k_BT_c = 2.7$, which is nearly the same as that deduced above for the less ideal crystal where there is a significant T^2 term in $\lambda_{ab}(T)$ due to the existence of defects. All these results consistently suggest that the pairing symmetry in electron-doped cuprates is the anisotropic s wave with no line nodes.

Polarized Raman scattering [56] has also shown that the symmetry of the order parameter in $Nd_{1.84}Ce_{0.16}CuO_{4-y}$ (NCCO) is consistent with an anisotropic s wave. More precisely, the tunneling spectra [57] are consistent with a gap function: $\Delta(\theta) = \Delta(s+\cos 4\theta)$ with $s > 1$. If we use $\Delta_M = 2.5T_c$ (Ref. [56]) and the minimum gap $\Delta_m = 1.4T_c$ [from the $\lambda_{ab}(T)$ data], we find $\Delta(\theta) = 1.15(3.52 + \cos 4\theta)$ meV for an electron-doped cuprate with $T_c = 24$ K. Therefore, three bulk-sensitive experiments consistently indicate an anisotropic s-wave pairing symmetry in electron-doped cuprates.

14.4. Phase-Sensitive Experiments Along the *C*-Axis Direction

The most reliable phase-sensitive experiment is the atomically clean BSCCO Josephson junctions between identical single-crystal cleaves stacked and twisted at an angle ϕ_0 about the c axis [12]. The quality of the junction is nearly the same as that of the intrinsic Josephson junctions made of the Bi-O insulating layers. Theoretically, it has been shown that the critical current I_c of the twist junction is [58]

$$I_c \propto \sum_l \eta_l \Delta_l \cos l\phi_0, \tag{14.4.1}$$

where $l = 0, 1, 2, \ldots$, and $\eta_l \ll \eta_0$ for $l \geq 1$. The above equation indicates that the s-wave component contributes to the critical current much more effectively. The experiment shows [12] that the I_c value is nearly independent of the twist angle ϕ_0, and the temperature dependence of I_c is consistent with the Ambegaokar-Baratoff model for an s-wave superconductor. This indicates that the s-wave component in this material must be significant compared with the other high angular momentum components. For slightly overdoped BSCCO, we have found that the gap function is $\Delta(\theta) = 17.75(\cos 4\theta + 0.46)$ meV for $T_c = 90$ K, and $\Delta(\theta) = 18(\cos 4\theta + 0.33)$ meV for $T_c = 86$ K. Then we have $\Delta_s = 6 - 8$ meV, which is not small compared with the g-wave component $\Delta_g = 18$ meV. Since $\eta_4 \ll \eta_0$ [58], the dominant contribution to the I_c should be the s-wave component, as observed [12]. From the magnitude of the s-wave component, we can calculate $I_c R_N = (\pi/2e)\Delta_s = 9 - 12$ mV. The measured $I_c R_N$ value is about 8 mV [12]. This is in quantitative agreement with the predicted value considering the fact that the strong coupling effect can reduce the $I_c R_N$ value by more than 20%.

Another reliable phase-sensitive experiment is the c-axis Pb/$YBa_2Cu_3O_{7-y}$ Josephson-junction experiment [13]. This junction can be described as $SINS'$ (where S = YBCO, S' = Pb, and I and N represent the insulating and normal-metal layers, respectively). Due to a very short coherent length ξ_c along the c-axis direction, the bulk gap will be strongly depressed at the SI interface; the depression factor is c/ξ_c (where c is the lattice constant along the c axis) [59]. From $\xi_c = \xi_{ab}/\gamma$ (where γ is the mass anisotropy parameter and equal to about 8 for optimally doped YBCO [60]), we get $\xi_c = 1.7$ Å by taking $\xi_{ab} = 14$ Å. Therefore, the gap size at the SI interface will be suppressed by a factor of about 7. Since the bulk s-wave component Δ_s in slightly overdoped YBCO is 3–5 meV (see above), this component at the SI interface should be reduced to 0.4–0.7 meV. Then the I_cR_N value is calculated to be 0.93–1.27 mV, in quantitative agreement with the measured one ($\sim$ 0.9 mV) [13].

Now we discuss another c-axis Josephson tunneling experiment in which a conventional superconductor (Pb) is deposited across a single twin boundary of a YBCO crystal [61]. Because Pb is an s-wave superconductor, the Pb counterelectrode couples only to the s-wave component of the YBCO order parameter. If YBCO were predominantly d wave, any small s-wave component added to the dominant d-wave component would change sign across the twin boundary. In this case, magnetic fields parallel to the boundary would produce a local minimum in I_c at $B = 0$, in agreement with the observation [61]. The experimental results thus appear to provide evidence for mixed d-and s-wave pairing symmetries in the bulk with a reversal in the sign of the s-wave component across the boundary. However, if the bulk OP symmetry in a single domain were $d + s$ or $d - s$, one would expect a nearly zero I_c in heavily twinned crystals. The fact that the observed I_cR_N in heavily twinned crystals [13] is nearly the same as the one in the single-domain crystal [61] rules out the bulk $d + s$- or $d - s$-wave OP symmetry in YBCO. Therefore the only possibility is that a half or fractional flux is trapped in the twin boundary. Also, this can naturally explain why I_c does not go to zero even for a symmetric junction with the same junction area in both sides of the twin boundary [61].

14.5. Phase-Sensitive Experiments Along the *ab* Planes

The phase-sensitive tricrystal experiments on both hole-and electron-doped cuprates [9, 10, 20] show that the OP symmetry is d wave, in contradiction with the above conclusion drawn from many bulk-sensitive experiments. In order to resolve the above discrepancy, one should notice that the tricrystal experiments are rather surface sensitive, so these experiments are probing the OP symmetry at the surface/interface, rather than in the bulk. Based on the Ginzburg-Landau free energy, Bahcall [18] has shown that the OP symmetry near surfaces/interfaces can be different from that in the bulk if the bulk OP is strongly suppressed at the surfaces. Experimentally, the observed I_cR_N values in all the tricrystal experiments are about two orders of magnitude smaller than the intrinsic Ambegaokar–Baratoff limit. For example, in the optimally doped $YBa_2Cu_3O_{7-y}$, the magnitude of the maximum gap $\Delta_M(0)$ is about 30 meV [8, 17]. Then the intrinsic I_cR_N value should be equal to the Ambegaokar–Baratoff limit $\pi\Delta_M(0)/2e = 47$ mV, which has been confirmed by a nearly ideal SIS break junction experiment [17]. However, the observed I_cR_N values in the tricrystal experiments on $YBa_2Cu_3O_{7-y}$ and $Tl_2Ba_2CuO_{6+y}$ (Refs. [9] and [10]) are about 1.8 mV and 0.5 mV, respectively. These values are about two orders of magnitude smaller than the intrinsic bulk values. Similarly, the observed I_cR_N value in the NCCO and PCCO tricrystal experiments is about 0.1 mV, as inferred from the measured critical current density $J_c = 6$ A/cm^2 (Ref. [20]) and the empirical relation between I_cR_N and J_c [62]. This I_cR_N value is also about two orders of magnitude smaller than the intrinsic bulk value, which is estimated to be ~8 mV with $\Delta_M(0) = 2.5T_c$ [56]. Therefore, the OP at the interfaces of the grain-boundary junctions must be strongly depressed in order to explain such small I_cR_N values. This strong depression in the order parameter ensures the condition under which the OP symmetry near surfaces/interfaces can be different from that in the bulk [18]. Hence, it is very likely that the tricrystal experiments are detecting the OP symmetry at the degraded interfaces, which may be different from the intrinsic one in the bulk.

Now the question arises: How can the bulk OP be so strongly depressed at the surfaces of the grain-boundary junctions? It is known that the coherent length in cuprates is very short due to a large superconducting gap and small Fermi velocity. The short coherent length in cuprates can lead to a large depression of the OP near the interfaces even within the conventional theory of the proximity effect [59, 63]. Alternatively, several groups [64–66] showed that there are possibly nonsuperconducting regions near the boundary of the junction due to hole depletion and/or strain, so that the critical current density can be reduced by several orders of magnitude compared with the intrinsic bulk value.

There is another way to explain the tricrystal experiments. As discussed above, the boundaries of the grain-boundary junctions are intrinsically underdoped superconductors or nonsuperconducting anti-ferromagnets due to hole depletion and/or strain [64–67]. For underdoped cuprates, the superconductivity mainly arises from the Bose-Einstein condensation of preformed pairs [68]. In this case, the symmetry of the superconducting condensate is different from the pairing symmetry; the former is d wave while the latter might be s-wave [68]. Since Josephson tunneling probes the symmetry of the superconducting condensate, the d-wave symmetry of the condensate is consistent with the tricrystal experiments.

14.6. Conclusion

In conclusion, the existing bulk- and nearly bulk-sensitive experiments consistently show that the dominant bulk pairing symmetry in hole-doped cuprates is of extended s wave with eight line nodes and of anisotropic s wave in electron-doped cuprates. The deduced extended s-wave pairing symmetry for hole-doped cuprates is also in quantitative agreement with the phase-sensitive experiments along the c axis for both $Bi_2Sr_2CaCu_2O_{8+y}$ and $YBa_2Cu_3O_{7-y}$. The proposed pairing symmetries do not contradict some surface- and phase-sensitive experiments that show a predominant d-wave pairing symmetry at the degraded surfaces.

Acknowledgments

The author would like to thank R. Prozorov, L. Alff, S. Kamal, and W. N. Hardy for sending their published data.

References

[1] W. N. Hardy, D. A. Bonn, D. C. Morgan, Ruixing Liang, and K. Zhang, *Phys. Rev. Lett.* **70** (1993), 3999.
[2] T. Jacobs, S. Sridhar, Q. Li, G.D. Gu, and N. Koshizuka, *Phys. Rev. Lett.* **75** (1995), 4516.
[3] S.-F. Lee, D. C. Morgan, R. J. Ormeno, D. Broun, R. A. Doyle, J. R. Waldram, and K. Kadowaki, *Phys. Rev. Lett.* **77** (1996), 735.
[4] A. Bhattacharya, I. Zutic, O. T. Valls, A. M. Goldman, U. Welp, and B. Veal, *Phys. Rev. Lett.* **82** (1999), 3132.
[5] M. Willemin, C. Rossel, J. Hofer, H. Keller, Z. F. Ren, and J. H. Wang, *Phys. Rev. B* **57** (1998), 6137.
[6] A. Sacuto, R. Combescot, N. Bontemps, P. Monod, V. Viallet, and D. Colson, *Europhys. Lett.* **39** (1997), 207.
[7] C. Kendziora, R. J. Kelley, and M. Onellion, *Phys. Rev. Lett.* **77** (1996), 727.
[8] J. Y. T. Wei, N.-C. Yeh, D. F. Garrigus, and M. Strasik, *Phys. Rev. Lett.* **81** (1998), 2542.
[9] C. C. Tsuei, J. R. Kirtley, M. Rupp, J. Z. Sun, L.-S. Yu-Jahnes, C. C. Chi, A. Gupta, and M. B. Ketchen, *J. Phys. Chem. Solids* **56** (1995), 1787.

[10] C. C. Tsuei, J. R. Kirtley, M. Rupp, J. Z. Sun, A. Gupta, M. B. Ketchen, C. A. Wang, Z. F. Ren, J. H. Wang, and M. Bhushan, *Science* **271** (1996), 329.
[11] J. R. Kirtley, C. C. Tsuei, and K. A. Moler, *Science* **285** (1999), 1373.
[12] Q. Li, Y. N. Tsay, M. Suenaga, R. A. Klemm, G. D. Gu, and N. Koshizuka, *Phys. Rev. Lett.* **83** (1999), 4160.
[13] A. G. Sun, D. A. Gajewski, M. B. Maple, and R. C. Dynes, *Phys. Rev. Lett.* **72** (1994), 2267.
[14] H. Ding, J. C. Campuzano, A. F. Bellman, T. Yokoya, M. R. Norman, M. Randeria, T. Takahashi, H. Katayama-Yoshida, T. Mochiku, K. Kadowaki, and G. Jennings, *Phys. Rev. Lett.* **74** (1995), 2784.
[15] R. J. Kelley, C. Quitmann, M. Onellion, H. Berger, P. Almeras, and G. Margaritondo, *Science* **271** (1996), 1255.
[16] I. Vobornik, R. Gatt, T. Schmauder, B. Frazer, R. J. Kelley, C. Kendziora, M. Grioni, M. Onellion, and G. Margaritondo, *Physica C* **317–318** (1999), 589.
[17] Ya. G. Ponomarev, B. A. Aminov, M. A. Hein, H. Heinrichs, V. Z. Kresin, G. Müller, H. Piel, K. Rosner, S. V. Tchesnokov, E. B. Tsokur, D. Wehler, K. Winzer, A. V. Yarygin, and K. T. Yusupov, *Physica C* **243** (1995), 167.
[18] S. R. Bahcall, *Phys. Rev. Lett.* **76** (1996), 3634.
[19] J. Ma, C. Quitmann, R. J. Kelley, G. Margaritondo, and M. Onel lion, *Solid State Commun.* **94** (1995), 27.
[20] C. C. Tsuei and J. R. Kirtley, *Phys. Rev. Lett.* **85** (2000), 182.
[21] N. Miyakawa, P. Guptasarma, J. F. Zasadzinski, D. G. Hinks, and K. E. Gray, *Phys. Rev. Lett.* **80** (1998), 157.
[22] Y. DeWilde, N. Miyakawa, P. Guptasarma, M. Iavarone, L. Ozyuzer, J. F. Zasadzinski, P. Romano, D. G. Hinks, C. Kendziora, C. W. Crabtree, and K. E. Gray, *Phys. Rev. Lett.* **80** (1998), 153.
[23] I. Kosztin and A. J. Leggett, *Phys. Rev. Lett.* **79** (1997), 135.
[24] K. Suzuki, K. Ichimura, K. Nomura, and S. Takekawa, *Phys. Rev. Lett.* **83** (1999), 616.
[25] A. Mourachkine. cond-mat/9901282, unpublished.
[26] V. M. Krasnov, A. Yurgens, D. Winkler, P. Delsing, and T. Claeson, *Phys. Rev. Lett.* **84** (2000), 5860.
[27] R. Prozorov, R. W. Giannetta, A. Carrington, P. Fournier, R. L. Greene, P. Guptasarma, D. G. Hinks, and A. R. Banks, *Appl. Phys. Lett.* **77** (2000), 4202.
[28] L. Buschmann, M. Boekholt, and G. Güntherodt, *Physica C* **203** (1992), 68.
[29] L. Ozyuzer, J. F. Zasadzinski, C. Kendziora, and K. E. Gray, *Phys. Rev. B* **61** (2000), 3629.
[30] I. Maggio-Aprile, Ch. Renner, A. Erb, E. Walker, and O. Fischer, *Phys. Rev. Lett.* **75** (1995), 2754.
[31] Y. Yagil, N. Hass, G. Desgardin, and I. Monot, *Physica C* **250** (1995), 59.
[32] S. Kamal, R. X. Liang, A. Hosseini, D. A. Bonn, and W. N. Hardy, *Phys. Rev. B* **58** (1998), R8933.

[33] D. Thelen, D. Pines, and J. P. Liu, *Phys. Rev. B* **47** (1993), 9151.
[34] I. Zutic and O. T. Valls, *Phys. Rev. B* **56** (1997), 11279.
[35] M. Chiao, R. W. Hill, C. Lupien, L. Taillefer, P. Lambert, R. Gagnon, and P. Fournier, *Phys. Rev. B* **62** (2000), 3554.
[36] A. Carrington, F. Manzano, R. Prozorov, R. W. Giannetta, N. Kameda, and T. Tamegai, *Phys. Rev. Lett.* **86** (2001), 1074.
[37] A. Kaminski, M. Randeria, J. C. Campuzano, M. R. Norman, H. Fretwell, J. Mesot, T. Sato, T. Takahashi, and K. Kadowaki, *Phys. Rev. Lett.* **86** (2001), 1070.
[38] A. Junod, B. Revaz, Y. Wang, and A. Erb, *Physica B* **284–288** (2000), 1043.
[39] F. Yu, M. B. Salamon, A. J. Leggett, W. C. Lee, and D. M. Gins-berg, *Phys. Rev. Lett.* **74** (1995), 5136.
[40] H. Aubin, K. Behnia, M. Ribault, R. Gagnon, and L. Taillefer, *Phys. Rev. Lett.* **78** (1997), 2624.
[41] X. K. Chen, J. C. Irwin, H. J. Trodahl, T. Kimura, and K. Kishio, *Phys. Rev. Lett.* **73** (1994), 3290.
[42] C. Panagopoulos, B. D. Rainford, J. R. Cooper, W. Lo, J. L. Tallon, J. W. Loram, J. Betouras, Y. S. Wang, and C. W. Chu, *Phys. Rev. B* **60** (1999), 14617.
[43] G. Deutscher, N. Achsaf, D. Goldschmidt, A. Revcolevschi, and A. Vietkine, *Physica C* **282–287** (1997), 140.
[44] B. Lake, G. Aeppli, T. E. Mason, A. Schröder, D. F. McMorrow, K. Lefmann, M. Isshiki, M. Nohara, H. Takagi, and S. M. Hay-den, *Nature (London)* **400** (1999), 43.
[45] T. E. Mason, A. Schröder, G. Aeppli, H. A. Mook, and S. M. Hay-den, *Phys. Rev. Lett.* **77** (1996), 1604.
[46] L. Alff, S. Meyer, S. Kleefisch, U. Schoop, A. Marx, H. Sato, M. Naito, and R. Gross, *Phys. Rev. Lett.* **83** (1999), 2644.
[47] R. Prozorov, R. W. Giannetta, P. Fournier, and R. L. Greene, *Phys. Rev. Lett.* **85** (2000), 3700.
[48] B. Giovannini and L. Weiss, *Solid State Commun.* **27** (1978), 1005.
[49] J. Halbritter, *Supercond. Sci. Technol.* **12** (1999), 883.
[50] J. Halbritter, *J. Appl. Phys.* **71** (1992), 339.
[51] A. F. Hebard, A. T. Fiory, M. P. Siegal, J. M. Phillips, and R. C. Haddon, *Phys. Rev. B* **44** (1991), 9753.
[52] B. Mühlschlegel, *Z. Phys.* **155** (1959), 313.
[53] C. C. Homes, B. P. Clayman, J. L. Peng, and R. L. Greene, *Phys. Rev. B* **56** (1997), 5525.
[54] D. H. Wu, J. Mao, S. N. Mao, J. L. Peng, X. X. Xi, T. Venkatesan, R. L. Greene, and S. Anlage, *Phys. Rev. Lett.* **70** (1993), 85.
[55] P. J. Hirschfeld and N. Goldenfeld, *Phys. Rev. B* **48** (1993), 4219.
[56] B. Stadlober, G. Krug, R. Nemetschek, R. Hackl, J. L. Cobb, and J. T. Markert, *Phys. Rev. Lett.* **74** (1995), 4911.

[57] S. Kashiwaya, T. Ito, K. Oka, S. Ueno, H. Takashima, M. Koyanagi, Y. Tanaka, and K. Kajimura, *Phys. Rev. B* **57** (1998), 8680.
[58] R. A. Klemm, C. T. Rieck, and K. Scharnberg, *Phys. Rev. B* **58** (1998), 1051.
[59] K. A. Müller. *Nature* (*London*) **377** (1995), 133; G. Deutscher and K. A. Müller. *Phys. Rev. Lett.* **59** (1987), 1745.
[60] M. Willemin, A. Schilling, H. Keller, C. Rossel, J. Hofer, U. Welp, W. K. Kwok, R. J. Olsson, and G. W. Crabtree, *Phys. Rev. Lett.* **81** (1998), 4236.
[61] K. A. Kouznetsov, A. G. Sun, B. Chen, A. S. Katz, S. B. Bahcall, J. Clarke, R. C. Dynes, D. A. Gajewski, S. H. Han, M. B. Maple, J. Giapintzakis, J. T. Kim, and D. M. Ginsberg, *Phys. Rev. Lett.* **79** (1997), 3050.
[62] S. Kleefisch, L. Alff, U. Schoop, A. Marx, R. Gross, M. Naito, and H. Sato, *Appl. Phys. Lett.* **72** (1998), 2888.
[63] M. Y. Kupryanov and K. K. Likharev, *IEEE Trans. Magn.* **27** (1991), 2400.
[64] J. Halbritter, *Phys. Rev. B* **46** (1992), 14861.
[65] J. Betouras and R. Joynt, *Physica C* **250** (1995), 256.
[66] A. Gurevich and E. A. Pashitskii, *Phys. Rev. B* **57** (1998), 13878.
[67] J. Mannhart and H. Hilgenkamp, *Physica C* **317–318** (1999), 383.
[68] A. S. Alexandrov, *Physica C* **305** (1998), 46.

Part V

Effect of Lattice

15
Block Model and Origin of Strong Anisotropy in High-T_c Superconductors

H. Zhang

Materials Physics Laboratory, State Key Laboratory for Mesoscopic Physics, Department of Physics, Peking University, Beijing, 100871, China

All the high T_c superconductors are structurally composed of two different blocks, the perovskite, and the rock salt. The perovskite block is the active block where the Cu(2)–O planes are located and the carriers concentrate. The rock salt block corresponds to the charge-reservoir block, which supplies the carriers to the Cu(2)–O planes. Based on this fact, we developed a model to calculate the interaction between the two blocks (marked by combinative energy) and to study their relationship with the value of T_c. Bi-system superconductors, (such as $Bi_2Sr_2CuO_y$, $Bi_2Sr_2CaCu_2O_y$ and $Bi_2Sr_2Ca_2Cu_3O_y$), Hg-system superconductors (such as $HgBa_2CuO_y$, $HgBa_2 CaCu_2O_y$, $HgBa_2Ca_2Cu_3O_y$, $HgBa_2Ca_3Cu_4O_y$ and $HgBa_2Ca_4Cu_5O_y$), Tl-system superconductors (such as $Tl_2Ba_2Ca_{n-1}Cu_nO_{2n+3}$ with 5 superconductive compounds and $Tl_2Ba_2Ca_{n-1}Cu_nO_{2n+4}$ with 4 superconductive compounds), Y-system superconductors with different oxygen deficiencies, and $La_{2-x}M_xCuO_4$ (M=Ba, Sr) with different M concentrations, were studied with our model, respectively. A close relationship between the superconducting transition temperature and the combinative energy in these superconducting systems is established, i.e., the higher the combinative energy is, the lower the T_c value is. Furthermore, it is discovered by X-ray analysis and Raman spectroscopy that the Cu(2)-O plane located in the boundary of the two blocks is very steady, i.e., the bond lengths and angles in this plane are hardly changed and form a so-called "fixed triangle". This fixed triangle may be in charge of the strong anisotropy of the high T_c superconductors. It is suggested that because of the very strong anisotropy, not only the energy, but the anisotropy of phonons should be taken into account. Under this strong anisotropy, the state of the carrier may be changed or polarized, and then determines the superconductivity. The strong anisotropy of phonons may be the main difference between the conventional superconductors and the high T_c superconductors. The results show that the lattice dynamics or electron–phonon interaction is still of great importance for high T_c superconductivity.

15.1. Introduction and Motivation

Because the high T_c superconductors demonstrate strong 2D characteristics, the coupling along the c direction in the crystals is of great importance, and this characteristic has been studied extensively [1]. In most of these studies, the high T_c superconductor is regarded as a multilayer with superconducting and non-superconducting layers, and the coupling between the superconducting layers is considered. However, the role of different layers to superconductivity is not very clear. As is well known, there are many layers along the c direction in the high T_c superconductors. If we consider the coupling between each plane, it will be difficult to deal with this problem; however, if we can divide these layers into a few blocks reasonably based on the structural characters and experimental facts, the problem will become easy. The aim of this study is based on this consideration.

A lot of work has been done on understanding the superconductivity from different ways. Among those researches, to study the relationship between cohesive energy and the superconductivity is an important way. Billesbach [2] calculated the lattice instability using a rigid-ion model. Torrance [3] studied the effect of Madelung energy on the hole conductivity in the high T_c superconductors. Muroi [4] calculated the cohesive energy as a function of different Cu-O planes, and found the cohesive energy had correlation with the hole concentration in the Cu-O planes. Zhang [5] used the cohesive energy to explain the change of T_c of Y-doped superconductors successfully. Ohta [6] studied the Madelung energy of Pb-system superconductor, and found some relations between the energy and carrier concentration. Mizuno [7] found some relationship between the Madelung energy and carrier in La-system superconductor. Mueller [8] reviewed that the study of Madelung energy was of great importance to high T_c superconductivity. Although much work has been done on this aspect, quantitative relationship between the cohesive energy and superconductivity has not been established within our knowledge.

All the high-T_c superconductors are structurally composed of two different blocks, the perovskite, and the rock salt [9] (see Fig. 15.1). The

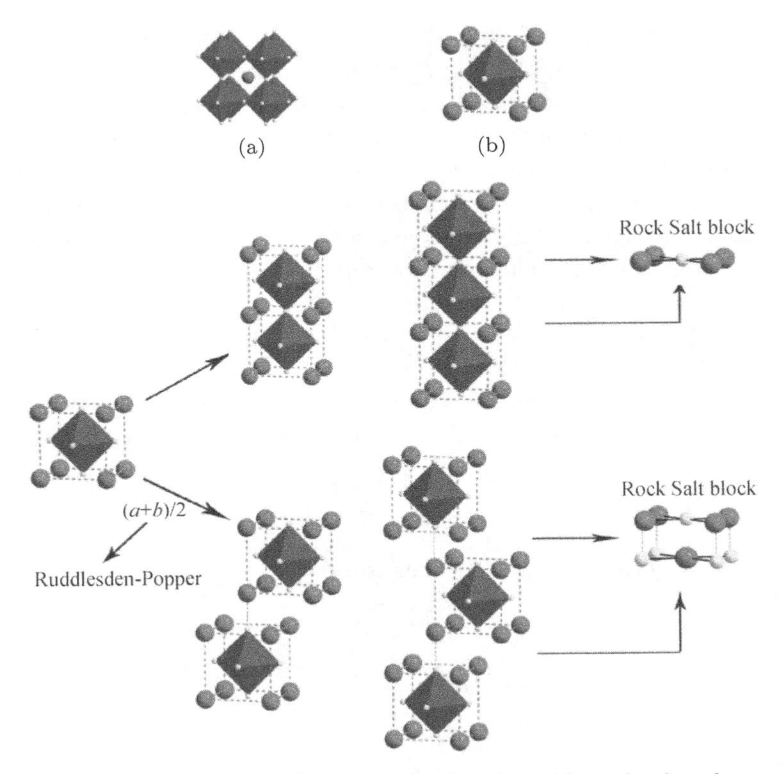

Fig. 15.1. Structure of the high-T_c superconductors, stacked by rock salt and perovskite blocks along c-direction. (a) and (b) are two different stacking ways.

perovskite block is the active block where the Cu(2)–O planes are located and the carriers concentrate. The rock salt block corresponds to the charge-reservoir block, which supplies the carriers to the Cu(2)–O planes. Without any one of the two blocks, it is not high-T_c superconductors or even non-superconducting. This fact demonstrates that the interaction of the two blocks is a key fact for the high-T_c superconductivity. Based on this fact, e believe that the lattice dynamics is still important for the high-T_c superconductivity, and developed a model to calculate the cohesive energy of different superconducting systems. When treating the cell as two blocks: perovskite and rock salt, it is found that the combinative energy between the two blocks is closely related with the value of T_c. This result supports our point of view that the interaction between the two blocks is very important to superconductivity.

Furthermore, it is discovered by X-ray analysis and Raman spectroscopy that the Cu(2)–O plane located in the boundary of the two blocks is very steady, and forms a so called "fixed triangle". This fixed triangle may be in charge of the strong anisotropy of the high-T_c superconductors. It is suggested that because of the very strong anisotropy, not only the energy, but the anisotropy of phonons should be taken into account. Under this strong anisotropy, the state of carriers may be changed or polarized, and then determines the superconductivity.

15.2. Model

According to the classical theory of crystals, the cohesive energy E_{n} is made up of Madelung energy E_{m}, repulsive energy of ions E_{r}, and electron affinity energy E_{a},

$$E_{\mathrm{n}} = E_{\mathrm{m}} + E_{\mathrm{r}} + E_{\mathrm{a}}, \tag{15.2.1}$$

which can be derived by the following formula:

$$E_{\mathrm{m}} = 1/2\alpha \sum e_i e_j / r_{ij}, \tag{15.2.2}$$

$$E_{\mathrm{r}} = a\mathrm{e}^{-r/\rho}, \tag{15.2.3}$$

$$E_{\mathrm{a}} = \sum \sum \varepsilon_{ij}. \tag{15.2.4}$$

Here, E_{m}, E_{r}, and E_{a} represent the Madelung energy, repulsive energy and electron affinity, respectively. e_i, e_j are the electric charges of different atoms in cell, ε_{ij} is the ith ionization energy of the jth atoms in cell, r is the distance between the positive and negative ions and a, α are the coefficients. We discard the electron affinity energy, because once the atom becomes ion, the ion has a closed outer shell then, and the electron affinity energy will not strongly affect other electrons or vacancies any more. We use ionic model to simplify this problem. Some authors [2–8] have demonstrated that the ionic model can be used to deal with the high-temperature superconductors. According to Pauling's rule [10], it is reasonable to consider that the high-T_c superconductors are ionic compounds. But, obviously, they have some covalent character. In the Cu(2)–O plane, Cu3d and O2p orbits hybridize, giving carriers. In order to compensate for the deficiency of the ionic model for the high-T_c superconductors, we directly put some holes on the Cu(2)–O plane, the number of the holes depends on the oxygen deficiency. The whole cell is kept electrically neutral. This method is consistent with the experimental fact that the holes are mainly concentrated on the Cu(2)–O plane. In this way, the covalence is approximately considered, which makes the calculation more precise and the model more reasonable.

To calculate the Madelung energy, we use the standard Evjen [11] method. In this way, the distribution of charges in a cell is balanced and the summation is highly convergent. In the calculation of the repulsive energy we use a Bohr approximation. To test the accuracy of this program we calculated several samples and found that the calculated results matched the experimental results very well [12]. We believe that the program is reliable.

Besides the calculation of the cohesive energy of whole cell, the energy of the different parts in a single cell was calculated for the consideration of the interaction of the different blocks mentioned above. To calculate the energy between the different parts in a single cell (Hereafter, in order to differentiate it from the cohesive energy of whole cell, it is called combinative energy.) there are two ways. The first way is to separate all the Cu–O planes out of the cell and leave the remains. This method is demonstrated by Fig. 15.2(a). In this way, all the Cu–O planes are considered equally. All the Cu–O planes are separated from the cell into independent plane,

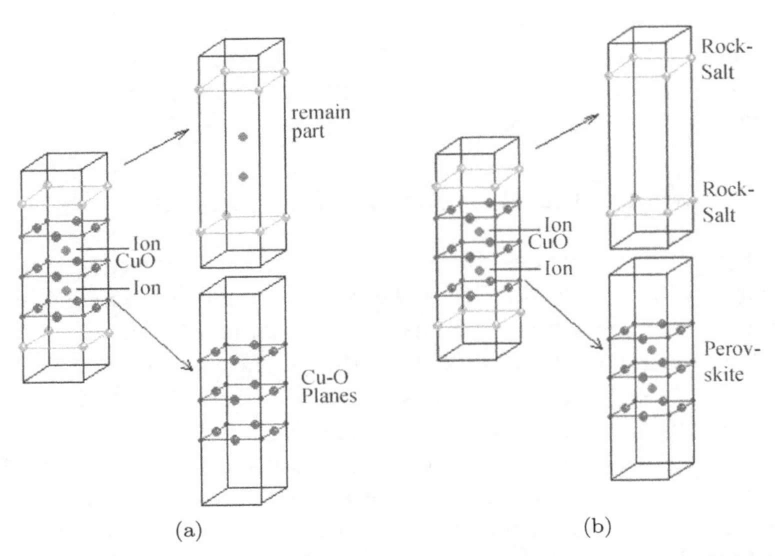

Fig. 15.2. (a) The structure is separated into layers; (b) the structure is divided into two blocks: perovskite and rock salt bocks.

leaving some discrete remaining parts. After calculating, we get the average combinative energy between each Cu–O plane and the remains. The combinative energy indicates the strength of interaction between each Cu–O plane and its adjacent planes.

Another way is to treat the cell as two different blocks, so-called perovskite and rock salt, instead of some independent planes. Figure 15.2(b) demonstrates this process. Unlike the first method, in this way the perovskite and rock salt block are considered as "packaged unit". Then the combinative energy calculated will mainly indicate the interaction between the two blocks in the cell. In fact, we do find something interesting in this way, which does not appear in the first method.

The relationship among these parameters is:

$$E_{\mathrm{n}} = E_{\mathrm{pb}} + E_{\mathrm{rb}} + E_{\mathrm{cb}}, \tag{15.2.5}$$

where E_{n} is the total cohesive energy of the cell, E_{pb} is the cohesive energy of perovskite black, E_{rb} is the cohesive energy of the rock salt block, and E_{cb} is the combinative energy between the two blocks.

Bi-system superconductors [12], Hg-system superconductors [12], Tl-system superconductors [13, 14], $Tl_2Ba_2Ca_{n-1}Cu_nO_{2n+3}$ and $Tl_2Ba_2Ca_{n-1}Cu_nO_{2n+4}$, with different numbers of Cu-O planes, Y-system superconductors [15] with different oxygen deficiencies, $La_{2-x}M_xCuO_4$ (M=Ba, Sr) [16] with different M concentrations have been studied with this model, respectively. The detailed results can be known from the related references. The following just gives some brief results.

15.3. Calculating Results and Discussions

With the first methods described in the model, i.e., separating the structure into layers, we calculated the cohesive and combinative energies of the Bi-system superconductors (such as $Bi_2Sr_2CuO_y$ (2201), $Bi_2Sr_2CaCu_2O_y$ (2212) and $Bi_2Sr_2Ca_2Cu_3O_y$ (2223)), Hg-system superconductors (such as $HgBa_2CuO_y$ (1201), $HgBa_2CaCu_2O_y$ (1212), $HgBa_2Ca_2Cu_3O_y$ (1223), $HgBa_2Ca_3Cu_4O_y$ (1234) and $HgBa_2Ca_4Cu_5O_y$ (1245)), Y-system superconductors with different oxygen deficiencies. The calculated results show that there is no obvious correlation among the cohesive energy, combinative energy, value of T_c and the number of the Cu-O planes. For example, in the Bi-system, the combinative energy is 46.83 eV for the 2201 phase, 38.28 eV for the 2212 phase, 51.37 eV for the 2223 phase. We can't find any regular

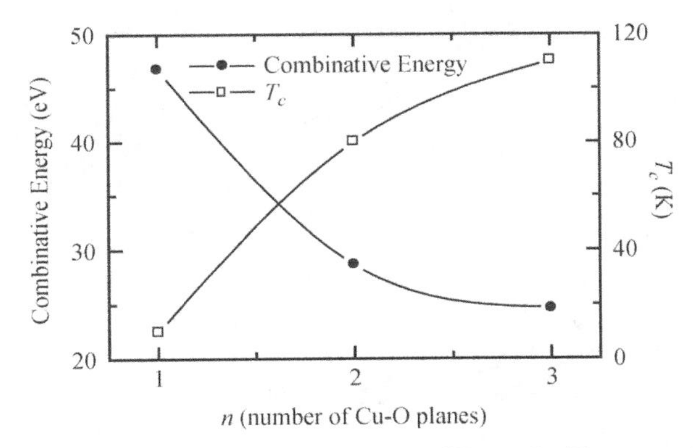

Fig. 15.3. The combinative energy and the value of T_c in the Bi-system. There exists a close relationship between the combinative energy and the value of T_c.

pattern among these parameters. These results show that separating the structure into different layers is not reasonable.

The second method shows different result from the first one. We also calculated the same parameters as we did on the first model. Figure 15.3 illustrates the relationship among the combinative energy between the two blocks, the value of T_c and the number of the Cu–O planes in the Bi-system. Clearly, there exists an obvious correlation between the combinative energy and the value of T_c. As the value of T_c gets the maximum in the three Cu–O planes, the combinative energy between the two blocks gets the minimum. The value of T_c and combinative energy demonstrate very good correspondence.

In order to confirm our calculation and the correlation among the value of T_c, combinative energy between the two blocks and the number of the Cu–O planes, the parameters described above were further calculated for the Hg-system superconductors by the two different methods. The Hg-system has five superconducting phases, 1201, 1212, 1223, 1234 and 1245. The relationship between the value of T_c and the number of the Cu–O planes shows a clear dome-ship. If the correlation as that in the Bi-system exists in the Hg-system, it will be more reliable and important. In the Hg-system, with the increase of the Cu–O planes, the value of T_c gets a maximum ($\sim$133 K) in the 1223 phase, i.e., the number of the Cu–O plane is three. Then it decreases with the increase of more Cu–O planes. For the 1234 and 1245 phases, they have more than three Cu–O planes, but the value of the T_c of them is not over 133 K. So far, there has not been a

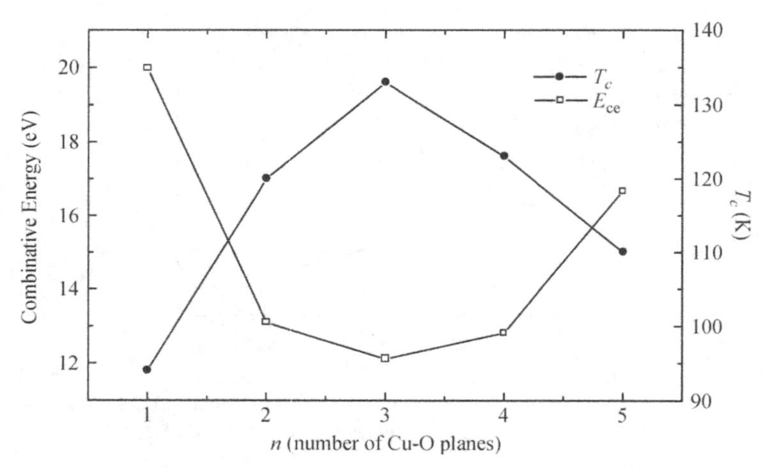

Fig. 15.4. The combinative energy E_{ce} and the value of T_c in the Hg-system with five Cu–O planes.

satisfying answer for the change of the value of T_c as the function of the number of the Cu–O planes. The following results may give a possible way to understand it.

Figure 15.4 illustrates the relationship among the combinative energy between the two blocks, the value of T_c and the number of the Cu–O planes in the Hg-system. Clearly, there exists an obvious correlation between the combinative energy and the value of T_c. As the value of T_c gets the maximum in the three Cu–O planes, the combinative energy between the two blocks gets the minimum. The value of T_c and combinative energy demonstrate very good correspondence. For the $TlBa_2Ca_{n-1}Cu_nO_{2n+3}$ series, the highest T_c value appear in $n = 4$, but for the $Tl_2Ba_2Ca_{n-1}Cu_nO_{2n+4}$ series, $n = 3$ (see Figs. 15.5 and 15.6). There is no corresponding relationship between the T_c and the number of n, but the relationship between the T_c value and the combinative energy is obviously corresponding.

For the Bi-system Hg-system and Tl-system, we calculated the systems with different superconducting phases. For $YBa_2Cu_3O_{7-\delta}$, one superconducting phase with the different δ value (from 0.07 to 0.62), was calculated. Figure 15.7 demonstrates the combinative energy calculated from the orth I, II, i.e., when $\delta < 0.35$, using the orth I; when $\delta > 0.35$, using the orth II, respectively. The difference between the three curves is very small. The results demonstrate a very close relationship between the T_c value and the combinative energy. As the T_c value increases, the combinative energy

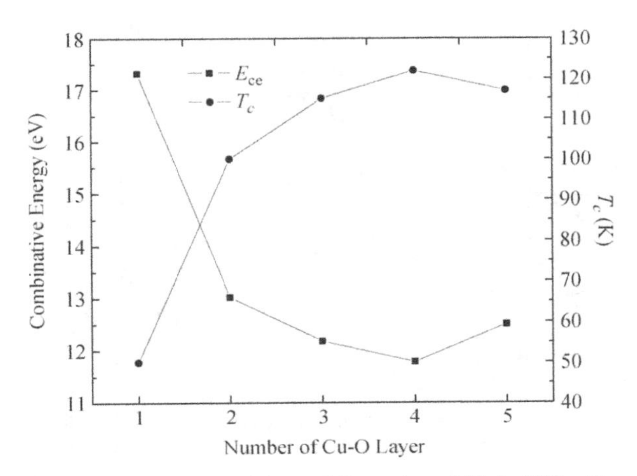

Fig. 15.5. The combinative energy E_{ce} and the value of T_c in $\text{TlBa}_2\text{Ca}_{n-1}\text{Cu}_n\text{O}_{2n+3}$ ($n = 1, 2, 3, 4, 5$).

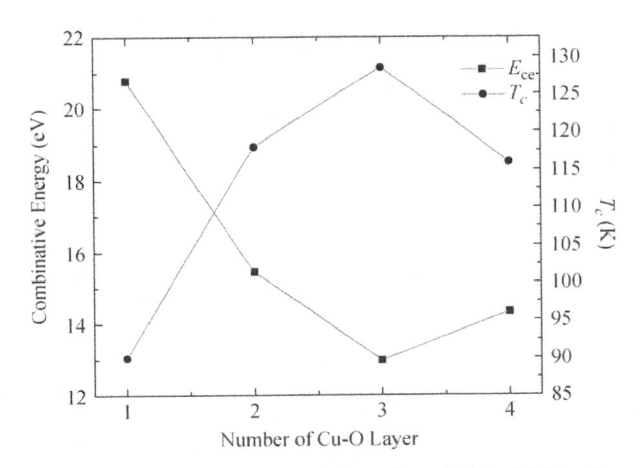

Fig. 15.6. The combinative energy E_{ce} and the value of T_c in $\text{Tl}_2\text{Ba}_2\text{Ca}_{n-1}\text{Cu}_n\text{O}_{2n+4}$ ($n = 1, 2, 3, 4$).

decreases. At about 60 K, the T_c value shows a plateau, and the combinative energy shows a plateau at about $\delta \sim 0.45$ too (corresponding to $T_c = 60$ K). This result further demonstrates that the interaction between the two blocks is of great importance for superconductivity.

Figures 15.8 and 15.9 shows the relationship between the T_c and combinative energy in $\text{La}_{2-x}\text{Sr}_x\text{CuO}_4$ and $\text{La}_{2-x}\text{Ba}_x\text{CuO}_4$ systems, respectively. In Fig. 15.8, it is easy to see that there is an obvious relationship

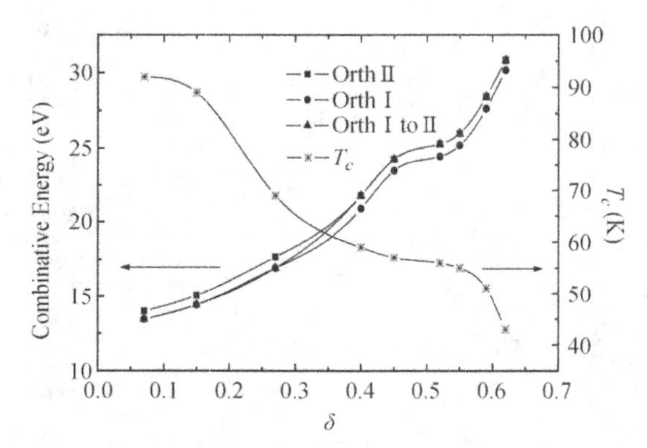

Fig. 15.7. The variation of T_c and the combinative energy with different oxygen deficiency δ. The results are obtained for different phases in $YBa_2Cu_3O_{7-\delta}$.

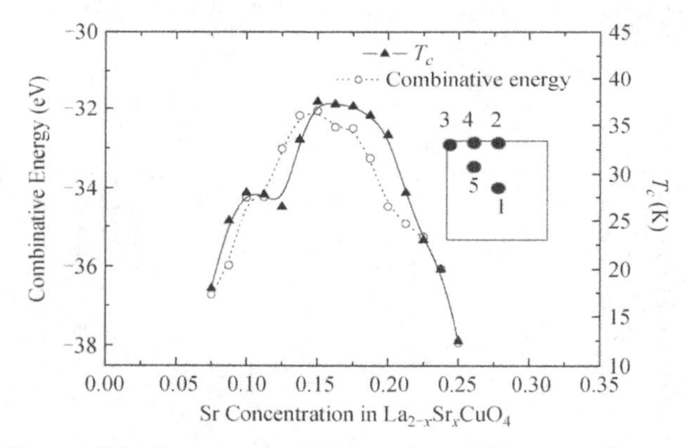

Fig. 15.8. The combinative energy and the value of T_c versus Sr concentration in $La_{2-x}Sr_xCuO_4$ with holes at position 1 (0.5, 0.5) in the CuO_2 planes. To guide the eyes, the scale of the combinative energy is overturned in Fig. 15.8 and 15.9, regarding to other figures about the relationship between the T_c and the combinative energy.

between the combinative energy and the T_c values versus Sr concentrations in $La_{2-x}Sr_xCuO_4$ when holes are at position 1 (The definition of the positions of the holes is shown in Fig. 15.8). The similar correlation also exists in $La_{2-x}Ba_xCuO_4$ as displayed in Fig. 15.9. When T_c gets the maximum, the combinative energy also reaches the maximum. It indicates that the slacking of the interaction between the two blocks does contribute to the change of the T_c in these systems. The combinative energy begins decreasing with increasing of Ba concentration when $0.08 < x < 0.125$, $0.17 < x < 0.24$ and of Sr concentration when $x > 0.15$ as the T_c decreases.

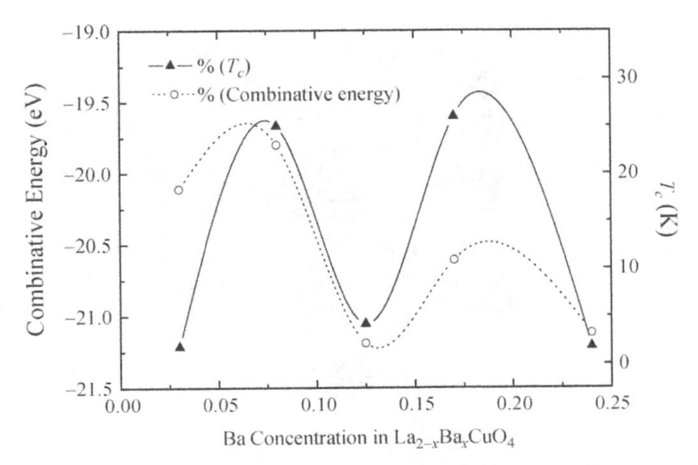

Fig. 15.9. The combinative energy and the value of T_c versus Ba concentration in $La_{2-x}Ba_xCuO_4$ with holes at position 1 (0.5, 0.5) in the CuO_2 planes.

The results indicate that the increasing of the interaction between the two blocks do have something to do with the suppression of superconductivity in these systems.

To conclude briefly, the method considers the cell as two relatively independent blocks and the combinative energy between those blocks mainly represents the strength of the interaction between the blocks. It seems that the interaction of these blocks (the perovskite and the rock salt) plays an important role in the superconductivity. This result supplies some important clues for understanding the mechanism of the high-T_c superconductivity. How the interaction between the blocks affects crystalline structure and then the superconductivity? The following result (part 5) about the stable Cu(2)–O plane or "fixed triangle" will give some interesting hints.

15.4. Another Way to Demonstrate that the Lattice Dynamics is Still of Great Importance for Superconductivity: Carrier Compensated System

According to the electronic phase diagram, the carrier concentration determines the superconductivity. Is the lattice dynamics still important to the superconductivity? We studied a compensated system to demonstrate it. In the system of $Y_{1-x}Ca_xBa_{2-y}La_yCu_3O_z$, partial Y^{3+} is replaced by

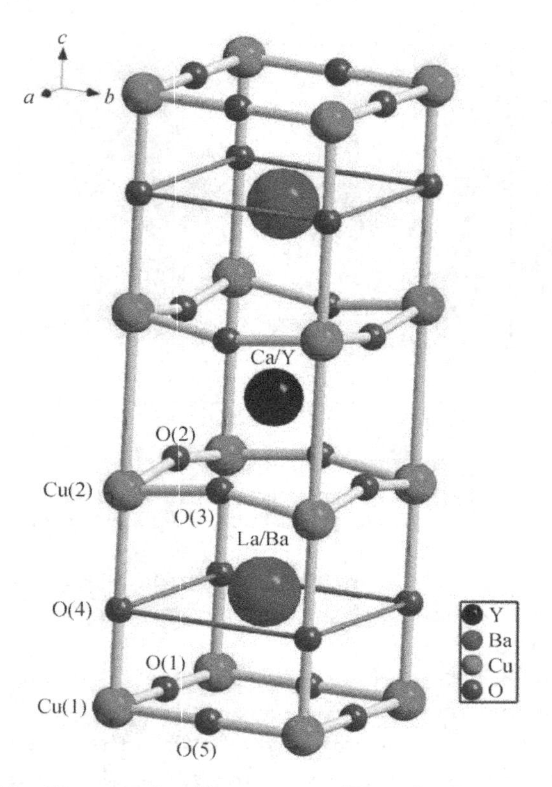

Fig. 15.10. The orthorhombic structure of $Y_{1-x}Ca_xBa_{2-y}La_yCu_3O_z$.

Ca^{2+}, and partial Ba^{2+} is replaced by La^{3+} (see Fig. 15.10). In this way, the carrier concentration in the system is not changed when $x = y$. If the T_c is changed, it is attributed to the change of the lattice.

$YBa_2Cu_3O_y$ (YBCO) system has been investigated by many research groups in the manner of individual calcium or lanthanum doping [17–30]. Generally speaking, cation doping in YBCO depresses the T_c. Partial substitution of Ca^{2+} at the Y^{3+} site will supply hole carriers, hence the carrier concentration or charge distribution in this system will regenerated. The Ca substitution also leads to the reduction of total oxygen content, little change in the effective copper valence, and the depression of the T_c. On the contrary, La^{3+} replacing Ba^{2+} will provide electronic carriers. The La substitution at the Ba site makes oxygen content increase and the effective copper valence slightly decrease, and the T_c depresses too.

The samples were synthesized by solid-state reaction method. Pure oxides of Y_2O_3, $CaCO_3$, $BaCO_3$, La_2O_3 and CuO were mixed according to the chemical formula of $Y_{1-x}Ca_xBa_{2-y}La_yCu_3O_z$ with both x and y from 0 to 0.5 in intervals of 0.1, respectively. Preparing process is similar to the preparation of self-compensating $Y_{1-x}Ca_xBa_{2-y}La_yCu_3O_z$ ($y = x$) system [31].

The crystalline structures of the samples were characterized by X-ray diffraction (XRD) on an X′pert MRD diffractometer with Cu K_α radiation. Detailed structural parameters were obtained by Rietveld refinement method, using the X′pert Plus software. The goodness of fit (GOF $= R_{wp}/R_{exp}$) of all the samples is less than 1.7, demonstrating that the refinement results are reliable. The T_c of each sample was determined by DC magnetization measurement at 10 Oe field in the temperature of 30–100 K, using Quantum Design MPMS system. The detailed result was published in Ref. [32].

To show clearly the T_c dependence on the content of Ca or La, respectively, the T_c is plotted as functions of both x and y for the $Y_{1-x}Ca_xBa_{2-y}La_yCu_3O_z$ system in Fig. 15.11, which is a 3D phase diagram. As an interesting phenomenon, the maximum T_c values are evidently along the diagonal of the basal plane, which correspond the optimum doping samples, i.e., the self-compensated samples. But it is clearly shown that even if the sample is the optimum doped or self-compensated, with the increase of the x and y values, the T_c is still depressed gradually.

The highest transition temperature is about 92 K for pure YBCO, which corresponds to the optimum oxygen content of 6.93 and the average copper valence of +2.28 [22]. If assuming that in the system, the average copper valence and the optimum oxygen content are both the same as that of pure YBCO (in fact, all the samples were fully oxygenated at 450° for 24 h in a flowing oxygen), the optimum doped condition is $y = x$ according to the electric neutrality condition. It is well consistent with the experimental result, i.e., the self-compensated $Y_{1-x}Ca_xBa_{2-y}La_yCu_3O_z$ ($y = x$) samples have higher T_c values compared with the samples un-compensated ($y \neq x$).

In the $Y_{1-x}Ca_xBa_{2-y}La_yCu_3O_z$ system, Ca^{2+} (x) replacing Y^{3+} will provide more hole carriers and depress the T_c in the manner of over-doping. But $La^{3+}(y)$ substituting for Ba^{2+} will supply electronic carriers, which neutralize part of holes caused by the Ca doping. When $y < x$ and the system remains overdoped, the T_c increases with the increase of La content

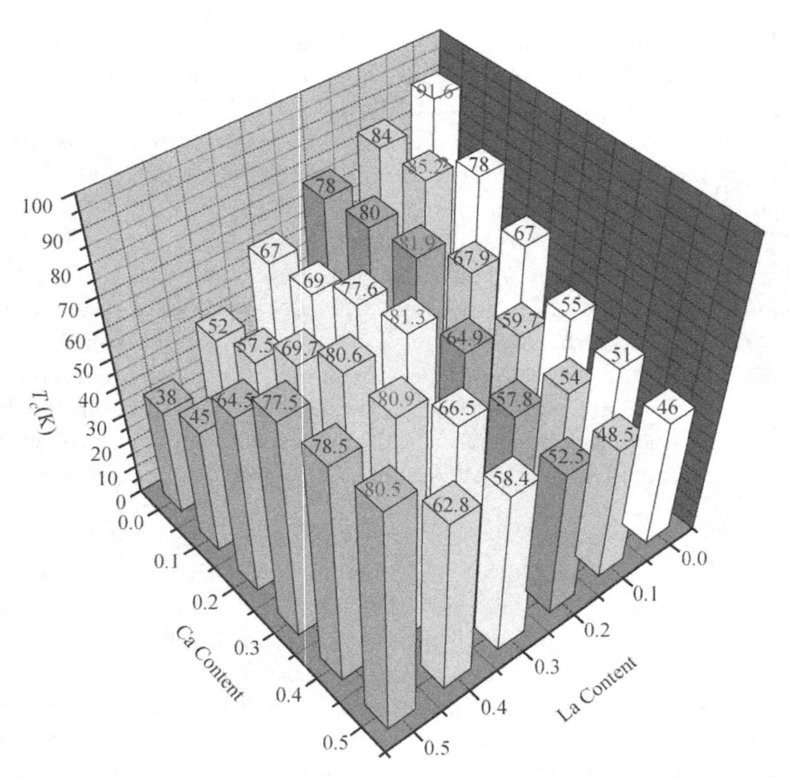

Fig. 15.11. 3D phase diagram, depicting the T_c dependence of the Ca and La content simultaneously in $Y_{1-x}Ca_xBa_{2-y}La_yCu_3O_z$ system.

at first. Once the content of La exceeds the optimum doping value, i.e., $y > x$, the excess electrons will neutralize more hole carriers and the sample becomes underdoped. As a result, the T_c decreases with the further addition of La. In other words, with x keeping constant and y increasing, the system $Y_{1-x}Ca_xBa_{2-y}La_yCu_3O_z$ undergoes a transition from overdoped to optimum-doped at $y = x$, and finally to underdoped. Accordingly, the T_c values exhibit a roof-shape in the 3D phase diagram, where the ridge represents the T_c of the optimum doped sample.

From Fig. 15.11, It is clearly shown that although in the self-compensated $Y_{1-x}Ca_xBa_{2-y}La_yCu_3O_z$ ($y = x$) samples, the T_c value is still depressed. It is reasonable to attribute this depression to lattice dynamics. Besides the interaction of the two blocks, this evidence demonstrates that the lattice is still of importance to superconductivity too.

Furthermore, the relationship between the T_c and the bond lengths of the Cu(2)-O(4) and the Cu(1)-O(4) is investigated and illustrated in Figs. 15.12 and 15.13. It is evident that the dependence of the T_c on the bond length of the Cu(2)-O(4) has better linearity than on that of the Cu(1)–O(4). It is quite interesting that the relationship between the T_c

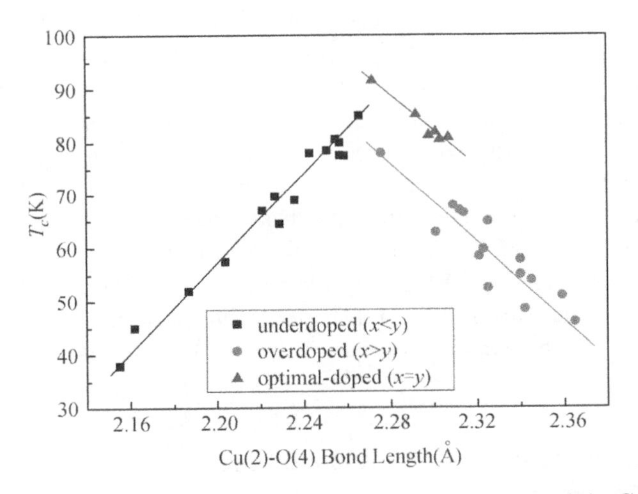

Fig. 15.12. The correlation between the T_c and the bond length of the Cu(2)–O(4).

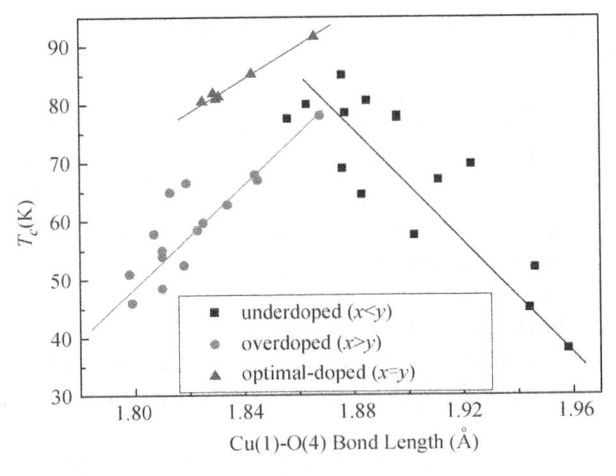

Fig. 15.13. The correlation between the T_c and the bond length of the Cu(1)–O(4).

and the bond lengths is different for the underdoped ($x < y$), optimal-doped ($x = y$) and overdoped ($x > y$) samples. In the underdoped region, the T_c increases linearly with the lengthening of the Cu(2)–O(4) bond and the constriction of the Cu(1)–O(4) bond. That may be attributed to the charge transfer mechanism. The variations of the two bond lengths indicate that the bridge atom O(4) moves far from the Cu(2)–O plane and accordingly the hole carriers transfer toward the Cu(2)–O plane. From the result of the electronic phase diagram, in the underdoped region, larger hole carrier concentration on the Cu(2)-O plane implies higher T_c. But after the highest T_c value, the behavior is completely opposite. It seems to show that the carrier concentration could not response to it. From this fact, although we can consider that the carrier transfer affects the T_c value, on the other hand, we also can think that the change of the bonds represents the change of the lattice. In the compensated region, it is obviously that the change of the bonds affects the superconductivity although the carrier concentration is kept constant. The structural change may be more basic than the carrier concentration. This hypothesis is being further studied.

15.5. Existence of Fixed Triangle by X-Ray Diffraction

Calculation of the bond angles and lengths demonstrates that Cu(2)–O(2), Cu(2)–O(3) and O(2)–O(3) form a stable triangle in $Y_{1-x}Ca_xBa_{2-y}La_yCu_3O_z$ system, called "fixed triangle" and independent of doping level in the system. The four fixed triangles form a stable Cu(2)–O plane. Figure 15.14 shows the results. The study in other superconducting systems also found the same phenomena. Figure 15.15 shows the results in $YBa_2Cu_{3-x}Al_xO_z$ and the $YBa_2Cu_{3-x}Zn_xO_z$ systems. As the content of the dopant changes, these bond angles and lengths in the Cu(2)–O plane almost don't change, but other bond angles and lengths obviously changed. For the sake of simplicity the bond lengths are just given, and the bond angles are not shown. We have studied nearly 10 doped systems of YBCO, and the fixed triangle or stable Cu(2)-O plane exists in all the systems. It can be concluded that the fixed triangle is a special local structure in the high-T_c superconductors.

Because the Cu(2)–O plane is just located in the boundary of the two blocks and it has been demonstrated above that the interaction between

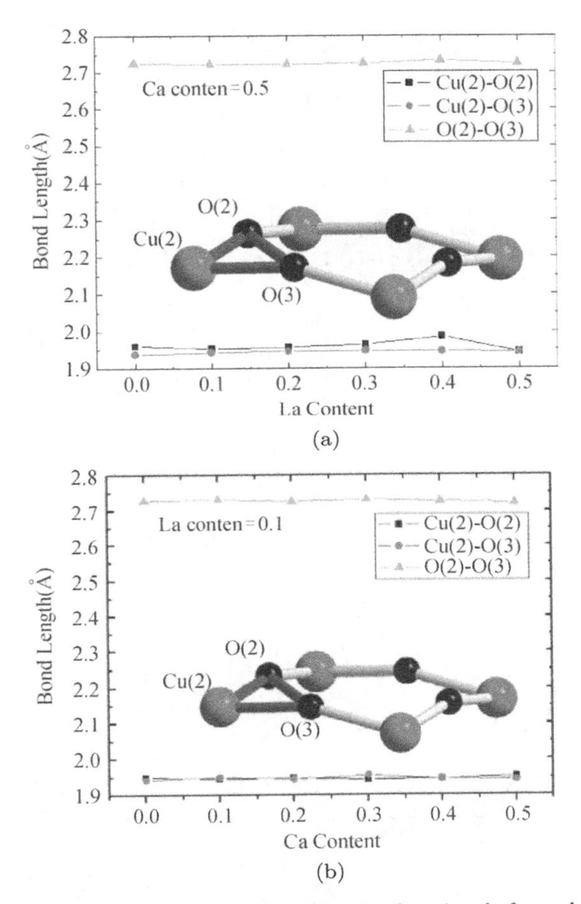

Fig. 15.14. The bond lengths between each other in the triangle formed by Cu(2), O(2) and O(3) versus the La content in $Y_{0.5}Ca_{0.5}Ba_{2-y}La_yCu_3O_z$ (a) and the Ca content in $Y_{1-x}Ca_xBa_{1.9}La_{0.1}Cu_3O_z$ (b).

the perovskite block and rock salt block has close relationship with the superconductivity, it is reasonable to think that the stable Cu(2)–O plane is resulted from the interaction of the two structural blocks. We [33] once elucidated that the interaction between the two blocks result in the strain between the two blocks, which can be derived from the combinative energy between them. The strain or combinative energy affects the superconductivity. How does it affect the superconductivity? We try to attribute it to make the Cu–O plane steady. Because the strain between

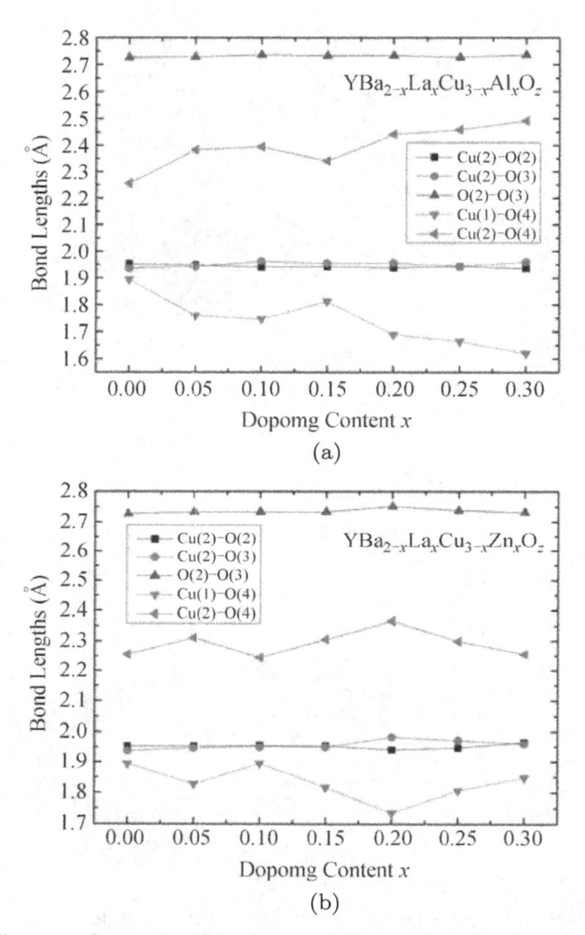

Fig. 15.15. Changes of some bond lengths with doping in $YBa_2Cu_{3-x}Al_xO_z$ (a) and the $YBa_2Cu_{3-x}Zn_xO_z$ (b) systems.

the two blocks has direction, the atoms in the boundary of the blocks can not move or vibration as the atoms in other positions. This may be the reason of the formation of the stable Cu(2)–O planes. This stable Cu(2)–O plane induces very strong anisotropy in the crystal. The phonons produce in the Cu(2)–O plane are very strong anisotropic too. In this case, the interaction between carriers and phonons becomes different from that in the conventional superconductors, and it then affects the superconductivity. This will be further discussed in the later section.

15.6. The Existence of the Fixed Triangle Demonstrated by Raman Spectroscopy

Someone may question that the existence of fixed triangle demonstrated by X-ray is not enough, and still needs other evidences. Here we use Raman spectroscopy to prove it.

The Raman spectra of the polycrystalline La-doped samples were obtained at room temperature with a Horiba–Jobin Yvon HR800 spectrometer equipped with a microscope and a CCD detector. The 785 nm laser line was used for Raman excitation. The laser was focused to a spot of approximately 2μm in diameter. The detailed results were published in Ref. [34].

As is pointed out by Cava *et al.* [35], the off-stoichiometry of the oxygen content in the La-doped YBCO system is slight (no more than 0.05 with $x \leq 0.4$) similar to our doping region except $x = 0.5$. Therefore, most structural and spectroscopic changes are due to the cation substitutions effects. Hence, we did no chemical analysis to get the exact oxygen content values of different samples.

Since the Cu(2)–O planes have already been widely recognized as the key ingredient for HTSC, the variation of the local structure about the Cu(2)-O planes with doping deserves more attention. Three bond lengths and three bond angles related to the Cu(2)–O(2)–O(3) triangle were calculated and illustrated in Fig. 15.16, respectively. It's evident that the "fixed triangle" reported previously by us [36, 37] is confirmed again in

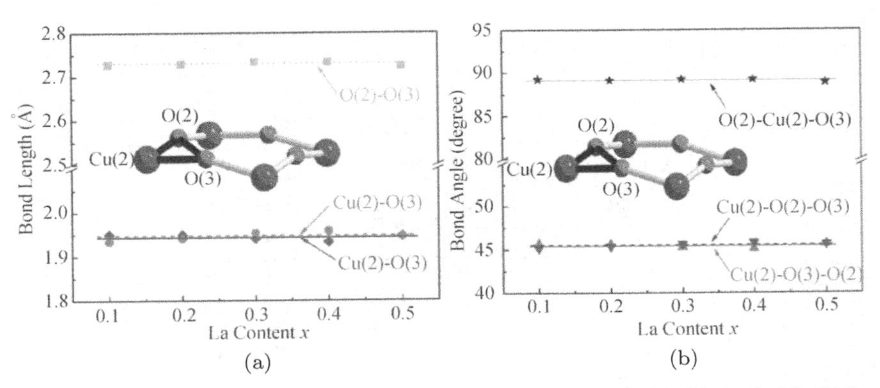

Fig. 15.16. The variation of 3 bond lengths (a) and 3 bond angles (b) in Cu(2)–O(2)–O(3) triangle with La content x in $YBa_{2-x}La_xCu_3O_z$ system.

this La-doped YBCO system, since all the six parameters constituting the Cu(2)-O(2)-O(3) triangle are relatively stable. They merely fluctuate in a small range around their average values, as is shown by the horizontal lines in Figs. 15.16(a) and 15.16(b). Our recent study indicates that the amplitude of fluctuations (denoted by δ, the average square deviation of three bond lengths) is closely related to the suppression of the T_c (denoted by ΔT_c) for different doped YBCO systems in the exponential form as $\Delta T_c = 1.09 + 3.08\exp(0.32\delta)$. In other words, the stability of the fixed triangle is favorable for the conservation of the HTSC.

The Raman spectra of the $YBa_{2-x}La_xCu_3O_z$ system with different content of La after background corrections are displayed in Fig. 15.17. As is known, 13 atoms in the Y123 unit cell give rise to a total of 39 phonon modes, 15 of which are Raman active with symmetries of $5A_g+5B_{2g}+5B_{3g}$ (orthorhombic notation) [38, 39]. It's evident that in our spectra of the La-doped system, four A_g modes corresponding to the c-direction vibrations of Ba, Cu(2), O(2)-O(3) (out-of-phase) and O(4), are clearly visible around 115 cm^{-1}, 150 cm^{-1}, 337 cm^{-1}, and 505 cm^{-1}, respectively. However, the c-direction O(2)+O(3) (in-phase) mode around 440 cm^{-1} is too week to be identify.

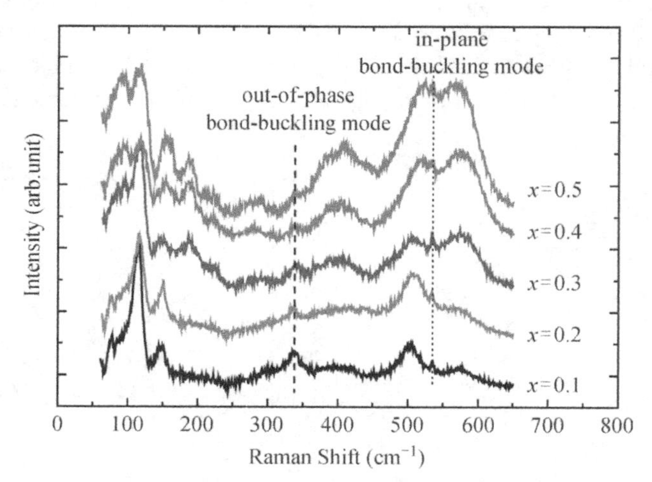

Fig. 15.17. The background-corrected Raman spectra of polycrystalline $YBa_{2-x}La_xCu_3O_z$ with $x = 0.1, 0.2, 0.3, 0.4$ and 0.5 at room temperature. The vertical dashed and dotted lines mark the O(2)–O(3) out-of-phase bond buckling modes and the in-plane Cu(2)–O(2) bond stretching modes, respectively.

Furthermore, several modes around 190 cm^{-1}, 210 cm^{-1}, 300 cm^{-1}, 390 cm^{-1}, 534 cm^{-1} and 575 cm^{-1} are discernible in the background-corrected spectra. The 210 cm^{-1} and 300 cm^{-1} mode are identified as the B_{2g} (along a axis) vibration and B_{3g} (along b axis) vibration of the apical O(4) atom, respectively [40, 41]. The 534 cm^{-1} mode is very week and sharp, supporting its origin as the in-plane Cu(2)–O(2) bond stretching B_{2g} mode [42]. In addition, at high doping levels, a mode below 100 cm^{-1} evolves, which could not be perfectly explained by Fano asymmetry. We speculate that it is caused by the vibration of La impurity since this mode lies near the A_g mode of Ba and its intensity increases dramatically with the La content.

The 390 cm^{-1} mode and 575 cm^{-1} mode are quite broad and their intensity increase dramatically with the addition of La, thus could be attributed to IR modes which become Raman active through the loss of local inversion symmetry associated with the increase doping level [43, 44]. The weak 190 cm^{-1} mode is not usually seen in the Raman spectra of YBCO system, however, it was once reported by Burns [45] and no assignment was given so far. We note that this frequency is in accord with the c-direction vibration of Y atom, which should be IR-active and of B_{1u} symmetry [46]. Thus it is possible that the disorder induced by the cation dopants mentioned above break the perfect centrosymmetry about Y which activates this mode into Raman-active.

The profiles of all the peaks could be well fitted with Lorentzian lineshape. For instance, the fitting of the $x = 0.1$ spectrum is illustrated in Fig. 15.18. The doping dependence of phonon frequencies of all the intrinsic (not disorder-induced) modes obtained by fitting are shown in Fig. 15.19. And the doping dependence of phonon linewidths of 4 A_g modes and the in-plane Cu(2)–O(2) bond stretching B_{2g} mode around 534 cm^{-1} (the other two in-plane modes around 210 cm^{-1} and 300 cm^{-1} are relatively flat and the fittings about their linewidths are not reliable) are plotted in Fig. 15.20. As Fig. 15.19 shows, the A_g phonon modes of Ba, Cu(2) and O(4) all harden evidently with the addition of La, which is naturally caused by the contraction of the c-axis since the bond length influences the corresponding phonon frequency greatly. The broadening of the three modes with the doping of La, as is shown in Fig. 15.20, indicates that the polycrystalline system becomes more disordered as a result of increasing La substitution.

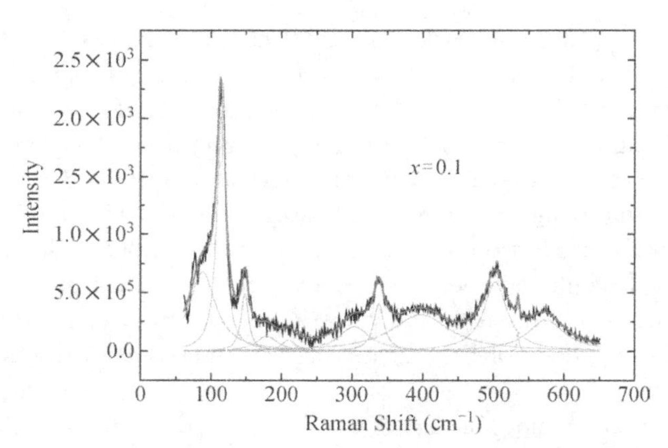

Fig. 15.18. Best fit of the Raman spectrum of $YBa_{1.9}La_{0.1}Cu_3O_z$. Black and red solid line are the raw and fitted profiles. Green solid lines represent different Lorentzian components.

In contrast, it's noticeable that the frequencies of the c-axis O(2)-O(3) out-of-phase bond buckling A_g mode around 335 cm^{-1} and the in-plane Cu(2)-O(2) bond-stretching B_{2g} mode around 534 cm^{-1} are almost independent of La content. And the linewidths of both modes represent similar behaviors. The insensitivity of these two modes related to the CuO_2 subunit to the doping level reflects the stiffness of the in-plane Cu(2)–O(2) bond and O(2)–O(3) bond, and thus provides direct evidence for the stability of the fixed triangle structure.

In addition, we notice that when the YBCO system is doped by Ca or Co impurities, there were similar observations from Raman spectroscopy that the frequency of the out-of-phase bond buckling mode around 335 cm^{-1} remains constant with increase doping [47, 48]. It is generally accepted that Ca substitutes into Y site, Co into Cu(1) site, and La into Ba site in our case. Therefore, the insensitivity of this phonon frequency and the stability of the CuO_2 triangle should be universal to the YBCO cuprate family, although the local environment around the CuO_2 plane might change significant.

Here, we give a brief conclusion about the Raman spectroscopy. It is discovered in high-T_c superconducting $YBa_{2-x}La_xCu_3O_z$ system by Raman spectroscopy that there exists a stable "fixed triangle" structure in Cu(2)–O planes. All chemical bonds and angles constituting the O(3)–Cu(2)–O(2) triangle are almost constant with doping. The frequencies and linewidths of the out-of-phase c-axis O(2)–O(3) buckling modes around 337 cm^{-1} and

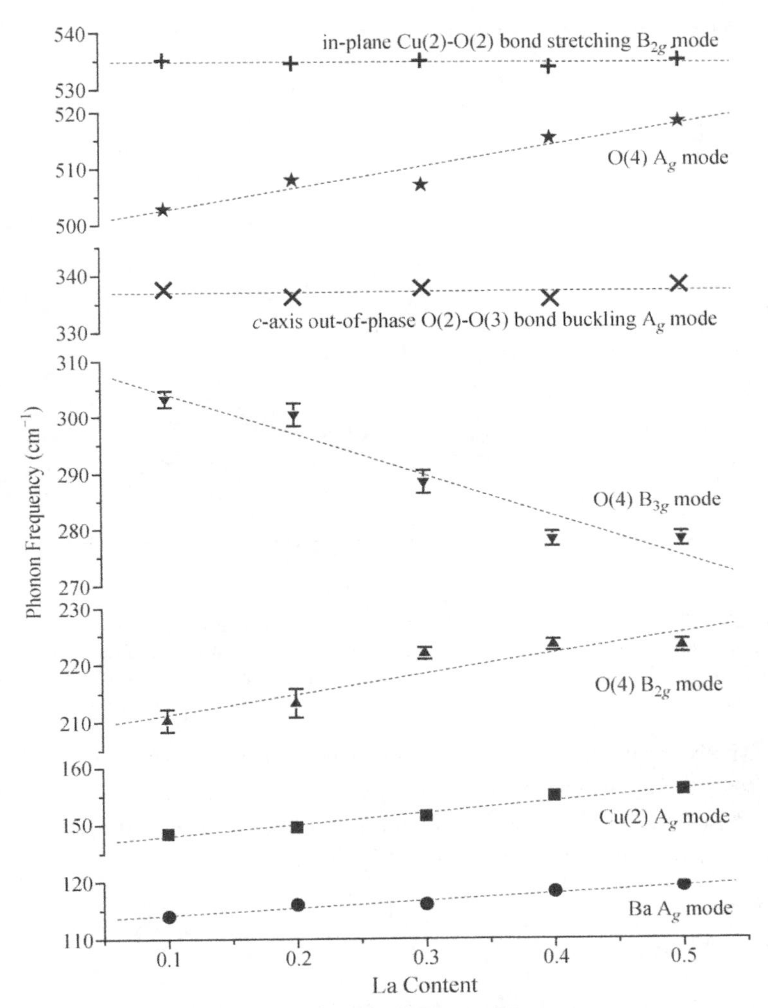

Fig. 15.19. The peak frequencies of 4 A_g (along c-aixs) phonon modes and 3 in-plane phonon modes appearing in the Raman spectra of the $YBa_{2-x}La_xCu_3O_z$ system as a function of the content of La. The error bars are denoted by the vertical solid lines. The dashed lines are linear fits to the data.

the in-plane Cu(2)–O(2) bond-stretching modes around 534 cm^{-1} are both independent of the doping level, providing direct evidence for the stability of this "fixed triangle". It was revealed previously that just these two phonons couple strongly with the anti–nodal and nodal electronic states,

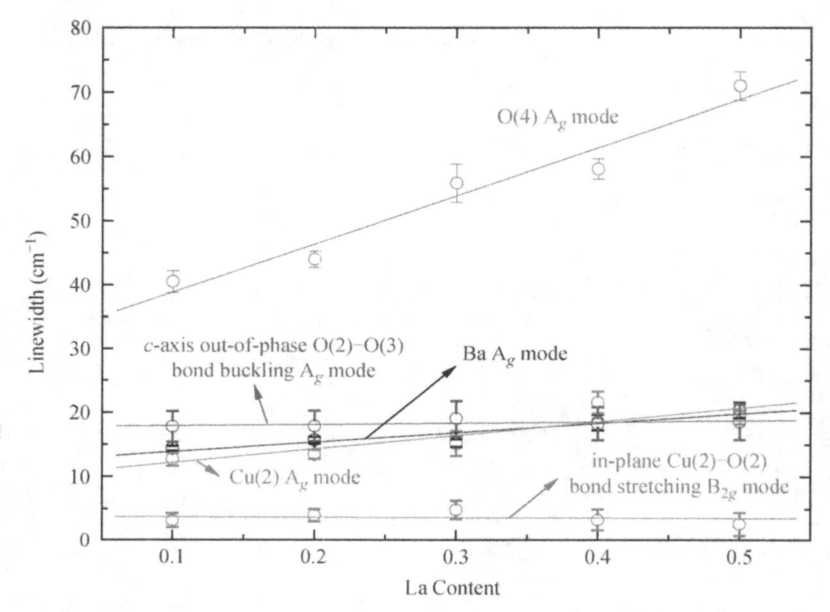

Fig. 15.20. The phonon linewidths of 4 A_g phonon modes and the in-plane Cu–O bond stretching mode around 534 cm^{-1} as a function of the content of La. The error bars are denoted by the vertical solid lines. The solid lines are linear fits to the data.

respectively, resulting in an anisotropic electron-phonon interaction in the cuprates [49, 50]. Anyway, the stability of the CuO_2 triangle should be paid more attention, its role in the electron–phonon interaction and in inducing the *d*-wave pairing might be of great significance.

15.7. Discussions

15.7.1. *Electronic-Phonon Interaction and Lattice Dynamics*

In early studies of the high-T_c superconductivity (HTSC), the role of electron–phonon interaction (EPI) in cuprates was questioned by researchers because some important properties of the cuprates could not be explained by conventional BCS theory in which the EPI is key and responsible for Cooper pairing [51]. For instance, the isotope effect in the cuprates is not evident as in conventional superconductors. Many theories were proposed to account for the mechanism of HTSC, such as resonating valence bond

(RVB) theory [52], strong coupling theory [53], exciton model [54], t–J model [55] and so on. But it was soon realized that these new theories were inadequate to explain the mechanism of HTSC. At the same time, substantial electron–phonon interaction became visible in the cuprates. In 2001, Lanzara *et al.* observed for the first time that there was an abrupt change of the electron velocity at 50–80 meV in different families of the cuprates using angle-resolved photoemission spectroscopy (ARPES), which was referred to as the "kink" in the electronic dispersion [56], and the most possible candidate resulting in this "kink" is strong interaction between electrons and phonons. Meevasana interpreted this "kink" as the result of the coupling between electrons and special phonons with some collective behavior [57]. Subsequently, it was found that the isotope effect in high-T_c superconductors was not negligible but significant in fact. For example, Khasanov directly observed evident oxygen-isotope ($^{16}O/^{18}O$) effect in the in-plane penetration depth of YBCO film [58]. And Iwasava found a distinct oxygen isotope shift near the electron–phonon coupling "kink" in the electronic dispersion of BSCCO system, which demonstrated the dominant role of the EPI in the cuprates [59]. Recently, some review articles emphasizing the important significance of the EPI in the mechanism of HTSC came forth [60, 61].

Of course, a lot of researchers did not agree with the EPI scenario. For instance, Allen thought the EPI in the cuprates could not induce their d-wave symmetry [62]. This was just the difference between the high-T_c superconductors and conventional superconductors, whose wave function was of s-wave symmetry. Whether or not were the d-wave symmetry induced by the EPI? Although those experiments mentioned above indicated the important significance of the EPI, the relationship between the EPI and the d-wave symmetry was not clearly clarified. In order to explain the mechanism of HTSC, one must answer this question inevitably.

Alexandrov proposed a bipolaronic model to explain how the EPI in the cuprates determines the HTSC [63, 64]. An interesting experimental result, which reveals that the velocity of longitudinal ultrasonic waves along $a-b$ plane is almost twice larger than that along c axis in a Bi-2212 single crystal [65], was cited to indicate the strong anisotropy of phonons in the cuprates. Although this model could not completely elucidate the EPI in the high-T_c superconductors in detail, it was suggested that quasi-2D charge carriers weakly coupled with the anisotropic phonons undergo a quantum phase transition from conventional s-wave symmetry to unconventional d-wave symmetry. Therefore, the anisotropic phonons and thereby the anisotropic EPI is responsible for the d-wave pairing in the cuprates.

In spite of this, the anisotropy of phonons must arise from the anisotropy of the crystalline structure. As is known, anisotropy is the essential characteristic of crystals because every crystal is more or less anisotropic. Among the conventional superconductors, there also are such oxides with strong anisotropy as $SrTiO_3$ and $BaPbBiO_3$. Then, why is the anisotropy of phonons in the cuprates so strong? What is the particularity of the phonons in the cuprates? One must find the answers from the crystalline structure. In this paper, we report our systematic structural and Raman studies on many superconducting samples, from which the existence of Cu(2)–O(2)–O(3) "fixed triangle" in the Cu(2)–O planes is confirmed. The new discovery of this special local structure might provide answers to the questions above.

On the other hand, by theoretically calculating the electron–phonon coupling constant, Devereaux *et al.* concluded that the out-of-phase O(2)–O(3) buckling mode and the in-plane Cu–O stretching mode couple strongly with the anti-nodal and nodal electronic states, respectively, resulting in the anisotropic EPI and corresponding band renormalizations observed in the ARPES [49, 66]. Strikingly, it is a coincidence that among all the phonons, the nodal and anti-nodal electrons tend to couple largely with the two phonons whose behaviors are insensitive to doping as shown by our Raman experiments, respectively. Combined with our recent discovery that the suppression of the T_c with doping in different doped-YBCO systems is closely correlated with the amplitude of fluctuations of the CuO_2 fixed triangle [67], it is reasonable to speculate that the stability of the Cu–O_2 triangle might be favorable for achieving the most appropriate anisotropic EPI between the electrons and the two phonons associated with this subunit, and thus favorable for keeping the superconductivity and sustaining a high-T_c. The stability of the Cu(2)–O plane should be paid more attention, its role in the electron–phonon interaction and in inducing the *d*-wave pairing might be of great significance.

15.7.2. *Hints from the Block Model*

According to above results, it is believed that the interaction between the perovskite block and the rock salt block is crucial for the high-T_c superconductivity. Without any one of the blocks, it is not high-T_c superconductors or even non-superconductors. Most recently, superconductivity was observed on the interface between insulating oxides [68]

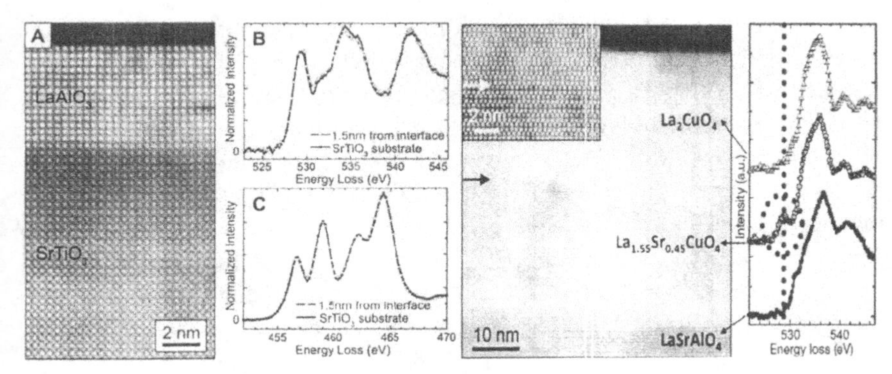

Fig. 15.21. Illustrations of the superconductivity in multilayer. Each single layer is not superconducting, but in the boundary of two layers, superconductivity exists (courtesy by I. Bozovic and D. Mueller. Refs. [68, 69]).

and the interface between metallic and insulating copper oxides [69]. The illustrations of the multilayer superconductivity are shown in Fig. 15.21. Each single layer is not superconducting, but in the boundary of two layers, superconductivity exists. These results clearly show that the interaction between the different crystalline structures can induce the superconductivity, which is a strong circumstantial evidence to our block model published 12 years ago [12].

The interaction of the blocks may result in the form of the fixed triangle and then a stable Cu(2)–O plane. This stable plane is responsible for the strong anisotropy of phonons, which actions are different from the phonons in the conventional superconductors because their anisotropy should be taken into account.

15.7.3. *Effect of the Local Structure of the Cu–O Plane on Superconductivity*

It is well-known that the coupling between electrons and phonons results in the formation of the Cooper pairs and conventional superconductivity. But the real mechanism remains elusive for the HTSC. Whether the interaction between electrons and phonons (or some other Boson mode) should be responsible for the HTSC is still under controversy. However, researchers have found some evidences for the electron–phonon coupling in the HTSC and it seems quite likely to be included in the mechanism of the HTSC. Cohen *et al.* [70] proposed a model that was capable of explaining the small oxygen isotope effect in $YBa_2Cu_3O_7$ within the framework of phonon-mediated electron pairing. Lanzara *et al.* [56] observed a dispersion anomaly ("kink") from ARPES experiment. They considered it was a direct evidence for ubiquitous strong electron–phonon coupling in HTSC, but some people did not agree with them [66]. Meevasana *et al.* found that the weight of the self-energy in the overdoped Bi-2201 system shifted to higher energies, which was related to a change in the coupling between electrons and *c*-axis phonons [57]. The above results indicate that it is hardly to exclude the electron–phonon coupling out of the HTSC. But how the coupling works in the HTSC and how much it effects on the HTSC are still very ambiguous.

From our experimental result, it is highly hypothesized that the mystery of the HTSC lies on the particularity of the phonon mode due to the structural specialty of the high-T_c cuprate, i.e., the stable Cu(2)–O plane formed by four fixed triangles.

It is intelligible that the atoms in the vital superconducting Cu(2)–O plane probably behaves collectively since Cu(2), O(2) and O(3) atoms form a "fixed triangle". Such an atom group with colossal effective mass is sure to represent a special kind of phonon mode different from a single atom. Furthermore, the coupling between electrons in the unit cell and that exceptional phonon might differ from that in the conventional superconductors. The vibration of the "fixed triangle" should be much weaker than individual atom. The interaction between electrons and it should also be much weaker, and the potential energy in the system should be lower. The fixed triangle is strongly coincident with the theory of Varma [71–73] in somewhere, who believes that there exists a loop current among Cu(2), O(2) and O(3) atoms and then highly determines the HTSC. The location of the loop is exactly at the fixed triangle (see Fig. 15.22).

Here, we give a guess about small isotopic effect by the fixed triangle. If the atoms in the fixed triangle vibrate together to some extent, like a "heavy atom", the small isotopic effect could be understood. Replacing copper or oxygen by their isotopes, it just changes a part of the mass of this heavy atom. In this case, the isotopic effect should be smaller than expected.

To better elucidate the possible significance of the fixed triangle, the electron-phonon interaction term is separated from the Hamiltonian of the system and expressed as:

$$H_{\mathrm{ep}} = H_{\mathrm{ep1}} + H_{\mathrm{ep2}}, \tag{15.7.1}$$

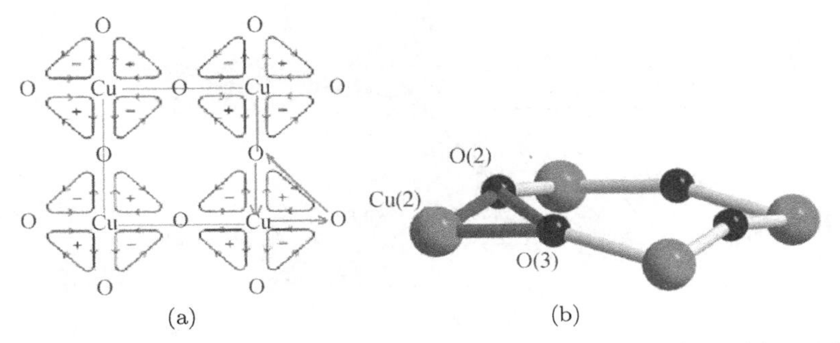

Fig. 15.22. Similarity in geometry of the loop current among Cu(2), O(2) and O(3) Atoms (a) and fixed triangle (b).

where H_{ep} is the total electron–phonon interaction Hamiltonian; H_{ep1} is the interaction Hamiltonian between electrons and the fixed triangle on the Cu(2)–O plane, where the state of the carrier may be changed and potential energy is slightly lowered; and H_{ep2} is the interaction Hamiltonian between electrons and other phonons in the crystal. The physical process of the interaction of electron–phonon in the YBCO unit cell is illustrated in Fig. 15.22(b). Assuming that the fixed triangle behaves as a phonon mode with a collective wavevector $\boldsymbol{q}$, and one electron (hole) with a wavevector $\boldsymbol{k}$ is scattered by it to another electronic state denoted by $\boldsymbol{k}'$. In this process, the energy of the system may be lowered, and forms a pseudogap (Fig. 15.22(a)). When it goes through the Cu(2)–O plane, it is scattered to the third electron state with a wavevector $\boldsymbol{k}''$ by absorbing other phonons (which are all included in a wavevector $\boldsymbol{q}'$) in the unit cell. In this process, it is likely to induce the formation of Cooper pairs, which may have something to do with the superconducting gap and the HTSC.

Here, we make the conjecture more clear. It is well known that there are pseudogap and gap in the HTSC [39], but so far, a last word about the two gaps has not been made. As illustrate in Fig. 15.22(a), the Fermi surface of the electrons is supposed to lower in two stages due to the lattice scattering. Firstly, it lowers (as the red curve shows) after the Hamiltonian H_{ep1} when scattered by the fixed triangle, and the pseudogap is formed. Secondly, it lowers further due to the Hamiltonian H_{ep2} when scattered by other phonons in the unit cell, and the superconductivity occurs. It should be mentioned that in the conjecture (Formula 6) the H_{ep1} could be due to the scattering by other phonons and the H_{ep2} could be due to the fixed triangle phonon.

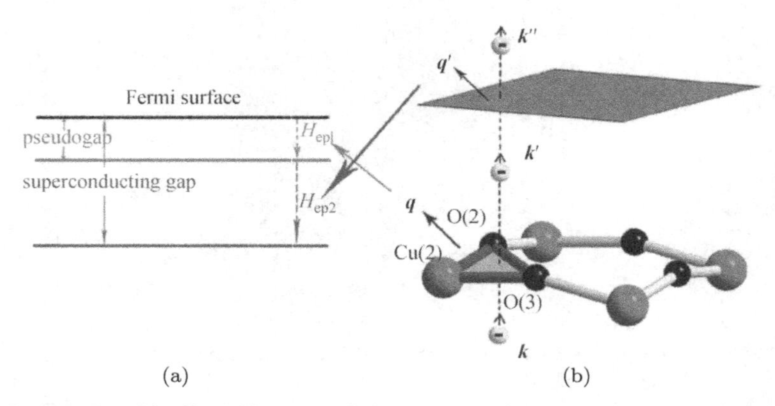

Fig. 15.23. Possible physical origin of the two-gap phenomenon associated with the lattice scattering (a). Lattice scattering of electrons in the YBCO unit cell (b). Green plane represents all the other phonons.

From that point of view, the final mechanism of the HTSC may possibly be approached. If the conjecture is correct, then the exceptional nature of the collective phonon mode and its coupling with electrons might be represented in various experimental measurements such as electronic quasi-particle dispersion expressed in ARPES, isotopic effect, the temperature dependence of the resistivity, Raman spectrum and so on.

Acknowledgments

During nearly past 20 years, many graduate students worked in my lab, and contributed to this work. Here I sincerely thank them for their works, although some of them have lost connection. They are: Qin X.C., Cheng L. L., Du X., Yang K., Tang X. T., Sun X. F., Wang S. X., Rui X. F., Zhang L., Yu J., Xu Y. Y., Jin W. T., Hao S. J., Guo C.Q. Among them, special thanks should be given to Qin X. C. and Jin W. T. for some important contributions in this work.

Most of the contents of this paper were once published in Phys. Rev. B, Inter. J. Mod. Phys. B., Physica C., New J. Phys., and Supercond. Sci. Technol.

References

[1] R. Kleiner, F. Steinmeyer, G. Kunkel, and P. Muller, *Phys. Rev. Lett.* **68** (1992), 2394.

[2] D. P. Billesbach and J. R. Hardy, *Phys. Rev. B* **39** (1989), 202.
[3] J. B. Torrance and R. M. Metzger, *Phys. Rev. B* **63**, (1989) 1515.
[4] M. Muroi and R. Street, *Physica C* **248** (1995), 290.
[5] H. Zhang, Y. Zhao, X. Y. Zhou, and Q. R. Zhang, *Phys. Rev. B* **42** (1990), 2253.
[6] Y. Ohta and S. Maekawa, *Phys. Rev. B* **41** (1990), 6524.
[7] Y. Mizuno, T. Tohyama, and S. Maekawa, *Physica C* **282** (1997), 991.
[8] K. A. Mueller, *Physica C* **185** (1991), 3.
[9] C. Park and R. L. Snyder, *J. Am. Ceram. Soc.* **78** (1995), 3171.
[10] L. Pauling, *Phys. Rev. Lett.* **59** (1987), 225.
[11] H. M. Evjen, *Phys. Rev.* **39** (1932), 675.
[12] H. Zhang, L. L. Cheng, X. C. Qin, and Y. Zhao, *Phys. Rev. B* **61** (2000), 1618.
[13] K. Yang, L. P. Yu, S. H. Han, H. Zhang, *Physica C* **357** (2001), 305.
[14] S. H. Han, J. Yu, Y. Y. Xu, and H. Zhang, *Inter. J. Mod. Phys. B* **21** (2007), 3048.
[15] H. Zhang, L. L. Cheng, X. C. Qin, and Y. Zhao, *Phys. Rev. B* **62** (2000), 13907.
[16] S. X. Wang and H. Zhang, *Phys. Rev. B* **68** (2003), 012503.
[17] D. H. Ha, S. Byon, and K. W. Lee, *Physica C* **340** (2000), 243.
[18] K. Hatada and H. Shimizu, *Physica C* **304** (1998), 89.
[19] V. P. S. Awana and A. V. Narlikar, *Phys. Rev. B* **49** (1994), 6353.
[20] R. Giri, V. P. S. Awana, H. K. Singh, R. S. Tiwari, O. N. Srivastava, A. Gupta, B. V. Kumaraswamy, and H. Kishan, *Physica C* **419** (2005), 101.
[21] R. Mohan, K. Singh, N. Kaur, S. Bhattacharya, M. Dixit, N. K. Gaur, V. Shelke, S. K. Gupta, and R. K. Singh, *Solid State Commun.* **141** (2007), 605.
[22] A. Sedky, A. Gupta, V. P. S. Awana, and A. V. Narlikar, *Phys. Rev. B* **58** (1998), 12495.
[23] N. L. Wang, M. C. Tan, J. S. Wang, S. Y. Zhang, H. B. Jiang, J. Sha, and Q. R. Zhang, *Supercond. Sci. Technol.* **4** (1991), S307.
[24] A. Tokiwa, Y. Syono, M. Kikuchi, R. Suzuki, T. Kajitani, N. Kobayashi, T. Sasaki, O. Nakatsu, and Y. Muto, *Jpn. J. Appl. Phys.* **27** (1998), L1009.
[25] K. Westerholt, H. J. Wueller, H. Bach, and P. Stauche, *Phys. Rev. B* **39** (1989), 11680.
[26] S. Mazumder, H. Rajagopal, A. Sequeira, R. Venkatramani, S. P. Garg, A. K. Rajarajan, L. C. Gupta, and R. Vijayaraghavan, *J. Phys. C: Solid State Phys.* **21** (1988), 5967.
[27] D. H. Ha, *Physica C* **302** (1998), 299.
[28] A. Manthiram, X. X. Tang, and J. B. Goodenough, *Phys. Rev. B* **37** (1988), 3734.
[29] H. J. Bornemann and D. E. Morris, *Phys. Rev. B* **44** (1991), 5322.
[30] X. S. Wu, F. Z. Wang, J. S. Liu, S. S. Jiang, and J. Gao, *Physica C* **320** (1991), 206.

[31] X. F. Sun, X. F. Rui, F. Wang, L. Zhang, and H. Zhang, *J. Phys.: Condens. Matter* **16** (2004), 2065.
[32] W. T. Jin, S. J. Hao, and H. Zhang, *New J. Phys.* **11** (2009), 113036.
[33] X. Du, L. L. Cheng, L. Zhang, and H. Zhang, *Physica C* **337** (2000), 204.
[34] W. T. Jin, S. J. Hao, C. X. Wang, C. Q. Guo, L. Xia, S. L. Zhang, A. V. Narlikar, and H. Zhang, *Supercond. Sci. Technol.* **25** (2012), 065004.
[35] R. J. Cava, B. Batlogg, R. M. Fleming, S. A. Sunshine, A. Ramirez, E. A. Rietman, S. M. Zahurak, and R. B. van Dover, *Phys. Rev. B* **37** (1988), 5912.
[36] J. Yu, Y. Zhao and H. Zhang, *Physica C* **468** (2008), 1198.
[37] J. Yu, W. T. Jin, Y. Zhao, and H. Zhang, *Physica C* **469** (2009), 967.
[38] C. Ambrosch-Draxl, H. Auer, P. Kouba, E. Y. Sherman, P. Knoll, and M. Mayer, *Phys. Rev. B* **65** (2002), 064501.
[39] R. Feile, *Physica C* **159** (1989), 1.
[40] T. Strach, T. Ruf, E. Schönherr, and M. Cardona, *Phys. Rev. B* **51** (1995), 16460.
[41] K. F. McCarty, J. Z. Liu, R. N. Shelton, and H. B. Radousky, *Phys. Rev. B* **41** (1990), 8792.
[42] R. Liu, C. Thomsen, W. Kress, M. Cardona, B. Gegenheimer, F. W. de Wette, J. Prade, A. D. Kulkarni, and U. Schröder, *Phys. Rev. B* **37** (1988), 7971.
[43] G. A. Kourouklis, A. Jayaraman, B. Batlogg, R. J. Cava, M. Stavola, D. M. Krol, E. A. Rietman, and L. F. Schneemeyer, *Phys. Rev. B* **36** (1987), 8320.
[44] P. Galinetto, L. Malavasi, and G. Flor, *Solid State Commun.* **119** (2001), 33.
[45] G. Burns, F. H. Dacol, F. Holtzberg, and D. L. Kaiser, *Solid State Commun.* **66** (1988), 217.
[46] M. Cardona, L. Genzel, R. Liu, A. Wittlin, and H. Mattausch, *Solid. State. Commun.* **64** (1987), 727.
[47] D. Palles, E. Liarokapis, T. Leventouri, and B. C. Chakoumakos, *J. Phys: Cond. Matter* **10** (1998), 2515.
[48] M. Kakihana, L. Borjesson, S. Eriksson, P. Svedlindh, and P. Norling, *Phys. Rev. B* **40** (1989), 6787.
[49] T. P. Devereaux, T. Cuk, Z. X. Shen, and N. Nagaosa, *Phys. Rev. Lett.* **93** (2004), 117004.
[50] S. Johnston, F. Vernay, B. Moritz, Z. X. Shen, N. Nagaosa, J. Zaanen, and T. P. Devereaux, *Phys. Rev. B* **82** (2010), 064513.
[51] J. Bardeen, L. N. Cooper, and J. R. Schrieffer, *Phys. Rev.* **108** (1957), 1175.
[52] P. W. Anderson, *Science* **235** (1987), 1196.
[53] R. Zeyher and G. Zwicknagl, *Z. Phys. B* **78** (1990), 175.
[54] S. M. Hayden, H. A. Mook, P. C. Dai, T. G. Perring, and F. Dogan, *Nature* **429** (2004), 531.
[55] F. C. Zhang and T. M. Rice, *Phys. Rev. B* **37** (1988), 3759.
[56] A. Lanzara *et al.*, *Nature* **412** (2001), 510.
[57] W. Meevasana *et al.*, *Phys. Rev. Lett.* **96** (2006), 157003.

[58] R. Khasanov *et al.*, *Phys. Rev. Lett.* **92** (2004), 057602.
[59] H. Iwasawa *et al.*, *Phys. Rev. Lett.* **101** (2008), 157005.
[60] V. Z. Kresin and S. A. Wolf, *Rev. Mod. Phys.* **81** (2009), 481.
[61] S. Johnston, W. S. Lee, Y. Chen, E. A. Nowadnick, B. Moritz, Z. X. Shen, and T. P. Devereaux, *Adv. Condens. Matter. Phys.* **2010** (2010), 968304.
[62] P. B. Allen, *Nature* **412** (2001), 494.
[63] A. S. Alexandrov, *Phys. Rev. B* **77** (2008), 094502.
[64] A. S. Alexandrov and A. M. Bratkovsky, *Phys. Rev. Lett.* **105** (2010), 226408.
[65] F. Chang, P. J. Ford, G. A. Saunders, L. Jiaqiang, D. P. Almond, B. Chapman, M. Cankurtaran, R. B. Poeppel, and K. C. Goretta, *Supercond. Sci. Technol.* **6** (1993), 484.
[66] B. A. Philip, *Nature* **412** (2001), 494.
[67] S. J. Hao, W. T. Jin, C. X. Wang, L. Xia, S. L. Zhang, and H. Zhang, unpublished results.
[68] N. Reyren *et al.*, *Science* **317** (2007), 1196.
[69] A. Gozar, G. Logvenov, L. F. Kourkoutis, A. T. Bollinger, L. A. Giannuzzi, D. A. Muller, and I. Bozovic, *Nature* **455** (2008), 782.
[70] R. E. Cohen, W. E. Pickett, and H. Krakauer, *Phys. Rev. Lett.* **64** (1990), 2575.
[71] C. M. Varma, *Phys. Rev. B* **55** (1997), 14554.
[72] C. M. Varma, *Phys. Rev. B* **73** (2006), 155113.
[73] V. Aji and C. M. Varma, *Phys. Rev. B* **75** (2007), 224511.

Index

CPSIA information can be obtained
at www.ICGtesting.com
Printed in the USA
LVHW041047210220
647558LV00003B/12